Springer-Verlag Berlin Heidelberg GmbH

Kermit L. Carraway • Coralie A. Carothers Carraway
Kermit L. Carraway III

Signaling and the Cytoskeleton

Springer

Kermit L. Carraway
Coralie A. Carothers Carraway

University of Miami
School of Medicine
Miami, Florida, U.S.A.

Kermit L. Carraway III

Harvard Medical School
Boston, Massachusetts, U.S.A.

ISBN 978-3-662-12995-1

Library of Congress Cataloging-in-Publication data

Carraway, K.L. (Kermit L.)
   Signaling and the cytoskeleton / Kermit L. Carraway, Coralie A. Carothers Carraway, Kermit L. Carraway III
      p. cm. — (Biotechnology intelligence unit)
   Includes bibliographical references and index.
   ISBN 978-3-662-12995-1   ISBN 978-3-662-12993-7 (eBook)
   DOI 10.1007/978-3-662-12993-7

   1. Cellular signal transduction. 2. Cytoskeletal proteins. 3. Cytoskeleton.
I. Carraway, C.A.C. (Coralie A. Carothers) II. Carraway, Kermit L., 1963- . III. Title.
IV. Series.
   [DNLM: 1. Signal Transduction—physiology. 2. Cell Communication—physiology. 3.
Cytoskeletal Proteins—physiology. QH 601 C313s 1997]
QP517.C45C37   1997   571.6—dc21
DNLM/DLC
for Library of Congress   97-34514
                                                                          CIP

© Springer-Verlag Berlin Heidelberg 1998
Originally published by Springer-Verlag Berlin Heidelberg New York in 1998

Typesetting: R.G. Landes Company Georgetown, TX, U.S.A.

SPIN 10672443   31/311 5 4 3 2 1 0 - Printed on acid-free paper

Acknowledgments

Our efforts on this book have been aided by many colleagues. Numerous individuals sent preprints and reprints, many of which have been cited in the text and references. In some cases we were unable to use contributed materials because of the organization we imposed on the book. Nevertheless, we are grateful to all who contributed to our learning process in this undertaking. We also want to thank those colleagues who sent illustrations which were used in preparing some of the figures and tables. We were assisted in these efforts by Maria Elena Carvajal and Maria Penton. Finally, and most importantly, we want to thank our colleagues here at the University of Miami who read and commented on portions of the manuscript: Drs. Robert H. Warren, Richard K. Assoian, Fulvia Verde and Greg Conner. Their suggestions were invaluable, but any errors or misinterpretations which appear in this work are our responsibility alone.

PREFACE

The interactions of cells with their environment and their responses to incoming signals from that environment are basic to all of cell and developmental biology. It follows that alteration in the capabilities of cells to respond to environmental signals in an appropriately regulated manner is of central concern in aberrant processes, ranging from developmental abnormalities to malignancy. The primary objective of this undertaking was to layout at the cellular and molecular levels, the basic concepts relating to two major classes of proteins, those involved in the transduction of signal and those involved in cytoskeletal structures and their organization. In the final analysis, the second category is a subset of the first, as exemplified throughout this treatise.

The development of this book was driven by the necessity to demonstrate the relationships between cell signaling and the cytoplasmic cytoskeleton in fundamental cell processes. The primary problem encountered in this effort is that there are literally hundreds of examples of these relationships which could be described. One approach would be to seek and catalog these examples. Although such a catalog might be useful to some researchers, it would undoubtedly fail to provide the biological rationales and mechanisms which are so important to understanding contemporary biological research on these cell behaviors. Instead of taking an encyclopedic approach, we have chosen to examine some of the major cell biological phenomena which involve both the cytoskeleton and signaling. The purpose is to use these processes to "teach by example" about the relationships between cell structure and signaling mechanisms. A major emphasis, which could not have been as effectively addressed in a catalog approach, was to provide an *integrative* analysis of subcellular systems and their regulatory mechanisms. A final objective was to attempt to provide a perspective which poses unanswered questions in major areas of cellular function and provokes ideas about the use of integrative approaches to addressing them.

The primary question of interactions between the signaling and cytoskeletal systems can be broken down into two reciprocal subsidiary questions:

1) How are the structure and function of the cytoskeleton affected by external signals which impinge on the cell?

2) How does the cytoskeleton influence the cellular signaling processes which determine cell behaviors?

Numerous studies have addressed the first question in a variety of systems, several of which we have described. Many of the observations are still fragmentary, but molecular mechanisms are beginning to emerge. Of particular importance is the limited number of mechanisms and types of components involved. Signals impinging on cells are mediated via receptors, and may be carried by such diverse vehicles as hormones, cytokines, extracellular matrix interactions and mechanical forces. The receptors may be either intracellular or extracellular, enzymic or nonenzymic, but their primary purpose is to transfer the signal into the cytoplasm and/or the nucleus. This signal transduction process frequently involves *multimolecular complexes associated with the cytoskeleton* and results in significant cytoskeletal perturbations, driven frequently by protein phosphorylation/dephosphorylation, lipid, cyclic nucleotide or Ca^{2+} second messengers or ionic changes. The primary mechanisms for cytoskeletal rearrangements are often regulated by small G-protein switches. Many of the components are known, but the specific mechanisms directing the cytoskeletal changes are often still unclear. Also unclear but developing is the relationship between these cytoskeletal changes and other signal transduction pathways important in cellular behavior, including the linkage to regulation of gene expession in the nucleus.

The second question has been much more difficult to approach, although an important paradigm is emerging. *Many important signaling events occur at sites of cytoskeleton attachment to membranes.* The proposed role of the cytoskeletal structure is in determining the *organization of the components in the multimeric complexes* at those sites. This organization allows not only the passage of signals down pathways, but also the integration of different pathways for regulating multifaceted responses to a single signal or inputs from different signals. Understanding this integration is one of the foremost challenges of modern cell and developmental biology. If a single statement can encompass our philosophy in approaching this research, it would be the following: *the cell is more than the sum of its parts; it is the product of its parts integrated over time and space.* Signaling mechanisms and the cytoskeleton are the elements which must provide that integration.

CONTENTS

ABBREVIATIONS

Ach	acetylcholine
AchR	acetylcholine receptor
ADF	actin depolymerizing factor
AKAP	A kinase anchor proteins
BPAG	bullous pemphigoid antigen
CamK	calmodulin-dependent kinase
cAMP	cyclic AMP
CAP	adenylyl cyclase-associated protein
cdk	cyclin-dependent kinase
cGMP	cyclic GMP
DAG	diacylglycerol
ECM	extracellular matrix
EGF	epidermal growth factor
ERK	extracellular signal-regulated kinase
ERM	ezrin/radixin/moesin family of proteins
FAK	focal adhesion kinase
FGF	fibroblast growth factor
GAP	GTPase-activating protein
GDI	guanine nucleotide dissociation inhibitor
GEF	guanine nucleotide exchange factor
IF	intermediate filament
IR	insulin receptor
IRS	insulin receptor substrate
KRP	kinesin-related protein
LPA	lysophosphatidic acid
MAP	microtubule associated protein
MAPK	mitogen-activated protein kinase
MARCKS	myristoylated alanine-rich C kinase substrate
MEK	mitogen-activated ERK-activating kinase
MEKK	MEK kinase
NDF	Neu differentiation factor
NGF	nerve growth factor
PE	phosphatidylethanolamine
PI3K	phosphatidylinositol 3-kinase
PICK	protein that interacts with C kinase
PIP	phosphatidylinositol phosphate
PKA	protein kinase A
PKC	protein kinase C
PLA	phospholipase A

ABBREVIATIONS

PLC	phospholipase C
PLD	phospholipase D
PTP	protein tyrosine phosphatase
RACK	receptors for activated C kinase
Rsk	ribosome specific kinases
SH2	Src homology 2
SH3	Src homology 3
TAM	tyrosine activation motif

CHAPTER 1

Cell Morphology and the Cytoskeleton

Introduction

Development of multicellular organisms can be viewed as an expression of information encoded in the nuclear genome to provide instructions for the synthesis of cell- and tissue-specific proteins. These proteins assemble into structures that allow the differentiated cell to perform its specialized functions. Specific cellular morphologies are as varied as their functions. The span of morphologies expressed by animal cells is astounding, ranging from the simple, spherical shape of the unperturbed lymphocytes to the spinal cord motor neuron which may extend meters in length. The morphology of differentiated cells is not necessarily static, since many cells are capable of radically altering their morphology according to changing conditions in their environment that require them to carry out different functions at different times. For example, neutrophils that circulate as spherical cells in the blood can, in response to signals produced by local tissue inflammation, adopt an "ameboid" migratory configuration to crawl between the lining cells of blood vessels, and home in on the site of inflammation. There they can reorganize their structure yet again to become phagocytic cells (1). The role and dynamics of morphology are not limited to cells of multicellular organisms. The budding yeast, *Saccharomyces cerevisiae,* responds to mating pheromone by extending a mating projection and by a rearrangement of its cytoskeleton and secretory apparatus (2). Such examples clearly demonstrate that cellular behavior is dependent on cell morphology, and that morphology is both dynamic and responsive to extracellular signals. These morphological changes are not limited to specialized cell functions. Many cells in the organism retain the ability to undergo cell division, often under the influence of signals from their environment or from other cells. This division involves massive internal morphological rearrangements.

The cytoskeleton is the fundamental structural system that organizes and maintains the varied and complex shapes of specialized cells. In one sense, it is the means by which the instructions encoded in the nuclear genome can be expressed as an organized structure in the cytoplasm. It provides internal scaffolding to hold the nucleus in position, to maintain internal membrane systems in proper relationships, to perform the transport of vesicular material between different membrane compartments of the cell, to elaborate specializations of the surface membrane and to anchor the cell through the membrane to extracellular matrix and to other cells. Recent advances in our understanding of the mechanisms of signal transduction have made clear that information flowing into the cell from its environment in

Signaling and the Cytoskeleton, by Kermit L. Carraway, Coralie A. Carothers Carraway and Kermit L. Carraway III. © 1998 Springer-Verlag and R.G. Landes Company.

Table 1.1. Signals which induce morphological changes

Mitogens
Cell-matrix adhesion
Cell-cell adhesion
Platelet activation
Chemoattraction
Immunoglobulins
Growth factors and cytokines
Cell surface proteases/proteolysis
Phosphatase inhibitors
Tyrosine kinase inhibitors
Stress responses

---

the form of signal ligands interacting with receptors can also have powerful effects upon cell structural organization and function. Changes in cell behavior, including rearrangements of cytoskeletal architecture and altered patterns of gene expression are commonly observed. Although many signaling pathways that mediate information transfer between the cell surface and the nucleus have now been defined. There has been speculation that the cytoskeleton, by virtue of its structural linkages between the cell surface and nucleus, can also play a direct role in signal transduction (3). Such speculation has been reinforced by observations that mechanical forces acting upon the cell surface can modify gene expression in the absence of any apparent extracellular biochemical signals (4).

It is clear from all of these considerations that there is good reason to view the cytoskeleton as a necessary crossroad in the flow of information from the cell to its environment and vice versa. It is therefore important to know which cytoskeletal elements define cell morphology, how their organization into filaments and filament assemblies is controlled by the cell and the environment, and how these assemblies themselves may mediate signals. In animal cells, which do not have rigid cell walls, cell morphology is primarily determined by two types of structures, cytoplasmic cytoskeletal elements and cell adhesions. As we shall describe, these structures work cooperatively to determine cell shape. Moreover, a large number of extracellular agents or processes, some of which are listed in Table 1.1, will promote changes in cell shape through perturbations in one or both of these structures. Thus, to understand cell morphology and its dynamics, we need to understand the organization of the cytoskeleton and cell adhesions and the mechanisms by which they respond to and transmit signals.

Actin and Microfilaments

Globally, the cytoskeleton consists of an interconnected network of three types of filaments: microfilaments, microtubules and intermediate filaments. Each of these filament types play a specific role in determining cell morphology, plasticity and cellular responses to signals. Each can be considered and studied in isolation. In fact, most of the information concerning the cytoskeleton comes from studies focused on properties or functions of the individual filament systems. However, for complex cellular behaviors, such as cell division, it must be remembered that the different filaments are interconnected and that their organizations are coordinately regulated.

Many of the responses of cells, such as spreading, motility, polarization and cytokinesis, depend specifically on reorganization of cellular microfilaments. This dynamic behavior often involves the cell surface, and microfilaments form a cortical meshwork underlying the plasma membrane (5). Thus, understanding microfilaments is critical to understanding the dynamics of cell morphology. This study is not a trivial problem since more than a hundred different actin-binding proteins may contribute to actin organization, depending upon the types of cells and cell processes. However, use of the combined powers of biochemistry, biophysics, molecular biology, immunology and genetics has made substantial progress in defining the proteins involved in microfilament organization and the mechanisms which contribute to microfilament dynamics. In contrast, much less is known about how these processes are regulated and how they are integrated with other cellular events.

The essential event in actin organization is the polymerization of the monomer (G-actin) to form polymer (F-actin). G-actin has a two-lobed protein structure containing a central cleft with nucleotide and metal binding sites (6,7). Under cellular conditions, binding of Mg^{2+} and ATP is favored to yield an activated monomer, which is polymerization competent. The rate-determining step in polymerization is nucleation (8). In an actin solution, activated monomers combine to produce a trimer which can then act as the nucleation site. In the cell cytoplasm with its multitude of actin binding components, other molecular species are undoubtedly able to induce nucleation (9). The F-actin formed by polymerization is usually described as a polar, two-start α-helix. The strongest interactions are between monomers in the same chain; weaker interactions form associations between the chains (6). The filament has polarity; the two ends have different properties. Polymerization is favored by about 6-fold at one end, traditionally called the plus end or barbed end. This latter designation results from the arrowhead appearance in electron micrographs of F-actin complexed with myosin II fragments. Another aspect of importance to polymer formation is the ability of the F-actin, but not G-actin, to act as an ATPase, converting the ATP associated with the polymer subunits to ADP. Since actin-actin interactions are weaker for ADP-actin, these subunits more readily dissociate from the polymer. In the absence of other factors to regulate these processes, the favorable addition of ATP-actin to the barbed end and dissociation of ADP-actin from the pointed (minus) end leads to treadmilling of the polymer.

The concentration of free G-actin in equilibrium with polymer is known as the *critical concentration*. In the test tube, this concentration depends primarily on the number of free barbed ends, and can be modulated by factors which either bind G-actin or filament barbed ends. Interestingly, in most cells G-actin is present at a level some 50 times the predicted critical concentration. Therefore, most of the G-actin must be in a sequestered form in the cell, making polymerization dependent primarily on two processes, desequestration of monomer and presence of nucleation sites. Two different types of sequestering proteins appear to play a role in polymerization, profilins and thymosins; some of their properties are listed in Table 1.2 (10, 11). Although profilin was originally described as a primary sequestering protein, its concentration in cells is incompatible with that role. Instead, thymosin-β4 appears to be the major G-actin sequestering protein in most cells which have been examined (12). The thymosin-β4-actin complex serves as a pool of actin which can be released when barbed ends are freed (11). However, β-thymosin

Table 1.2. Properties of actin monomer binding proteins

Properties	Profilin	Thymosin	ADF/Cofilin
Molecular weight (kDa)	12-15	5	15-19
Stoichiometry	1:1	1:1	1:1
K_d ATP-actin (µM)	0.3-0.5	0.6	0.2
K_d ADP-actin (µM)	3.7	100	1.3
Platelet concentration (µM)	40	580	<140
Localization	ruffles membranes	cytoplasm ruffles	cytoplasm ruffles
Effect on nucleotide exchange	increase	decrease	decrease
Competitors	barbed end cappers	–	tropomyosin, myosin
Effects of			
PPIs	inhibit	none	none
Phosphorylation	none	none	inhibit
pH	none	none	yes

Modified from ref. 11 (Sun H-Q, Kwiatkowska K, Yin HL. Curr Opin Cell Biol 1995; 7: 102-110), courtesy of Current Biology Ltd.

is not simply a monomer buffering protein. At higher concentrations it can interact with F-actin and other actin binding proteins (13,14).

Profilin acts as a modulator of the polymerization process (15), lowering the critical concentration >10-fold to promote the transfer of actin subunits in an energy-dependent manner from the pool of sequestered actin to filaments. This transfer results from the addition of profilin-actin complexes to filament barbed ends (11). Profilin also facilitates ATP for ADP exchange in the monomer, which may be important for rapid activation of the monomer during periods of filament depolymerization and reorganization. Filament barbed-end capping proteins play important roles in regulating actin polymer dynamics in cells (16). By blocking the barbed end, they raise the critical concentration of actin to that of the pointed end. Barbed-end capping proteins also reduce dissociation from the barbed end. There are two classes of barbed end-capping proteins, examples of which are listed in Table 1.3. Some capping proteins such as gelsolin and its analogs (17), sever filaments, creating more ends for polymerization. Others simply bind to the barbed ends to block polymerization. Since nucleation is critical to polymerization, a combined severing/uncapping mechanism provides a powerful impetus for increasing F-actin (18). Alternatively, proteins such as cofilin, a monomer binding protein (Table 1.2), can sever filaments without binding to the ends (19), thus creating additional ends and nucleation sites, and facilitating the depolymerization and repolymerization required for microfilament reorganizations. Interestingly, knock-outs of ADF/cofilin are lethal in all organisms tested, but knock-outs of gelsolin in mice only impair mobility of cells (20). Such experiments suggest that *cofilin plays a critical role in microfilament dynamics.*

Organizational changes of actin in cells are both temporally and spatially regulated. Thus, the recent finding that G-actin is nonsymmetrically distributed in cells

Table 1.3. Barbed end capping proteins

Protein	Mol wt kDa	Major sources
Severing		
gelsolin	83	widespread in tissues, plasma form
villin	92	intestinal microvilli
severin	40	*Dictyostelium*
fragmin	42	*Physarum*
adseverin	74	adrenal medulla
scinderin	80	chromaffin cells
Nonsevering		
MCP	39	macrophage, fibroblast
CapZ	32/35	*Dictyostelium*, muscle, yeast
aginactin	70	*Dictyostelium*
radixin	69	liver adherens junctions
insertin	39	smooth muscle

Modified from ref. 17 (Weeds A, Maciver S. Curr Opin Cell Biol 1993; 5: 63-69), courtesy of Current Biology Ltd.

is of great interest. Three different forms of nonfilamentous actin have been described: freely diffusable monomer, diffusible complex with sequestering factor and nondiffusible actin in punctate foci (21). How this localized nonfilamentous actin contributes to polymerization processes is still unclear. Much of the dynamic activity of actin, including polymerization, is localized near the plasma membrane. However, microfilaments are associated with the membrane at their plus ends, raising the question of how rapid polymerization can occur at a membrane-associated plus end. As noted in chapter 4, this is one of the fundamental questions in understanding the protrusive activity at cell surfaces that is critical to cell motility.

Although the exact mechanism for this behavior has not been explained, an interesting model for this mechanism is the actin-polymerization-driven motility of intracellular bacteria (22, 23). Certain classes of pathogenic bacteria can invade mammalian cells by inducing phagocytosis, then escaping the phagocytic vacuole by lysis. Inside the cell they associate with microfilaments which become organized into a tail as the bacterium moves. The tails are composed of dynamic, short microfilaments, formed by polymerization of actin at the bacterial surface and random depolymerization within the tail. Polymerization requires only a single transmembrane, cell surface bacterial protein, ActA in *Listeria monocytogenes*. Two separable domains of ActA are required for interaction with the actin cytoskeleton; one involved in nucleation and the other associated with filament dynamics. The second is a proline-rich region which binds profilin. An open question is how polymerized actin generates motility. One possibility is that a motor protein from the cytoplasm is involved. An alternative suggestion invokes a kind of rachet mechanism (24), in which thermal fluctuations permit forward motion, and the addition of actin to the membrane-proximal filament plus end prevents backward motion.

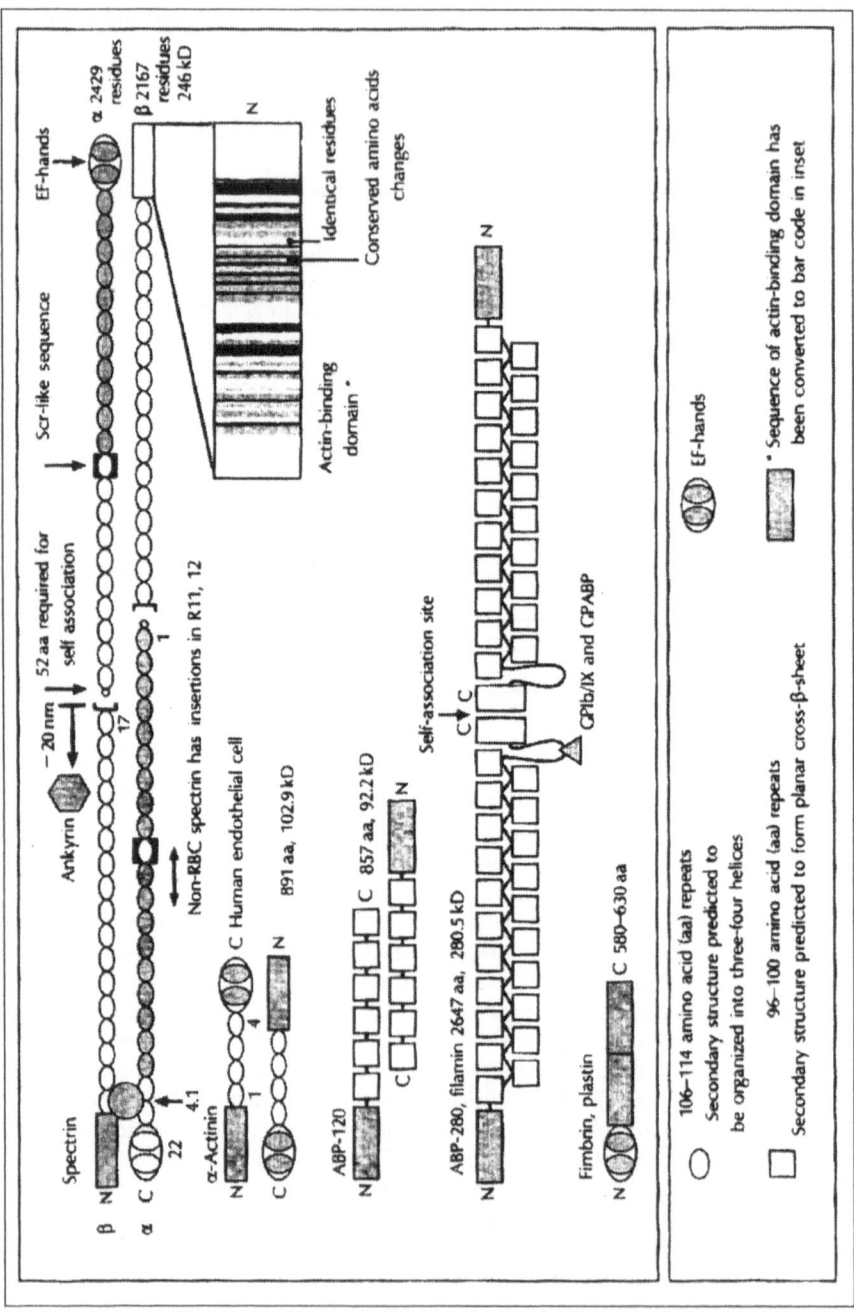

Fig. 1.1. Domain structures of some actin crosslinking proteins with common actin binding domains. Figure from ref. 28 (Hartwig JH, Kwiatkowski DJ. Curr Opin Cell Biol 1991; 3: 87–97), courtesy of Current Biology Ltd.

A similar mechanism may drive membrane protrusion in cell motility (see Fig. 4.4).

Microfilament-Containing Structures in Cells

Actin distribution in cells is also determined by proteins which associate with intact microfilaments. Two types of effects are particularly important: stabilization of the filaments and crosslinking of filaments into defined structural arrays (25). Stabilizing proteins, such as tropomyosin (26), bind along the sides of filaments to multiple subunits to prevent fragmentation and block severing. Crosslinking proteins organize filaments into bundles or meshworks. The type of structure formed is dependent on not only the specific crosslinking protein, but also its concentration relative to filaments (27). Tight bundles are formed by monomers containing two actin binding sites, such as fimbrin (Fig. 1.1) (28). Most other filament crosslinking proteins are parallel or antiparallel, homo- or hetero-dimers whose binding sites are at the ends of the dimers. Thus, they can form either loose bundles or networks of parallel or antiparallel filaments. An important exception is filamin, which forms orthogonal networks by crosslinking filaments at right angles (29, 30).

In considering higher order organizations of microfilaments, it is useful to discuss the types of morphological elements to which microfilaments contribute, particularly those which are involved in cell shape and dynamics. *Stress fibers (actin cables)* have received substantial attention because they are one of the most readily observable structures in cells in culture; they are composed of loose bundles of microfilaments. Stress fibers are relatively stable compared to other microfilament structures (31), probably due to the association of tropomyosin along considerable portions of their individual microfilaments. However, it should be remembered that all microfilaments have appreciable turnover rates compared to many other cellular components. The loose bundles comprising stress fibers appear to be held together primarily by α-actinin crosslinks, although other crosslinking proteins may also contribute (27). Two features of the stress fiber bundles are particularly important to their functions:

1) Filaments are arranged in both parallel and antiparallel fashion.
2) Filament bundles are inserted into focal adhesion sites at the plasma membrane, described in more detail in chapter 4. This structural organization is similar to that of the muscle sarcomere, and with the addition of myosin II, permits the stress fibers to act as a contractile unit which can contribute to cell motility (chapter 4).

The association of myosin ATPases with microfilaments provides a motor for several types of cellular motions (32). At least 13 different families of myosins are known (33, 34). All known myosins have a heavy chain with an N-terminal motor domain containing the ATPase activity, a regulatory neck domain to which light chains bind and a class-specific tail domain. The tail domains probably dictate cellular localization and function. The best characterized myosins are I, II and V. Myosin II is the conventional myosin of muscle contraction and drives the shape changes which occur in nonmuscle cells during cell division (35). Myosin II molecules are two-headed structures associated by their coiled-coil helical tails into bipolar thick filaments, as found in muscle. The contractile force in muscle is provided by calcium-promoted changes in the interaction of the myosin heads with actin filaments to produce an ordered displacement (sliding) of the myosin filaments relative to the actin filaments. Similar displacement mechanisms have been

Table 1.4. Microfilament-containing cell surface structures

Structure	Crosslinking protein	Source
Filopodia	α-actinin filamin	animal cells
Lamellipodia	α-actinin filamin fimbrin	animal cells
Microvilli	fimbrin villin	enterocyte
Microvilli	α-actinin	tumor cells

suggested for contractility along fibroblast stress fibers and the contractile rings of dividing cells. However, no thick myosin filaments are observed in these cases and other mechanisms are possible.

Myosin I differs from II and V in having only a single head domain. Myosins I and V contain membrane binding domains and are probably involved in membrane-microfilament interactions and intracellular membrane trafficking (33, 36). Loss of myosin I function results in changes in pseudopod formation, microfilament organization, endocytosis and secretion, depending on the cells involved (34). Studies of myosin V mutants suggest that it acts as a vesicle motor (chapter 6). Myosins I and V are also related in having calmodulin as their light chain. Myosins are nominally regulated by calcium and phosphorylation. Myosin II in most species is regulated by phosphorylation of its light chains (37). However, amoeba heavy chain phosphorylation also contributes to changes in motility. The heavy chain kinases are related to the PAK family of kinases which are activated by the small G-proteins, Rac and cdc42, as discussed in chapter 5. Complete sequencing of the budding yeast genome shows that there are five myosins in this organism, two each of classes I and V and one of class II (38). Thus, only three classes of myosin are needed for the survival and function of this organism.

Many of the most dynamic structures of cells are those at the cell periphery, which include a number microfilament-containing protrusions or processes which contribute to cell behavior (5, 39). The organization of these structures is determined by various actin binding proteins (Table 1.4). Filopodia, pseudopodia and lamellipodia are protrusions extended at the advancing end of motile cells (see chapter 4) (40); they differ in their shape and organization of actin (41). *Filopodia* are needle-shaped and contain a tight bundle of long actin filaments. *Lamellipodia* are thin veils which contain a meshwork of actin filaments and a ruffling membrane. *Pseudopodia* are thick processes containing a crosslinked mesh of filaments, having an organization between that of filopodia and lamellipodia. *Microvilli* are membrane-bounded cylinders containing bundled actin filaments and are prominent cell surface features of unattached cells and the dorsal surfaces of attached cells (5). In most cells, the microvilli are dynamic and interconvert with cell surface ruffles or blebs (42). The function of the microvilli in these cells is not clear, but they may provide a reservoir of membrane and actin for dynamic cell surface

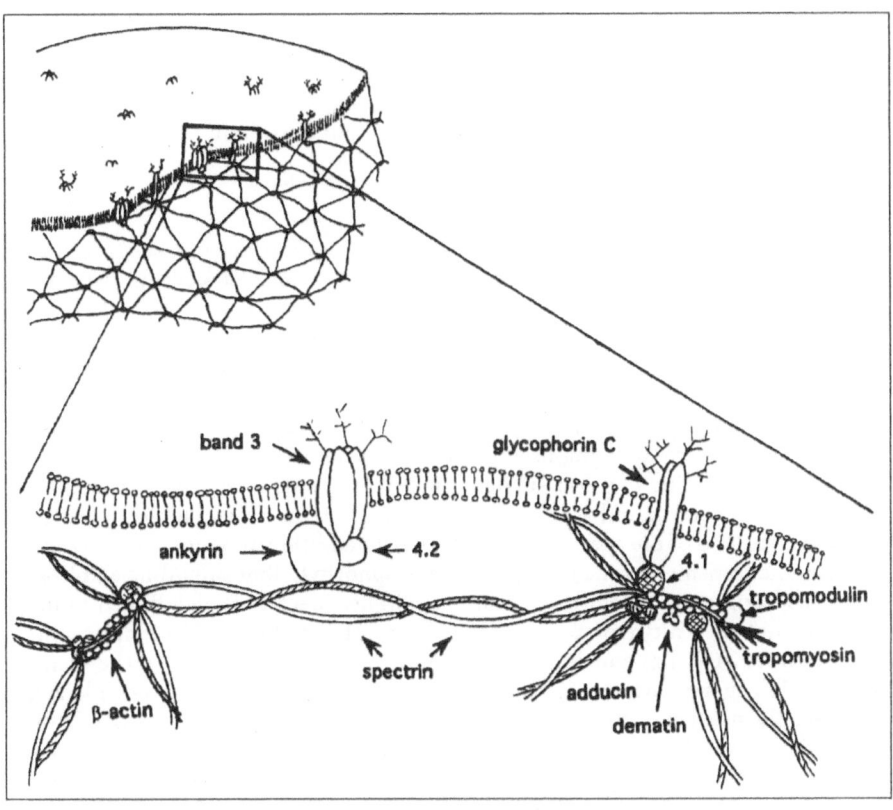

Fig. 1.2. Model of the organization of proteins in membrane skeleton of erythrocyte. Figure from ref. 49 (Bennett V, Gilligan DM. Annu Rev Cell Biol 1993; 9:27-66), with permission, from the Annual Review of Cell Biology, Vol 9, 1993, by Annual Reviews Inc.

reorganizations (42). In the specialized case of intestinal absorptive cells (see chapter 3), the microvilli are more highly structured and less dynamic. Their primary function is to increase the cell surface area containing hydrolases and permeases responsible for moving nutrients from the intestinal lumen into the cells (43). The microfilaments of these microvilli are tightly bundled by the monomeric crosslinking proteins villin and fimbrin (43). In microvilli of ascites tumor cells, the microfilaments are not so highly organized and appear more like those in cytoplasmic meshworks. Like these meshworks, the crosslinking protein in the ascites microvilli is α-actinin (44). Another specialized class of cells, the auditory hair cells of the inner ear, exhibit large microvilli called *stereocilia*, which are the primary sensors of sound (chapter 3). They are highly stable and long-lived, containing tight bundles of microfilaments (45). Membrane *ruffles* are dynamic structures which may be involved in rapid responses and interconversions among different types of membrane structures. However, it should be recognized that the term is a morphological description which may encompass different molecular structures regulated by different types of mechanisms (46). In summary, different classes of cell protrusions are variable in both structure and dynamics, but they all share the common characteristics of a membrane stabilized by internal microfilaments. The roles of these membrane-microfilament associations include stabilization of the membrane and participation in both inside-out and outside-in signal transduction processes.

Membrane Skeletons

Membrane skeletons are loosely defined as the supporting cytoskeletal structure of specialized membrane regions (47). The first membrane skeleton to be described and the best defined is the *erythrocyte cytoskeleton* or *membrane skeleton*, which was originally defined operationally as the insoluble material resulting from Triton extraction of erythrocyte membranes (ghosts) (48). The erythrocyte membrane skeleton is composed of a meshwork of filamentous structures in hexagonal or pentagonal arrays (Fig. 1.2) (49). The fibers of the meshwork are spectrin tetramers. The vertices are composed of protofilaments of actin, stabilized by tropomyosin, which bind to a domain near the end of spectrin (Fig. 1.2). The interaction between spectrin and actin is enhanced by proteins 4.1 and adducin, which bind near the end of the spectrin tetramer (50). In addition, protein 4.1 can associate with the transmembrane glycoprotein glycophorin C, thus providing a linkage of the spectrin-actin meshwork to the membrane. Spectrin is also linked to the membrane via ankyrin, which binds near the middle of the spectrin/ heterodimer (Fig. 1.2) comprising the tetramer. Ankyrin is linked to the membrane anion channel band 3. Protein 4.2 associates with both band 3 and ankyrin, presumably stabilizing their interaction. This complex membrane-cytoskeleton network provides both the toughness and flexibility necessary for the erythrocyte to survive its tortuous journey through the circulation. Genetic abnormalities leading to deficiencies or defects in spectrin subunits, ankyrin, band 3, protein 4.1 or 4.2 results in an increased fragility of the erythrocytes (51), substantiating the importance of these proteins in the network. Most of these proteins are also found in other tissues. In fact, the presence of spectrin associated with membranes has been used, probably inappropriately in the absence of other supporting evidence as a secondary definition for membrane skeletons.

Platelets are anucleate blood cells, or more appropriately, cell fragments which can respond rapidly to extracellular signals as part of the clotting process (chapter 4) (52). The disc shape of the unactivated platelet results from an internal scaffolding of microfilaments, a peripheral microtubule coil and a membrane skeleton at the inner surface of the plasma membrane (53). *Platelet membrane skeletons* can be observed morphologically after Triton extraction (54, 55, 56) and isolated from these lysates by differential centrifugation (54). However, because the platelet skeletons are less stable than erythrocyte membrane skeletons, defining their composition and structure has been more difficult. One important element of the platelet skeleton is short actin filaments linked to a membrane glycoprotein complex GPIb-IX by the actin crosslinking protein filamin (ABP-280) (Fig. 1.1). Depending on the lysis and centrifugation conditions, spectrin, talin, vinculin and the platelet integrins GPIIb-IIIa may also co-sediment. However, these latter components can be differentially sedimented and may be part of a separate network (see chapter 4) (53). Interestingly, signaling molecules such as Src, Yes and GAP are associated with this spectrin-rich complex (57). Changes in the platelet cytoskeleton with activation will be considered in chapter 4.

Membrane skeletons have also been implicated in specific structural and functional attributes of other, more complex cells, including neuromuscular junctions (chapter 3), epithelial cells (chapter 3) and nodes of Ranvier (47), though these have not been as well characterized. A specific isoform of spectrin (β-spectrin) is present in the neuromuscular junctions and appears to play a role in the localization of acetylcholine receptors. The junction is associated with actin filaments, but the molecular organization of the linkage of the receptors to the filaments is still unclear (58). A 43 kDa cytoplasmic protein appears to be most closely associated with the receptors. A 58 kDa protein is present in similar amounts; both are more tightly associated with the membrane than β-spectrin. Dystrophin, a spectrin analog, is also found in small amounts in the junctions, tightly associated with the 58 kDa protein, but its function is as yet uncertain. Ion channels other than the acetylcholine receptor are also clustered in the postsynaptic membrane, but the mechanisms involved in their concentration have not been characterized. The formation of neuromuscular junctions and their signaling complexes is discussed further in chapter 3.

Membrane skeletons provide linkage sites for association of the cytoskeleton with membrane proteins. The mobility of these linked proteins is thus highly restricted. Mobility of other membrane proteins can also be retarded by these interactions. One explanation for this phenomenon is the "fence" model, in which membrane skeleton linkages act as "posts" in the fence, restricting the movement of unlinked proteins in the plane of the membrane primarily to enclosed domains (59). How these components are organized at the molecular level is unclear except for the erythrocyte. Plasma membrane preparations from many types of cells have associated actin (42), some of which may be in an oligomeric form (60). In some instances the membrane actin appears to be directly linked to membrane glycoproteins (50, 61). In other cases, such as the focal adhesions, actin is linked to the membrane via actin-binding proteins in complex arrays. Both types of organization probably contribute to the reduced mobility of membrane proteins.

Microfilament-Associated Cell Adhesions

Cell morphology in animal cells is also dependent on extracellular constraints, such as cell-matrix and cell-cell interactions. Although the molecular details of these structures will be described in later chapters, certain features of their organization are common and critical to understanding their roles in cell morphology and their relationship to the cytoskeleton. *First*, cell adhesions are *transmembrane* structures (62). *Cell-cell interactions* may be mediated through several different classes of transmembrane molecules including cadherins, immunoglobulin-like cell adhesion molecules and selectins (62, 63). Molecular aspects of these adhesions are addressed in chapter 3. In the case of *cell-matrix interactions* the transmembrane glycoproteins are commonly integrins (64), though proteoglycans may be involved in some cases (62). *Second*, the extracellular domains of the transmembrane glycoproteins of cell adhesion sites interact specifically with components of the extracellular matrix or extracellular domains of cell surface components of other cells. For example, integrins interact by heterophilic interactions with a variety of extracellular matrix components (64). By contrast, cadherins form calcium-dependent homodimeric complexes which bridge two cells (65). *Third*, the cytoplasmic domains of the transmembrane cell adhesion molecules associate with cytoplasmic components to form complexes which are linked to the cytoskeleton. Interestingly, cytoplasmic domains of members of three different classes of cell-cell

Table 1.5. Actin-binding proteins regulated by calcium

Protein	Activities modified
Direct effect of calcium	
α-actinin	F-actin binding and crosslinking
gelsolin	Filament severing, nucleation of polymerization
villin	
severin	
fragmin	
adseverin	
scinderin	
villin	Barbed end uncapping
MCP/gCap-39	actin monomer binding, barbed end capping
annexins	F-actin/phospholipid binding
Effect of calcium/calmodulin	
MARCKS	F-actin binding
HSP100	F-actin binding
caldesmon	F-actin binding, inhibition of myosin/gelsolin effects
spectrin/4.1	microfilament crosslinking

Modified from ref. 68 (Janmey PA. Annu Rev Physiol 1994; 56: 169-191), with permission, from the Annual Review of Physiology, Volume 56, 1994, by Annual Reviews Inc.

adhesion molecules, integrins, immunoglobulin-like and selectins, have been observed to bind directly to the microfilament-binding protein α-actinin (66).

Since morphology is regulated by interactions of the cytoskeleton with cell surface proteins, the adhesions become primary determinants of cell morphology by providing plasma membrane-cytoskeleton interaction sites. *These complexes also often contain specific elements of signal transduction pathways involved in triggering cellular responses* (67). The consequence of these sets of interactions is that cell adhesions, whether cell-matrix or cell-cell, play a major role in determining the morphologies of cells by providing sites through which extracellular signals can be transmitted to the cytoskeleton to regulate various aspects of cell behavior. Recent studies have shown that other signaling elements, such as growth factor receptors, are localized to adhesion complexes. *These results suggest that the adhesion sites serve as integrators of signals from different pathways that affect cell behavior.*

Regulation of Microfilament Organization

Some cells can respond to external signals by doubling their concentrations of F-actin in 30 sec (9). Such dramatic changes require rapid formation of polymerization nuclei and rapid monomer desequestration. Nuclei can be formed either from pre-existing microfilaments, by severing proteins as described above, or by increased nucleating ability of cellular structures. Desequestration, as noted

Table 1.6. Actin-binding proteins regulated by phosphoinositides

Protein	Activity modified
profilin	monomer binding
gelsolin	severing, barbed end capping, monomer binding
villin	
severin	
adseverin	
cofilin	monomer binding
gCap39	monomer binding, barbed end capping
α-actinin	crosslinking
filamin	
MAP2C	
myosin I	F-actin binding to phospholipid bilayer
talin	
CapZ	barbed end capping
Cap 32/34	
Cap 100	

Modified from ref. 68 (Janmey PA. Annu. Rev. Physiol. 1994; 56: 169-191), with permission, from the Annual Review of Physiology, Volume 56, 1994, by Annual Reviews Inc.

above, involves profilin. Four different types of cell signaling components have been prominently implicated in changing microfilament organization Ca^{2+}, phospholipids, small G-proteins and protein kinases/phosphatases. Ca^{2+} has been shown to have effects on a number of proteins which alter microfilament assembly (Table 1.5) (68). The general pattern is that Ca^{2+} activates proteins which disrupt filament organization, such as the barbed end blocking and severing protein gelsolin. In addition, the binding of some crosslinking proteins, such as specific forms of α-actinin, can be inhibited or displaced by Ca^{2+}. Many of these proteins exhibit high dissociation constants for the ion (69). Thus, the effects are probably transient and local, rather than global. This localization of effects in time and space can help to explain morphological changes which appear to require simultaneous filament assembly and degradation in different parts of the same cell (9). Finally, the complex effects of Ca^{2+} on microfilaments must be integrated with its effects on other signaling pathways, such as the phospholipid pathways (68).

Phospholipids, particularly phosphoinositides, are important factors in many signaling pathways and cellular responses, as described in chapter 2. Furthermore, specific phosphoinositides have been shown to interact directly with actin binding proteins involved in microfilament organization (Table 1.6) (70). *First*, phosphatidylinositol phosphate (PIP) and PIP_2 can specifically dissociate the profilin/actin complex (71). The binding of profilin to these phosphoinositides can prevent their hydrolysis by PLC (72), a key event in some phosphoinositide regulatory pathways (see chapter 2). However, EGF stimulation of its receptor can trigger the phosphorylation of PLC, making the phosphorylated enzyme no longer inhibited by profilin (73). By these combined effects on phosphoinositides, profilin may

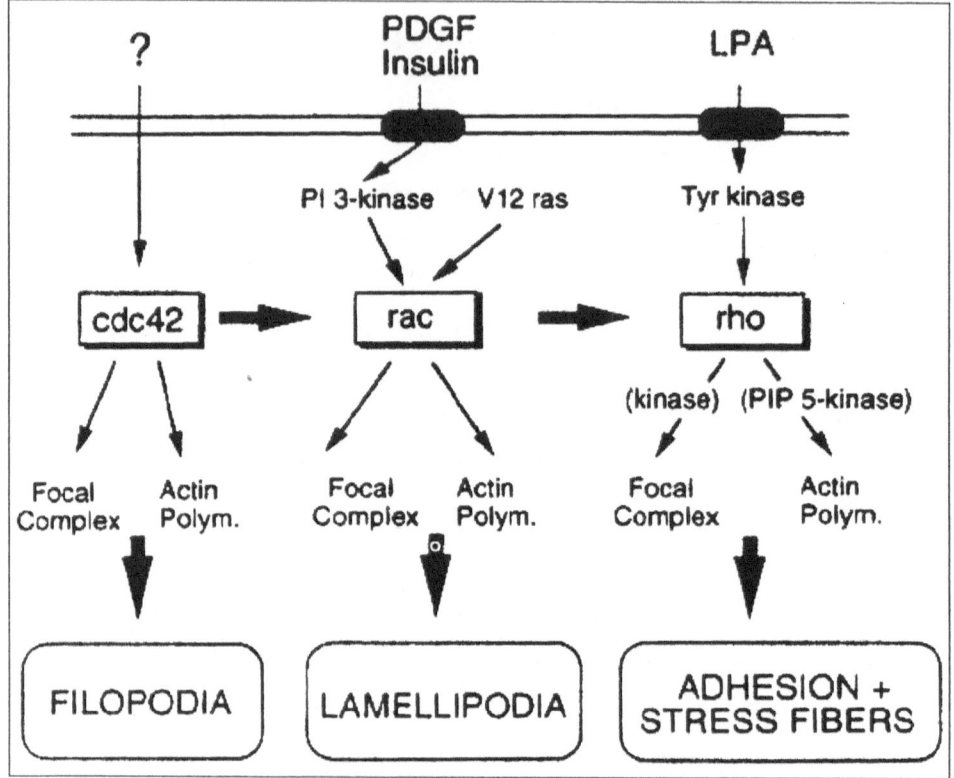

Fig. 1.3. Rho, Rac and cdc42 signaling pathways proposed to be involved in microfilament organization. Figure from ref. 86 (Nobes CD, Hall A. Cell 1995; 81: 53-62), courtesy of Cell Press.

play a dual role in regulating membrane signal transduction and microfilament organization. How these two effects are coordinated or whether they are even spatially compatible remains to be determined. In another type of mechanism, profilin may act on an adenylyl cyclase-associated protein (CAP) (74, 75) which binds G-actin and inhibits polymerization, an activity which is blocked by PIP$_2$ (76). CAP is located near the leading edge of chemotactically stimulated *Dictyostelium*, suggesting an involvement in migration. Since a CAP analog is also present in pig platelets, a similar pathway may be present in mammals (77). *Second*, PIP and PIP$_2$ inhibit the activities of severing proteins such as gelsolin (68). Severing activity is more susceptible to inhibition than nucleating or barbed end-capping (78). The phosphoinositides will block severing activity of both capping and noncapping classes of severing proteins (79). *Third*, phosphoinositides will release barbed-end capping proteins which do not sever filaments (80). *Fourth*, PIP and PIP$_2$ enhance the binding of some crosslinking proteins such as α-actinin to filaments (81), while inhibiting the crosslinking of others, such as gizzard filamin (82).

Several general observations about the phosphoinositide interactions with actin binding proteins are important:

1) Binding of the lipids is facilitated by lipid clustering (83), which probably imposes a spatial control on the effects of the lipids in cells.

2) Protein-lipid binding has been detected at sites, such as the Z-band of the sarcomere, which appear not to contain phospholipid bilayers. Since phosphoinositides co-purify with some of these proteins through stringent procedures, protein-lipid complexes may be present in the cytoplasm of some cells, as well as in membranes.

3) Proteins which bind phosphoinositides can inhibit their degradation by phospholipases, as described above for profilin. Since phospholipid hydrolysis is a key element in some signaling pathways (see chapter 2), these interactions may regulate or integrate signaling pathways which impinge on both the cytoskeleton and other cell functions.

4) The effects of phosphoinositides on microfilament networks are generally opposite to those of Ca^{2+}. Thus, the spatial and temporal cycling of Ca^{2+} and the phosphoinositides may provide the impetus for the reorganization of the microfilament networks during cell morphological transitions. Interestingly, diacylglycerol, a product of growth-factor stimulated phospholipases which hydrolyze inositides (see chapter 2), has been shown to stimulate microfilament nucleating activity of plasma membranes even in the absence of membrane actin (84). This case provides one example of how extracellular signals may be linked to cytoskeletal changes.

Extracellular signals, such as those provided by growth factors and hormones, can induce rapid changes in cell morphology and microfilament organization (Table 1.1). Mechanisms for transmission of signals to the cytoskeleton have been elusive, but some insights have been gained by recent studies on small G-proteins (85). These G-proteins, described further in chapter 2, act as molecular switches to control signaling pathways which regulate numerous cellular functions. The prototype for this family is the proto-oncogene Ras, which is a central element in many signaling pathways. Mutations in Ras lead to changes in microfilament organization. However, the specific mechanisms of the cellular morphological changes appear to involve three other small G-proteins, Rho, Rac and cdc42 (Fig. 1.3) (86). Microinjection studies on fibroblasts have shown that Rho regulates the assembly of focal adhesion sites and stress fibers (87). In contrast, Rac is required for a second signaling pathway which leads to the formation of lamellipodia and ruffles (88). Finally, cdc42 triggers the formation of filopodia (86). In general, Rho is coupled to heterotrimeric G-protein receptors and Rac to tyrosine kinase growth factor receptors. However, the detailed mechanisms by which these signals are transmitted from the cell surface to the small G-proteins and then onward to the cytoskeleton are still unclear, though cascades of both small G-proteins and of protein kinases may be involved (89). These will be discussed further in chapters 2 through 5.

Several candidate kinases from different segments of cytoskeleton-regulating pathways have been implicated in microfilament reorganizations. The inhibition of Rho-induced focal adhesion assembly by the tyrosine kinase inhibitor genistein implies the involvement of a tyrosine kinase (90). Candidates include focal adhesion kinase (FAK) and Src family members, tyrosine kinases are described further in chapters 2 and 4. However, an alternative candidate is the PI-4-P 5-kinase which synthesizes PIP_2 and is regulated by Rho (91). Rac and cdc42 bind a ubiquitous serine/threonine kinase p65[PAK] (92), but appears not to be involved in morphology changes (93). Recent studies suggest that one of the most important proteins involved in microfilament reorganizations is ADF/cofilin (20). Interestingly, *phosphorylation of ADF/cofilin inhibits its microfilament-severing activity directly* (19). Moreover, site-directed mutagenesis has shown that the phosphorylated form is concentrated at membrane ruffles and cleavage furrows, whereas an unphosphorylatable form is diffusely distributed in the cytoplasm (94). Thus, *phosphorylation may provide a localization signal which facilitates the participation of cofilin in microfilament reorganization.* The rapid dephosphorylation of ADF/cofilin in thrombin-treated platelets probably contributes to the reorganization of the actin cytoskeleton during platelet activation (95).

Another actin binding protein whose activity is influenced by phosphorylation is MARCKS (myristoylated alanine-rich C kinase substrate), a filament crosslinking protein also regulated by Ca^{2+}/calmodulin (Table 1.5) (96). Unphosphorylated MARCKS is associated with membranes and can crosslink filaments. Phosphorylation by protein kinase C (PKC) releases the protein from the membrane. The phosphorylated form is unable to crosslink filaments, though it still binds F-actin. MARCKS also loses its crosslinking ability in the presence of Ca^{2+}/calmodulin, but remains associated with the membrane and able to bind F-actin. MARCKS has been found in focal contacts and presynaptic junctions and is implicated in motility and secretion. The microfilament binding protein cortactin is a substrate for Src (97). Src may also influence microfilament organization through its phosphorylation of Rho-GAP (98), a GTPase-activating protein which modulates the activity of Rho, as described in chapter 2. Its presence in the cortical networks of ruffles and lamellipodia suggest an involvement in cell surface dynamics. Though fragmented, observations suggest multiple kinases and pathways can contribute to microfilament organization.

Tubulin and Microtubules

Microtubules are long, hollow cylinders, 25 nm in diameter, composed of the protomer tubulin. As the largest and most rigid of the cytoskeletal filaments, they are resistant to compression, approximately 300-fold stiffer than microfilaments (99). In one model for cell structure (*tensegrity model*), they serve as the internal struts that maintain cell morphology against the contractile microfilament network (100). Cytoplasmic microtubules in most cells emanate from a microtubule organizing center, the *centrosome*, which appears to be important in determining the overall organization of cell structure (101). For example, the microtubules determine the cellular localization of membrane organelles such as the endoplasmic reticulum (102) and Golgi apparatus (103), as described further in chapter 6. The role of cytoplasmic microtubules is not limited to the organizational or structural support. They also act as cellular "highways" or "guideways" for the movements of intracellular vesicles and organelles (chapter 6), e.g., cytoplasmic vesicles (104) or granules (105) and chromosomes (see chapter 5). The identification and characterization of microtubule-binding motor proteins involved in these dynamic processes has been a major accomplishment of recent advances in molecular cell biology. Microtubules form the internal structures of cilia and eukaryotic flagella, motile cell surface processes involved in either cell movements or movements of fluids near the cell surface (106). Microtubules also undoubtedly play a role in cell polarization (chapter 3), even in relatively unpolarized cells such as fibroblasts (107). In more highly polarized cells, such as neurons or epithelial cells, they provide the structural elements necessary for stable, long range polarization (108).

Table 1.7. Stability of different types of microtubules. Structures are listed in order of increased stability from top to bottom

Structure	Role
mitotic spindle	chromosome segregation
interphase network	intracellular organization
dendrites	neural morphogenesis
	postsynaptic densities
axon	neural morphogenesis
	presynaptic densities
	axonal transport
cilia, flagella	motility
basal bodies, centrioles	
axonemes	

Microtubule stability generally increases from top to bottom of table. Modified from ref. 113. Reprinted by permission of the publisher from Avila J. Microtubule functions. Life Sci 1992; 50: 327-334 copyright 1992 by Elsevier Science Inc.

Dynamics of Microtubules

The subunit of microtubules is the tubulin heterodimer, arranged in a cylinder of 13 linear protofilaments. As in the case of microfilaments, the polarity of the subunit imposes a polarity on the filament which has plus (fast-growing) and minus ends. The process of formation of microtubules is similar to that of microfilaments, the rate-limiting step being nucleation (99). Nucleotide hydrolysis by the polymer also plays a role in microtubule stability, but the nucleotide associated with polymerizing tubulin is GTP, rather than ATP. The hydrolysis of GTP on the microtubule and the association of GDP with tubulin subunits of the minus microtubule end lead to instability and depolymerization. This process is called *dynamic instability*, in which individual microtubules alternate between polymerization and depolymerization (109). In the dynamic instability models the depolymerization and polymerization phases are denoted as the *catastrophe* and *rescue* events, respectively. An intermediate state with no elongation or shortening is known as a *pause* (110). When unpolymerized tubulin is in excess, GTP-tubulin will add to microtubule ends until the *critical concentration* (equilibrium concentration) is reached. However, as with actin and ATP, the hydrolysis of GTP on microtubules lags polymerization (111). Thus, all microtubules will have a depolymerization-resistant "GTP cap" at the plus end until the polymerization rate slows as the process approaches equilibrium. At equilibrium, addition will continue to microtubules which have GTP tubulin at the plus end, but depolymerization (catastrophe) will occur when the GTP at the plus end of the filament has been depleted by hydrolysis to GDP.

Though dynamic instability is a property of tubulin, in vitro changes in microtubules are less rapid than those in cells. Comparisons of purified sea urchin tubulin with sea urchin extracts showed that the extracts exhibited a 6-fold faster elongation rate, 20-fold increase in catastrophes, 2-fold increase in rescues and no minus

Table 1.8. Microtubule binding proteins

Protein	M_r kDa	Source	Binding domain
MAP-1A,B,C	350	CNS, other	4-amino acid repeats
light chains	34,28,18	cell types	
MAP-2A,B	270	CNS	18-amino acid repeats
C	70		
MAP-4	190-240	widespread	18-amino acid repeats
tau	55-62	CNS, other	18-amino acid repeats
		cell types	

Modified from ref. 116 (Maccioni RB, Cambiazo V. Physiol Rev 1995; 75: 835-864), courtesy of American Physiological Society.

end assembly (112). These results indicate that microtubule dynamics is regulated by factors in the cells, particularly by factors which increase elongation, catastrophe and rescue at the plus end and inhibit polymerization at the minus end. Thus, as with microfilaments, cellular microtubules are more dynamic than those made from purified tubulin.

The dynamics of specific cellular microtubules is dependent on many factors, including the state of the cell cycle, cell differentiation and the localization of the microtubules. Table 1.7 describes the relative stability levels of different types of cellular microtubules (113). In undifferentiated interphase cells, about 80% of the microtubules are relatively dynamic (108), though less so than in cells undergoing mitosis. However, the remainder of the microtubules are more resistant to depolymerization. These microtubules can be identified by the presence of two posttranslational modifications of the tubulin in the microtubules: detyrosination and acetylation of the α-subunit (107). The former is catalyzed by a specific carboxypeptidase, leaving Glu as the C-terminal residue (Glu-tubulin). Acetylation occurs at Lys 40 of α-tubulin. The modifications are independent, occur only on microtubules, not tubulin, result from stochastic events, and appear not to play a direct role in microtubule stability. They are a consequence rather than a cause of microtubule stability (108). One possible function of these modifications is to modulate the binding of microtubule binding proteins which affect microtubule dynamics. For example, an additional C-terminal modification of the α-subunit, polyglutamylation with 1-6 glutamic residues (114), regulates the binding of microtubule-associated protein tau (115).

Microtubule Binding Proteins

Two different classes of microtubule binding proteins are known to promote assembly: high M_r (200-300 kDa or larger) *microtubule associated proteins*, most of which are designated *MAPs*, and *tau proteins* (55-62 kDa). Individual proteins of these classes are listed in Table 1.8 (116). Both classes have two domains, one which binds microtubules, while the other probably links the microtubules to other cellular components. The presence of multiple tubulin binding sites on these proteins may facilitate nucleation. Likewise, these multiple sites inhibit dissociation from microtubule ends, stabilizing the tubule by binding to several protomers of the filament, much as tropomyosin stabilizes the microfilament. This regulation of dynamic instability by MAP and tau proteins has been observed by in vitro studies using purified proteins, by microinjection of labeled tubulin into cells and by transfection of cells with microtubule binding proteins (116, 117). Microtubule binding proteins can also be classified according to their binding sites for tubulin. Tau, MAP-2 and MAP-4 contain 3-4 repeats of 18 amino acids which bind near the tubulin carboxy terminus. MAP-1 proteins contain 21 repeats with the core sequence KKEE or KKE(I/V).

In the cell, the centrosome clearly plays an important role in microtubule assembly (101). When cellular microtubules are depolymerized by drug treatments and allowed to repolymerize, the new microtubules grow from the centrosome from their minus ends to re-establish the original distribution. Centrosome structure and assembly are important to microtubule organization (118). A key component is γ-tubulin (119) which binds the β-subunit of the $\alpha\beta$ dimer, thus establishing the polarity of the microtubule (120). Other components involved in centrosome assembly are α- and β-tubulin, actin, the molecular chaperone hsp70 and pericentrin, a 220 kDa protein. Recruitment of γ-tubulin to the centrosome is required for nucleation at the minus ends. This nucleation function is highly conserved as human γ-tubulin will function when introduced into yeast (121). Recent studies indicate that the centrosome nucleation site is composed of a complex containing a ring of γ-tubulin molecules (122).

Functions of the individual microtubule binding proteins are under investigation. Much of the work has been done on brain or brain fractions because microtubules are abundant in the brain, particularly in axons or dendrites, described further in chapter 3, which carry electrical signals between cells (123). Microtubules of these neural processes differ from cytoplasmic microtubules in two important aspects:

1) Most of them are not associated with the centrosome.
2) They are packed into parallel arrays, suggesting mechanisms of association or bundling by microtubule binding proteins. One of the primary findings concerning the organization of microtubules in these processes is that MAPs are specifically localized, MAP-2A and MAP-2B in axons, tau in dendrites' and MAP-1A and MAP-1B in both (117). These proteins can serve as region-specific crosslinks between microtubules and between microtubules and other filaments or organelles in the organization and function of these structures. Moreover, they undoubtedly provide stability necessary for the long term survival of the nerve cell interconnections. Interestingly, the orientation of the microtubules relative to the cell body differs between the two types of processes. Dendrites have a mixed orientation of plus ends, while the axonal microtubules are uniform in orientation. Obviously, orientation is critical to the direction of axonal transport by specific motor proteins.

Bundled microtubules are also found in the circumferential marginal bands which underlie the membrane in nucleated erythrocytes and platelets (124). Since the marginal bands are not found in nonnucleated mammalian erythrocytes which have a different shape, the microtubules presumably play a role in establishing and maintaining cell morphology. MAP-2 and tau are both associated with the marginal bands. One issue of interest is how platelet bands differ from those of the erythrocytes, since morphological changes in platelets are much more pronounced than those required of the erythrocyte.

Observations on a number of cell types suggest that microtubule binding proteins play an important role in morphogenesis. Investigations of their behavior are still in the early stages. Transfection of tau or MAP-2 into fibroblasts changed their microtubule distribution, causing a disappearance of the microtubule organizing center and formation of microtubule bundles (117). Transfection of Sf9 insect cells with tau, MAP-2 and MAP-2C induced formation of neurite-like processes. However, the structural organization of the microtubules was different in the processes induced by MAP-2C from those induced by tau and MAP-2 (125). The involvement of these proteins in functional differentiation of neural cells is also supported by antisense studies in which antisense oligonucleotides for tau (126) and MAP-2 (127) blocked neuritogenesis. However, tau-null transgenic mice are viable, suggesting that other microtubule binding proteins can replace tau for this function (128). An important consideration of the involvement of microtubule binding proteins in neuritogenesis is the mechanism of localization. In situ transcript localization shows that tubulin RNA is confined to cell bodies, but MAP-2 RNA extends into dendrites and tau RNA is present in the proximal portion of the axon (129). Thus, RNA sorting appears to contribute to the localization of specific cytoskeletal proteins involved in neural morphogenesis. Transfection studies have also been used to study the functional domains of microtubule binding proteins. Surprisingly, the domain requirements for bundling of microtubules by tau was found to be cell-type specific. Four-repeat constructs were required in Chinese hamster ovary and 3T3 cells, but three-repeat constructs sufficed in L cells (130).

Three- and four-repeat forms of tau are generated from the tau gene by alternative splicing (116). Six tau isoforms, which differ in the presence of a 29-amino acid N-terminal domain, as well as the number of repeats, are developmentally regulated in the brain. Fetal brain which is undergoing extensive neuritogenesis, preferentially expresses the three-repeat forms, while adult brain, which requires more stable microtubules, has the four-repeat form. Tau splicing is also regulated by thyroid hormone (131). Alternatively spliced MAP-2 isoforms are developmentally regulated. A four-repeat variant of MAP-2C appears late in development and is expressed in glial cells (132).

Regulation of Microtubule Organization

All of the major microtubule binding proteins can be phosphorylated by cytoplasmic protein kinases, including protein kinase A, protein kinase C, calmodulin-dependent protein kinase, casein kinase II, MAP-2 kinase and even tyrosine kinases Src and insulin receptor (108, 116). A number of studies indicate that phosphorylation of some of the proteins reduces their binding and ability to contribute to the assembly of microtubules. Defining these functional modifications is complicated by the number and promiscuity of the cellular enzymes, the multiplicity of potential phosphorylatable sites and whether phosphorylation events are required for functional changes. For example, proline-directed kinases, such as MAPK, will phosphorylate tau and MAPs, but their effects on microtubule stability are relatively benign (133). In contrast, the protein kinase p110[mark] phosphorylates sites in the internal repeats and promotes MAP dissociation and increased dynamic instability. Moreover, specific phosphatases may be important in determining the level and positions of phosphorylation of individual binding proteins (134). Further studies are required to delineate the roles of specific enzymes and phosphorylation sites in particular cellular behaviors.

Other microtubule binding proteins may also contribute to the microtubule organization in different types of cells or cellular functions. An activity catalyzing microtubule severing has been detected in *Xenopus* mitotic extracts (135). Severing microtubules requires the breaking of at least 13 tubulin-tubulin bonds, a more complicated process than microfilament severing (136). A number of studies have suggested different types of interactions of microtubules with membranes. Gephyrin, a protein which copurifies with the glycine receptor and is localized to postsynaptic membranes binds with high affinity to microtubules, suggesting a role in organizing the postsynaptic membrane (116). A family of proteins which link cytoplasmic vesicles to microtubules is related to the *Drosophila* gene *Glued*, which is necessary for viability (137). Dynactin, the protein product of this gene, appears in a complex with cytoplasmic dynein that activates vesicle motility (138) (see chapter 6). Tubulin has also been linked to regulation of G-protein signaling which occurs at membranes (see chapter 2). Specific G-protein subunits can bind tubulin and transfer GTP under conditions where transfer from the medium is impossible (139), possibly linking microtubule dynamics and G-protein cascades. Finally, a class of cytoplasmic linker proteins has been identified which links membranes to microtubules (140). One of their proposed functions is in the organization of cellular organelles by linking them to microtubules.

Microtubule Motor Proteins

Dyneins and kinesins are motor proteins which use the energy from ATP hydrolysis for microtubule-based intracellular movements (32). Structurally, they both contain globular head groups which contain the motor function and tails which provide binding specificity (141). Both have been implicated in the transport of organelles and vesicles along microtubules. Dyneins are found in two forms: cytoplasmic dyneins which are involved in vesicle transport, and axonemal dyneins which drive movements of cilia and flagella (141). Regulation of the latter by cAMP and calcium has been reviewed by Thaler and Haimo (142). Cytoplasmic dyneins are huge molecules containing two heavy chains ($M_r > 400,000$) and a number of accessory chains, which target and regulate their activities (143). They direct particles toward the minus end of microtubules (144, 145). Fractionation and localization studies indicate that cytoplasmic dyneins are localized to a number of cellular structures, including late endosomes, lysosomes and elements of the Golgi apparatus (146). Membrane linkage is achieved through association with a soluble complex termed dynactin (see chapter 6).

Kinesins (147), generally direct particles toward the plus end, but some kinesins move toward the minus end (146). Kinesins are also important in mitosis for chromosome distribution (147). They are composed of two heavy ($\approx$110-130 kDa) and two light ($\approx$60-70 kDa) chains, and are defined by a 350 amino acid conserved sequence in the motor domain. Diversity arises from variations in the tail domains which can provide specificity for moving different types of intracellular particles (141). The tail domain also acts as an intramolecular inhibitor of motor activity whose inhibition is relieved when the kinesin is bound through its tail to "cargo" (146). Kinesin subpopulations have been localized to intracellular organelles such as the Golgi, ER-Golgi intermediate compartment and spindle membranes (146). Membrane association appears to be mediated by the kinesin-binding protein kinectin, whose sequence and behavior suggest that it is an integral membrane protein. However, kinectin is not observed in all systems in which kinesin appears

Table 1.9. Properties and distribution of the major mammalian intermediate filament proteins

IF protein	Sequence type	M_r (x 10^{-3})	Polypeptides	Tissue distribution
Keratin	I	40-56.5	15	Epithelia
Keratin	II	53-67	15	Epithelia
Vimentin	III	57	1	Mesenchymal cells
Desmin	III	53-54	1	Myogenic cells
GFAP	III	50	1	Glial cells and astrocytes
Peripherin	III	57	1	Peripheral neurons
Neurofilament proteins				Neurons of central and peripheral nerves
NF-L	IV	62	1	
NF-M	IV	102	1	
NF-H	IV	110	1	
Lamin proteins				All cell types
Lamin A	V	70	1	
Lamin B	V	67	1	
Lamin C	V	60	1	
Nestin	VI	240	1	Neuronal stem cells

Modified from ref. 156 (Albers K, Fuchs E. Int Rev Cytol 1992; 134: 243-279), courtesy of Academic Press.

to move vesicles and may not be the only component involved in linking kinesins to membranes (148). Interestingly, kinectin may also provide a membrane binding site for dyneins through an association with dynactin (146).

Microtubules and Filament Networks

As mentioned previously, interactions among microfilaments, microtubules and intermediate filaments are involved in the integrated structure of the cytoplasm. Both tau and MAP-2 have been implicated in the interactions between microfilaments and microtubules (116), an association that appears to be regulated by MAP phosphorylation (149). Antimitotic drug and antitubulin antibody disassembly of microtubules results in a collapse of the intermediate filament network of cultured cells (150, 151). Moreover, microinjection of fibroblasts with kinesin causes collapse of the vimentin network without breaking down the microtubular network, suggesting a role for kinesin in the organization and interaction of these cytoskeletal components (142). Consistent with the importance of microtubules in neural processes, biochemical analyses have indicated an association of neurofilaments and microtubules (116).

Intermediate Filaments

Intermediate filaments (IFs) are the most stable of the three filament types and are believed to contribute to the mechanical stability and organization of cells (152). Individual IFs often appear to be attached to the outer nuclear envelope membrane or nuclear pore complexes (153) and extend radially toward the cell surface where they may form associations with the cytoplasmic surfaces of plasma membranes, particularly at sites containing adhesion complexes such as desmosomes or hemidesmosomes (see chapter 3). They are so named because they are intermediate (8-12 nm) in size between microfilaments ($\approx$6 nm) and microtubules ($\approx$25 nm). IFs are the most highly diverse of the major cytoskeletal filaments and appear to be specific to multicellular species, possibly only to animals (154, 155, 156). IFs can be classified into five different structural types according to sequence homologies or tissue specific expression (Table 1.9) (156). The common feature of all intermediate filaments is a central α-helical domain of the IF subunit, composed of heptad repeats which forms a coiled-coil dimer (154). The central core is subdivided into four smaller regions (1A, 1B, 2A, 2B) by short linker sequences which interrupt both the heptad repeats and helix continuity (157). Hydrophobic amino acid residues are commonly found as the first and fourth residues of the heptads, stabilizing the coiled-coil structure. Cytoplasmic IFs contain $\approx$310 amino acids in the helical domain; nuclear lamins have an additional 42 amino acid insert. Regular periodic distributions of charged residues are also found in the central domain, providing electrostatic interactions which can further stabilize filament structures. The central helical domain is flanked by variable amino and carboxy terminal regions which protrude from the surface of the filament, and likely facilitate specific interactions for different filament types.

Types of Intermediate Filaments

In contrast to actin and tubulin which are highly evolutionarily conserved, intermediate filament subunits exhibit cell-type specific and often complex patterns of expression (158). There are about 50 known human IF subunits differentially expressed in different tissues. Keratins, which are primarily found in epithelia, comprise the largest group of these and include as many as 15 forms each of acidic (type I, 40-57 kDa) and basic (type II, 53-67 kDa) subunits (156). Individual members within each of the two classes have >50% sequence identity in their helical domains, while those of the opposite type are only about 30% identical in these domains. Keratins are obligate heteropolymers incorporating acidic and basic subunits in a 1:1 ratio to form filaments (159). Subunit incorporation is promiscuous, but filaments generated from the different keratin subunits have different physical properties, suggesting that IFs from different cell types have evolved to satisfy the particular requirements of those cells for IF function. Thus, one of the important aspects of keratins is the diversity of forms allowing the classification of differentiated epithelia on the basis of their keratin isoforms (160). Furthermore, keratin expression changes with neoplasia, as expected for a differentiation marker.

Type III is a broad group of structurally related IFs which include vimentin, desmin, glial fibrillary acidic protein (GFAP) and peripherin (156). Although they exist largely as homopolymers in specific tissues, individual subunits can form heteropolymers with other members of the class (161, 162) and with NF-L, a type IV neurofilament protein (163), but not with keratins. In cells expressing both type III and keratin subunits, two distinct networks are formed (164). Vimentin is widely expressed in mesenchymal cells and in less well differentiated transformed cell lines and tumors (165). In contrast, desmin is found in muscle, e.g., concentrated at the Z disks. GFAP is expressed in glial cells and astrocytes; peripherin is in the peripheral nervous system localized to neurons of dorsal root ganglion, sympathetic ganglia, cranial nerves and ventral motor neurons (166).

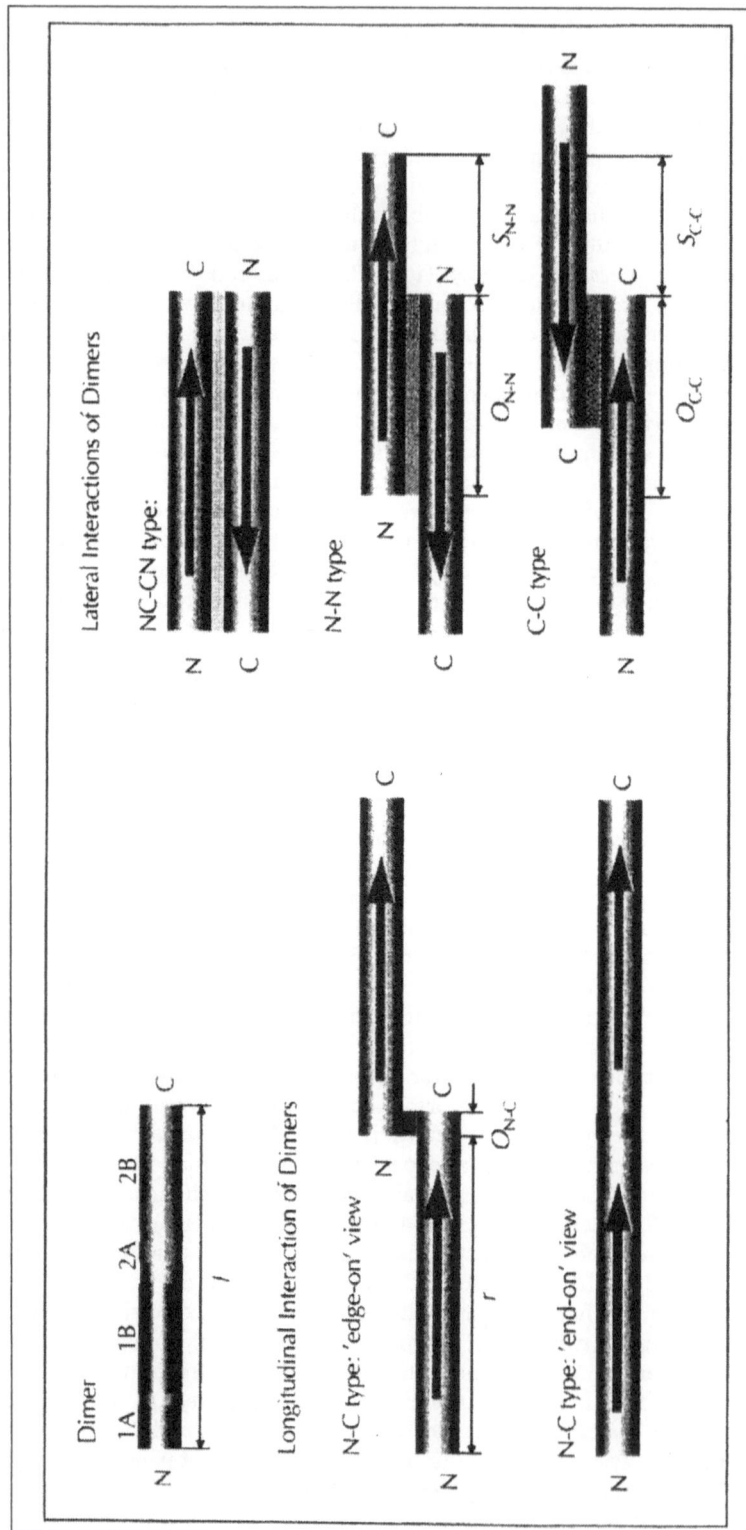

Fig. 1.4. Molecular models for intermediate filament organization. Figure from ref. 174 (Heins S, Aebi U. Curr Opin Cell Biol 1994: 6: 25-33), courtesy of Current Biology Ltd.

Table 1.10. *In vitro disassembly of IF proteins by phosphorylation*

IF	Kinase	Disassembly
Vimentin	A kinase	+
	C kinase	+
	CaMKII	+
	cdc2 kinase	+
	G kinase	-
	Casein kinase I	-
	Casein kinase II	-
GFAP	A kinase	+
	C kinase	+
	CaMKII	+
	cdc2 kinase	-
Desmin	A kinase	+
	C kinase	+
	cdc2 kinase	+
Keratin 8/keratin 18 complex	A kinase	+
	C kinase	+
	CaMKII	+
α-Internexin	A kinase	+
NF-L	A kinase	+
	C kinase	+
	cdc2 kinase	-
Lamin B	cdc2 kinase	-
	C kinase	+/-
	A kinase	-
	MAP kinase	+

Modified from ref. 177 (Inagaki M, Matsuoka Y, Tsujimura, K, Ando S, Tokui T, Takahashi T, Inagaki N. BioEssays 1996; 18: 481-487), courtesy ICSU Press.

Table 1.11. *In vivo regulation of IFs by phosphorylation*

IF	Kinase	Effects
Vimentin	cdc2 kinase	Bundle formation and fragmentation
	A kinase	Bundle formation and fragmentation
	C kinase	Bundle formation and fragmentation
	CaMKII	Bundle formation and fragmentation
Keratin	Unknown	Fragmentation and disappearance
Lamin	cdc2 kinase?	Nuclear lamina disassembly
NF	Unknown	Increase in axon caliber

Modified from ref. 177 (Inagaki M, Matsuoka Y, Tsujimura, K, Ando S, Tokui T, Takahashi T, Inagaki N. BioEssays 1996; 18: 481-487), courtesy of ICSU Press.

The type IV neurofilament proteins NF-L (62 kDa), NF-M (102 kDa) and NF-H (110 kDa) are coexpressed in axons, dendrites and perikarya (167). α-Internexin (66-70 kDa) is also found in neurons and appears to play a role in embryonic development (168). Except for α-internexin, type IV, IFs appear to be obligate heteropolymers in which NF-L forms the filament backbone, while NF-M and NF-H form filament crossbridges (169). Transient transfection studies indicate that the type III protein vimentin can substitute for NF-L in forming a backbone for NF-M (163).

The nuclear lamins (type V) form a meshwork at the inner surface of the nuclear membrane which may interconnect nuclear pore complexes and facilitate chromatin organization (156). Early vertebrate embryos express only B-type lamins. More highly differentiated cells may also express A-type lamins (70 kDa) and its truncated splice variant C-type (60 kDa). All lamins contain nuclear localization signals and CAAX C-terminal sequences specifying isoprenylation and carboxymethylation modifications (170) which facilitate their association with the nuclear membrane. The association of lamins with chromatin suggests a role in DNA replication. Their involvement in mitosis will be discussed in chapter 5.

Two proteins have been described which do not readily fit into the five types: nestin and filensin/phakinin (158). Both have structures intermediate between type III and IV subunits. Nestin has sometimes been listed as a type VI protein (156) and is found in proliferating stem cells of the developing mammalian nervous system (171). Filensin is expressed during differentiation of vertebrate lens epithelia (172). It is unable to polymerize separately, but forms a 1:3 co-polymer with phakinin (173).

Assembly, Disassembly and Turnover

Intermediate filaments are assembled from dimers which are parallel and in-register left-handed coils formed from two right-handed α-helices. Most IF proteins, including keratins can form homodimers, but keratin homodimers do not appear to participate in filament assembly (158). Instead, dimers associate to form stable tetramers. From crosslinking studies, three different types of tetramers have been demonstrated (Fig. 1.4):

1) antiparallel dimers in register.
2) approximately half-staggered, antiparallel dimers with overlapping C-termini.
3) approximately half-staggered, antiparallel dimers with overlapping N-termini (174).

A fourth model for tetramers has short head-to-tail overlaps. Tetramers further assemble into protofibrils and filaments. IFs commonly have about four protofibrils and 32 subunits contributing to the ≈10 nm width. Helical rod ends appear to be important for network formation, but the molecular mechanisms involved in this assembly process are not well understood (158). In contrast to microfilaments and microtubules, IFs exhibit a kind of polymorphism in which the number of dimers per filament cross-section varies between filaments in a cell and within a filament (175). Biophysical measurements indicate that filaments can incorporate and exchange subunits along their entire surface (158, 176). Moreover, dominant negative mutants of IF subunits expressed in cells can disrupt IF networks even though the cellular pools of soluble precursors are very small (159). Thus, intermediate filaments are dynamic but their dynamics differs from that of other cytoskeletal filaments and is not well understood.

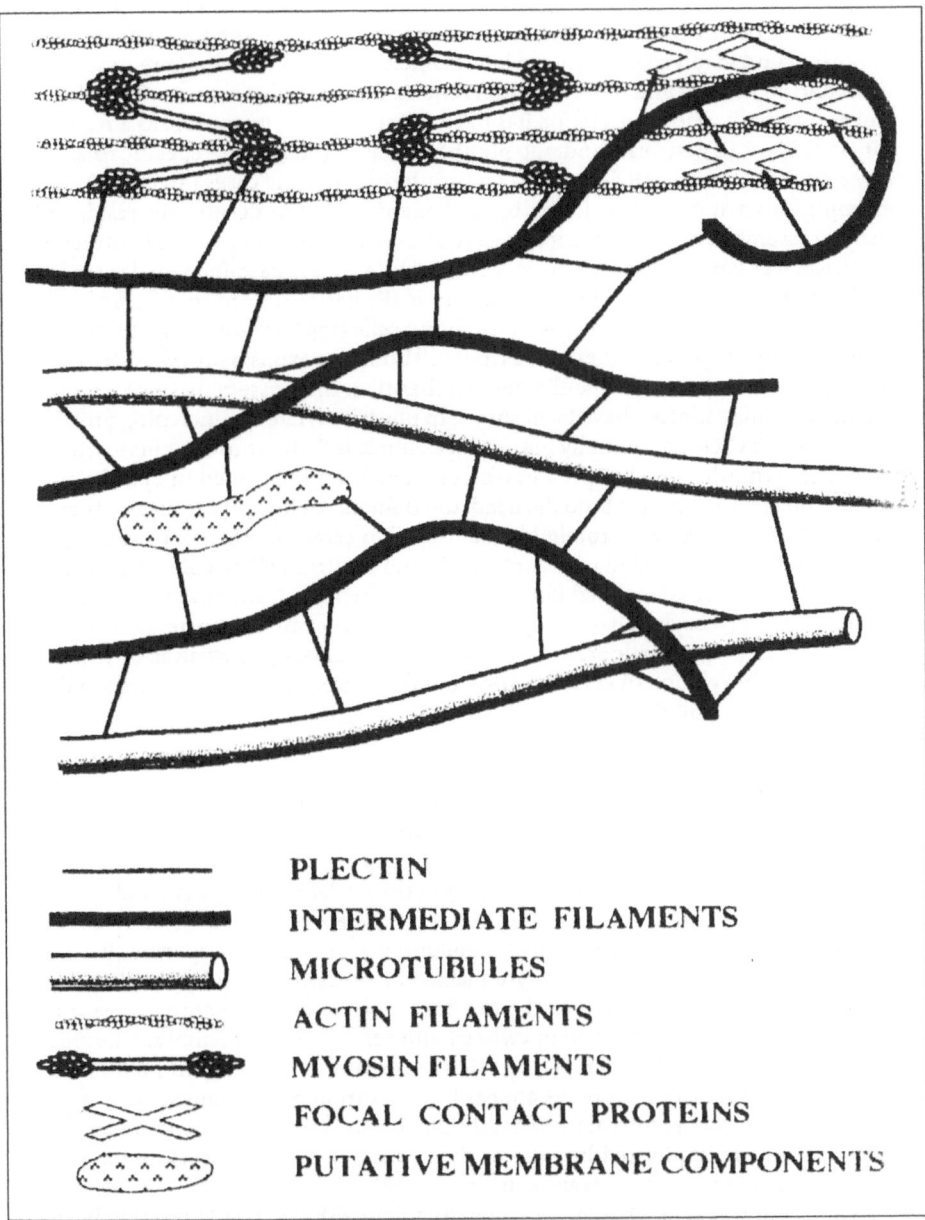

— PLECTIN

INTERMEDIATE FILAMENTS

MICROTUBULES

ACTIN FILAMENTS

MYOSIN FILAMENTS

FOCAL CONTACT PROTEINS

PUTATIVE MEMBRANE COMPONENTS

Fig. 1.5. Model for plectin integration of cytoplasmic filaments. Figure from ref. 193 (Svitkina TM, Verkhovsky AB, Borisy GG. J Cell Biol 96; 135:991-1007). Reproduced with permission from The Journal of Cell Biology, 1996; 135:991-1007, © The Rockefeller Press.

The variable head and tail regions contribute to regulating assembly, the head modulating both end-to-end and lateral interactions, while the tail appears less important and involved primarily in lateral interactions (159). Cellular reorganizations during mitosis and differentiation require disassembly and assembly of IFs. Not surprisingly, phosphorylation has been found to play a major role as a regulator (177). Most IF networks undergo phosphorylation (Table 1.10); specific roles of these modifications in IF dynamics can be inferred in some cases (Table 1.11). For example, in vitro phosphorylation of the head domains of desmin IFs results in their disassembly; assembly competence can be restored by phosphatase treatments (178). Nuclear lamina disassembly is regulated by a cdc2 kinase-mediated cascade, as described in chapter 5. Phosphorylation of IFs has also been implicated in IF breakdown at the cleavage furrow of dividing cells (179). Neurofilament tails are heavily phosphorylated as they move from cell bodies to axons, a process that probably affects interactions of the filaments rather than their assembly (180). Several serine/threonine kinases have been shown to phosphorylate IFs, including protein kinases A and C. One potentially important example is PKN, a protein kinase which associates with the small G-protein Rho and has been implicated in cytoskeletal regulation (181). It can also bind the head-rod domains of both neurofilaments and vimentin. Interestingly, a protein kinase C-epsilon catalytic fragment is reported to be specifically associated with keratins K8 and K18 from HT29 colon carcinoma cells (182). Other IF-associated kinases have been reported, but have not been defined (159). O-glycosylation of IFs by cytoplasmic glucosaminyl transferase has also been reported (174). Mitotic arrest of HT29 cells increases O-glycosylation on keratin 18 and phosphorylation of keratin 8 (183), but the role of these modifications in the cell cycle, if any, is ill defined.

Associations

The tensegrity model of cell structure suggests that IFs serve as a link between the plasma membrane and nucleus (184). Cytoplasmic IFs bind nuclear membranes, apparently through their tail domains (156). It has been proposed that initiation of filament assembly may occur preferentially at the nuclear envelope (185), although microinjection experiments have shown that nuclei are not required for assembly (158). Associations of IFs with plasma membranes are cell type specific. For example, vimentin binds to erythrocyte plasma membranes via ankyrin (186). The most familiar IF-plasma membrane interaction sites, desmosomes and hemidesmosomes, are involved in cell-cell and cell-basement membrane interactions, respectively (187) (chapter 3). In these structures, observed with both keratin and type III IFs, the filaments appear to loop through the membrane plaques. The molecular associations of the IFs at the plaques are unknown, but desmoplakin-like proteins are implicated because of their structures and co-recruitment with IFs to artificial junctions in transfected cells (188).

Other intermediate filament-associated proteins (IFAPs) (189) appear to be involved in the linkage of IFs to membranes, the most thoroughly studied being *plectin* (190). Plectin is a 300 kDa IFAP found in the cytoplasm of a wide range of cell types where it is partially co-distributed with IFs. It is also found at cell junctions and membrane attachment sites of IFs and microfilaments. Plectin binds several IF subunits including the neurofilament proteins, GFAP and some keratins, via the IF rod domain. Its ability to self-associate and to associate with spectrin, MAPs and lamin B suggest that it may function as a general linker among the cytoskeletal systems and also link IFs to membranes (189). A C-terminal homology to desmoplakin supports the latter proposal, and plectin has been localized to desmosomes and hemidesmosomes (191). Plectin binding to IFs may be modulated by phosphorylation since its association with lamin B and vimentin is decreased after PKA or PKC treatments (176, 189). Immunoelectron microscopic observations of glioma cell extracts suggest that plectin links IFs and microfilaments (192). Plectin binding to microtubules can be observed in cells from transgenic mice lacking IFs and was reversed by injection of exogenous vimentin (193). A model for the plectin-mediated integration of the cytoplasmic skeleton is presented in Figure 1.5.

A neural form of bullous pemphigoid antigen-1 has been implicated in a functional IF-microfilament interaction (194). The neural, but not epidermal isoform contains an actin binding domain, while both forms bind IFs. Transfection experiments indicate that it can induce co-alignment of actin and neurofilament networks. Moreover, BPAG1 null mice exhibit abnormal axonal and neurofilament architecture, suggesting that the NF-microfilament interactions stabilized by this crosslink are important for neuron structure. Another potential crosslinking protein is the 14-3-3 protein described in chapter 2, which is known to bind numerous signaling proteins and interacts with hyperphosphorylated keratin K18 during mitosis (176). How this protein may contribute to the dramatic changes of the IF network during mitosis needs to be clarified.

Another potentially important aspect of IF dynamics is the possible susceptibility of the IF networks to disrupting proteins. Three viral proteins have been shown to bind and disrupt IF networks: EBV-LMP (195), adenovirus E1B (196) and human papilloma HPV-16 E1-E4. Adenovirus E1B protein disrupts both cytoplasmic vimentin and nuclear lamin networks (196). In contrast, expression of E1-E4 in transfected cells causes the collapse of keratin networks without disrupting vimentin, microtubule or actin filaments (197). *Xenopus* oocytes undergoing meiotic maturation contain a keratin filament severing activity that temporally correlates with preferential phosphorylation of type II keratins (198). The mechanism of these severing activities is of considerable interest in understanding IF dynamics.

Functions

A major question in all studies of IFs is how their properties contribute to functions in cells. However, defining functions for the various IF types has proved difficult. Cytoplasmic IFs are not necessary for cell viability in vitro since several cultured cell lines do not have them (199, 200). Furthermore, knock-out mice lacking vimentin demonstrated no obvious phenotypic changes (201). The most widely accepted function for cytoplasmic IFs is that they contribute to mechanical stability of the cells. This function is exemplified by studies on transgenic mice and from human genetic diseases in which cells lacking an intact IF network become susceptible to mechanical stress (158, 159). Additional evidence for a mechanical function was obtained by a microinjection of peptides from helix initiation domain 1A into live fibroblasts (202). Rapid disassembly of the cellular IF network was observed with concomitant cell shape alterations and destabilization of microtubule networks and stress fibers. These studies emphasize not only the role of IFs in cell morphology, but their dynamics and associations with microtubule and microfilament organization. Both cytokeratins and vimentin have been implicated in the migration and invasiveness of tumor cells (203), though how they contribute is uncertain. Neurofilaments which are preferentially localized in axons may play a similar stabilizing role (173) and are proposed to determine the caliber of the axon (204). Given the extreme diversity of gene products, it seems likely that the individual IFs also have more specific functions in particular cells or tissues. Most of those functions remain to be defined. However, an intriguing possibility is that IFs play a role in gene regulation (205). Antisense inhibition of desmin expression inhibits myogenic differentiation, consistent with a regulatory role (206), but the mechanism of this effect remains to be defined. Recently, intermediate filaments have been implicated in the mechanical connections between integrins on the cell surface and the nucleus, suggesting a direct role for cell surface mechanical perturbations in the regulation of gene expression via the intermediate filaments (207).

Summary

The cytoskeleton is composed of three types of filaments: microfilaments, microtubules and intermediate filaments which form the skeletal framework of the cell. Together with cell-matrix and cell-cell adhesion sites they establish the morphology of cells which do not have defined cell walls. Microfilaments are particularly important to the organization and dynamics of structures at the cell periphery, such as filopodia, lamellae, microvilli, ruffles and cell adhesion sites. The contractility of actomyosin provides the motor which drives cellular movements such as cell motility and cytokinesis. Microtubules are highly rigid but dynamic structures. As such, they play an important role in establishing and maintaining the organization of the cell and its internal cell structures such as the Golgi apparatus. They also provide structural elements for the transport of membrane vesicles within the cell chromosome movements during mitosis and flagellar-driven cell motility, all powered by the molecular motors dynein and kinesin. Intermediate filaments are the least dynamic of the filament types and provide mechanical stability to the cell structure and connections between the nucleus and plasma membrane. They may also play a role in expression of some genes. Although each filament type can be studied in isolation, the interactions among them involving filament binding proteins are important to both cell structure and dynamics. Only in the context of the intact cell can one understand the importance of the cytoskeleton to cell behavior and its roles in, and responses to cell signaling.

References

1. Edwards SW. Biochemistry and Physiology of the Neutrophil. Cambridge, UK: Cambridge University Press.
2. Drubin DG, Nelson WJ. Origins of cell polarity. Cell 1996; 84:335-344.
3. Ingber DE, Dike L, Hansen L, Karp S, Liley H, Maniotis A, McNamee H, Mooney D, Plopper G, Sims S, Wang N. Cellular tensegrity: exploring how mechanical changes in the cytoskeleton regulate cell growth, migration, and tissue pattern during morphogenesis. Int Rev Cytol 1994; 150:173-224.
4. Ben-Ze'ev A. Animal cell shape changes and gene expression. BioEssays 1991; 13: 207-212.
5. Bretscher A. Microfilament structure and function in the cortical cytoskeleton. Annu Rev Cell Biol 1991; 7:337-374.
6. Ampe C, Vanderkerckhove J. Actin-actin binding protein interfaces. Sem Cell Biol 1994; 5:175-182.
7. Kabsch W, Holmes KC. The actin fold. FASEB J 1995; 9:167-174.
8. Pollard TD, Cooper JA. Actin and actin-binding proteins: a critical evaluation of mechanisms and functions. Ann Rev Biochem 1986; 55:987-1035.
9. Theriot JA. Regulation of the actin cytoskeleton in living cells. Sem Cell Biol 1994; 5:193-199.

10. Fecheimer M, Zigmond SH. Focusing on unpolymerized actin. J Cell Biol 1993; 123:1-5.
11. Sun H-Q, Kwiatkowska K, Yin HL. Actin monomer binding proteins. Curr Opin Cell Biol 1995; 7:102-110.
12. Nachmias VT. Small actin-binding proteins: the β-thymosin family. Curr Opin Cell Biol 1993; 5:56-62.
13. Carlier M, Didry D, Erk I, Lepault J, van Troys ML, Vandekerckhove J, Perelroizen I, Yin H, Doi Y, Pantaloni D. Tβ$_4$ is not a simple G-actin sequestering protein and interacts with F-actin at high concentration. J Biol Chem 1996; 271: 9231-9239.
14. Sun H-Q, Kwiatkowska K, Yin HL. β-thymosins are not simple actin monomer buffering proteins: insights from overexpression studies. J Biol Chem 1996; 271:9223-9230.
15. Carlier MF, Pantaloni D. Actin assembly in response to extracellular signals: role of capping proteins, thymosin beta 4 and profilin. Sem Cell Biol 1994; 5:183-191.
16. Schafer DA, Cooper JA. Control of actin assembly at filament ends. Annu Rev Cell Dev Biol 1995; 11:497-518.
17. Weeds A, Maciver S. F-actin capping proteins. Curr Opin Cell Biol 1993; 5: 63-69.
18. Hartwig JH. Mechanisms of actin rearrangements mediating platelet activation. J Cell Biol 1992; 118:1421-1442.
19. Moon A, Drubin DG. The ADF/cofilin proteins: stimulus responsive modulators of actin dynamics. Mol Biol Cell 1995; 6:1423-1431.
20. Welch MD, Nallavarapu A, Rosenblatt J, Mitchison TJ. Actin dynamics in vivo. Curr Opin Cell Biol 1997; 9:54-61.
21. Cao L-g, Fishkind DJ, Wang Y-l. Localization and dynamics of nonfilamentous actin in cultured cells. J Cell Biol 1993; 123:173-181.
22. Theriot JA. The cell biology of infection by intracellular bacterial pathogens. Annu Rev Cell Dev Biol 1995; 11: 213-239.
23. Lasa I, Cossart P. Actin-based bacterial motility: towards a definition of the minimal requirements. Trends Cell Biol 1996; 6:109-114.
24. Peskin CS, Odell GM, Oster GF. Cellular motions and thermal fluctuations-the Brownian rachet. Biophys J 1993; 65: 316-324.
25. Craig SW, Pollard TD. Actin-binding proteins. Trends Biochem Sci 1982; 7: 88-92.
26. Lees-Miller JP, Helfman DM. The molecular basis for tropomyosin isoform diversity. BioEssays 1991; 13:429-437.
27. Matsudaira P. Actin crosslinking proteins at the leading edge. Sem Cell Biol 1994; 5:165-174.
28. Hartwig JH, Kwiatkowski DJ. Actin-binding proteins. Curr Opin Cell Biol 1991; 3:87-97.
29. Stossel TP. From signal to pseudopod: how cells control cytoplasmic actin assembly. J Biol Chem 1989; 264:18261-18264.
30. Hartwig JH, Shevin P. The architecture of actin filaments and the ultrastructural location of actin-binding protein in the periphery of lung macrophages. J Cell Biol 1986; 103:1007-1020.
31. Theriot JA, Mitchison TJ. Actin microfilament dynamics in locomoting cells. Nature 1991; 352:126-131.
32. Vale RD. Switches, latches, and amplifiers: common themes of G proteins and molecular motors. J Cell Biol 1996; 135: 291-302.
33. Mooseker MS, Cheney RE. Unconventional myosins. Annu. Rev. Cell Dev Biol 1995; 11:633-675.
34. Titus MA. Unconventional myosins: new frontiers in actin-based motors. Trends Cell Biol 1997; 7:119-123.
35. Maciver SK. Myosin II function in non-muscle cells. BioEssays 1996; 18:179-182.

36. Ostap EM, Pollard TD. Overlapping functions of myosin-I isoforms? J Cell Biol 1996; 133:221-224.

37. Brzeska H, Korn ED. Regulation of class I and class II myosins by heavy chain phosphorylation. J Biol Chem 1996; 271:16983-16986.

38. Brown SS. Myosins in yeast. Curr Opin Cell Biol 1997; 9:44-48.

39. Allred LE, Porter KR. Morphology of normal and transformed cells. In: Hynes RO, ed. Surfaces of normal and malignant cells. New York: John Wiley and Sons, 1979; 21-61.

40. Condeelis J. Life at the leading edge: the formation of cell protrusions. Annu Rev Cell Biol 1993; 9:411-444.

41. Mitchison TJ, Cramer LP. Actin-based cell motility and cell locomotion. Cell 1996; 84:371-379.

42. Carraway KL, Carraway CAC. Membrane-cytoskeleton interactions in animal cells. Biochim Biophys Acta 1989; 988: 147-171.

43. Heintzelman MB, Mooseker MS. Assembly of the intestinal brush border. Curr. Topics Devel Biol 1992; 26:93-122.

44. Carraway KL, Huggins JW, Cerra RF, Yeltman DR, Carraway CAC. α-Actinin-containing branched microvilli isolated from an ascites adenocarcinoma. Nature 1980; 285:508-510.

45. Tilney LG, Tilney MS, DeRosier DJ. Actin filaments, stereocilia, and hair cells: how cells count and measure. Annu. Rev Cell Biol 1992; 8:257-274.

46. Ridley AJ. Membrane ruffling and signal transduction. BioEssays 1994; 16:321-327.

47. Pumplin DW, Bloch RJ. The membrane skeleton. Trends Cell Biol 1993; 3: 113-117.

48. Yu J, Fischman DA, Steck TL. Selective solubilization of proteins and phospholipids from red blood cell membranes by nonionic detergents. J Supramol Struct 1973; 1:233-248.

49. Bennett V, Gilligan DM. The spectrin-based membrane skeleton and micron-scale organization of the plasma membrane. Annu Rev Cell Biol 1993; 9:27-66.

50. Luna EJ, Hitt A. Cytoskeleton-plasma membrane interactions. Science 1992; 258:955-964.

51. Palek J, Sahr KE. Mutations of the red blood cell membrane proteins: from clinical evaluation to detection of the underlying genetic defect. Blood 1992; 80:308-330.

52. Frojmovic MM, Milton JG. Human platelet size, shape, and related functions in health and disease. Physiol Rev 1982; 62:185-261.

53. Fox JEB. The platelet cytoskeleton. Thromb Haemostasis 1993; 70:884-893.

54. Fox JEB, Boyles JK, Berndt MC, Steffen PK, Anderson LK. Identification of a membrane skeleton in platelets. J Cell Biol 1988; 106:1525-1538.

55. Bearer EL. Platelet membrane skeleton revealed by quick-freeze deep-etch. Anat Rec 1990; 227:1-11.

56. Hartwig JH, DeSisto M. The cytoskeleton of the resting human blood platelet: structure of the membrane skeleton and its attachment to actin filaments. J Cell Biol 1991; 112:407-425.

57. Fox JEB, Lipfert L, Clark EA, Reynolds CC, Austin CD, Brugge JS. On the role of the membrane skeleton in mediating signal transduction: association of GP IIb-IIIa, pp60[src], pp62[yes], and the p21[ras] GTPase-activating protein (GAP) with the membrane skeleton. J Biol Chem 1993; 268:25973-25984.

58. Froehner SC. Regulation of ion channel distribution at synapses. Annu Rev Neurosci 1993; 16:347-368.

59. Kusumi A, Sako Y. Cell surface organization by the membrane skeleton. Curr Opin Cell Biol 1996; 8:566-574.

60. Carraway KL, Cerra RF, Jung G, Carraway CAC. Membrane-associated actin from microvillar membranes of ascites tumor cells. J Cell Biol 1982; 94: 624-630.

61. Carraway CAC, Fang H, Ye X, Juang S-H, Liu Y, Carvajal M, Carraway KL. Membrane-microfilament interactions in ascites tumor cell microvilli. Identification and isolation of a large microfilament-associated glycoprotein complex. J Biol Chem 1991; 266:16238-16246.

62. Abelda SM, Buck CA. Integrins and other cell adhesion molecules. FASEB J 1990; 4:2868-2880.

63. Ruoslahti E, Obrink B. Common principles in cell adhesion. Exp Cell Res 1996; 227:1-11.

64. Hynes RO. Integrins: versatility, modulation, and signaling in cell adhesion. Cell 1992; 69:11-25.

65. Takeichi M. Cadherin cell adhesion receptors as a morphogenetic regulator. Science 1991; 251:1451-1455.

66. Heiska L, Kantor C, Parr T, Critchley DR, Vilja P, Gahmberg CG, Carpen O. Binding of the cytoplasmic domain of intercellular adhesion molecule-2 (ICAM-2) to α-actinin. J Biol Chem 1996; 271:26214-26219.

67. Carraway CAC, Carraway, KL. In: Hesketh HE, Pryme IF, eds. Treatise on the Cytoskeleton, Greenwich, CT: JAI Press, 1996:207-238.

68. Janmey PA. Phosphoinositides and calcium as regulators of cellular actin asembly and disassembly. Annu Rev Physiol 1994; 56:169-191.

69. Lamb JA, Allen PG, Tuan BY, Janmey PA. Modulation of gelsolin function: activation at low pH overides Ca^{2+} requirement. J Biol Chem 1993; 268:8999-9004.

70. Isenberg G. Actin-binding protein-lipid interactions. Cell Motil Cytoskel 1991; 12:136-144.

71. Lassing I, Lindberg U. Specific interaction between phosphatidylinositol 4,5-bisphosphate and profilin. Nature 1985; 314:472-474.

72. Goldschmidt-Clermont PJ, Machesky LM, Baldassare JJ, Pollard TD. The actin binding protein profilin binds to PIP_2 and inhibits its hydrolysis by phospholipase C. Science 1990; 247:1575-1578.

73. Goldschmidt-Clermont PJ, Janmey PA. Profilin, a weak CAP for actin and RAS. Cell 1991; 66:419-421.

74. Vojtek A, Haarer B, Field J, Gerst J, Pollard TD, Brown S, Wigler M. Evidence for a functional link between profilin and CAP in the yeast *S. cerevisiae*. Cell 1991; 66:497-505.

75. Machasky LM, Goldschmidt-Clermont PJ, Pollard TD. The affinity of human platelet and *Acanthamoeba* profilin isoforms for polyphosphoinositides account for the relative abilities to inhibit phospholipase C. Cell Reg 1991; 1:937-950.

76. Gottwald U, Brokamp R, Karakesisoglou I, Schleicher M, Noegel AA. Identification of acylase-associated protein (CAP) homologue in *Dictyostelium discoideum* and characterization of its interaction with actin. Mol Biol Cell 1996; 7:261-272.

77. Gieselmann R, Mann K. ASP-56, a new actin-sequestering protein from pig platelets with homology to CAP- an adenylate cyclase-associated protein from yeast. FEBS Let 1992; 298:149-153.

78. Janmey PA, Stossel TP. Modulation of gelsolin function by phosphatidylinositol 4,5-bisphosphate. Nature 1987; 325: 362-364.

79. Yonezawa N, Nishida E, Iida K, Yahara I, Sakai H. Inhibition of the interactions of cofilin, destrin, and deoxyribonuclease I with actin by phosphoinositides. J Biol 1990; 265:8382-8386.

80. Heiss SG, Cooper JA Regulation of CapZ, an actin capping protein of chicken muscle, by anionic phospholipids. Biochemistry 1991; 30:8753-8758.

81. Fukumi K, Furuhashi K, Inagaki M, Endo T, Hatano S, Takenawa T. Requirement of phosphatidylinositol 4,5-bisphosphate for α-actinin function. Nature 1992; 359:150-152.

82. Furuhashi K, Inagaki M, Hatano S, Fukami K, Takenawa T. Inositol phospholipid-induced suppression of F-actin-gelating activity of smooth muscle filamin. Biochem Biophys Res Commun 1992; 184:1261-1265.

83. Janmey PA, Stossel TP. Gelsolin-polyphosphoinositide interaction. Full expression of gelsolin-inhibiting function by polyphosphoinositides in vesicular form and inactivation by dilution, aggregation, or masking of the inositol head group. J Biol Chem 1989; 264:4825-4831.

84. Shariff A, Luna EJ. Diacylglycerol-stimulated formation of actin nucleation sites at plasma membranes. Science 1992; 256: 245-247.

85. Hall A. Small GTP-binding proteins and the regulation of the cytoskeleton. Annu Rev Cell Biol 1994; 10:31-54.

86. Nobes CD, Hall A. Rho, Rac, and Cdc42 GTPases regulate the assembly of multimolecular focal complexes associated with actin stress fibers, lamellipodia, and filopodia. Cell 1995; 81:53-62.

87. Ridley AJ, Hall A. The small GTP-binding protein rho regulates the assembly of focal adhesions and actin stress fibers in response to growth factors. Cell 1992; 70: 389-399.

88. Ridley AJ, Paterson HF, Johnston CL, Diekmann D, Hall A. The small GTP-binding protein rac regulates growth factor-induced membrane ruffling. Cell 1992; 70:401-410.

89. Chant J, Stowers L. GTPase cascades choreographing cellular behavior: movement, morphogenesis, and more. Cell 1995; 81:1-4.

90. Ridley AJ, Hall A. Signal transduction pathways regulating rho-mediated stress fiber formation: requirement for a tyrosine kinase. EMBO J 1994; 13:2600-2610.

91. Chong LD, Traynor-Kaplan A, Bokoch GM, Schwartz MA. The small GTP-binding protein rho regulates a phosphatidylinositol 4-phosphate 5-kinase in mammalian cells. Cell 1994; 79:507-513.

92. Manser E, Leung T, Salihuddin H, Zhao Z, Lim L. A brain serine/threonine protein kinase activated by cdc42 and rac1. Nature 1994; 367:40-46.

93. Lamarche N, Tapon N, Stowers L, Burbelo PD, Aspenstrom P, Bridges T, Chant J, Hall A. Rac and cdc42 induce actin polymerization and G1 cell cycle progression independently of p65PAK and the JNK/SAPK MAP kinase cascade. Cell 1996; 87:519-529.

94. Nagaoka R, Abe H, Obinata T. Site-directed mutagenesis of the phosphorylation site of cofilin: its role in cofilin-actin interaction and cytoplasmic localization. Cell Motil Cytoskel 1996; 35: 200-209.

95. Davidson MM, Haslam RJ. Dephosphorylation of cofilin in stimulated platelets: roles for a GTP-binding protein and Ca^{2+}. Biochem J 1994; 301:41-47.

96. Aderem A. Signal transduction and the actin cytoskeleton: the roles of MARCKS and profilin. Trends Biochem Sci 1992; 17:438-442.

97. Wu H, Parsons JT. Cortactin, an 80/85-kilodalton pp60^{src} substrate, is a filamentous actin-binding protein enriched in the cell cortex. J Cell Biol 1993; 120: 1417-1426.

98. Chang J-H, Sill S, Settleman J, Parsons SJ. c-Src regulates the simultaneous rearrangement of actin cytoskeleton, p190RhoGAP, and p120RasGAP following epidermal growth factor stimulation. J Cell Biol 1995; 130:355-368.

99. Mitchison TJ. Compare and contrast actin filaments and microtubules. Mol Biol Cell 1992; 3:1309-1315.

100. Ingber DE. Cellular tensegrity: defining new rules of biological design that govern the cytoskeleton. J Cell Science 1993; 104:613-627.

101. Karsenti E, Maro B. Centrosomes and the spatial distribution of microtubules in animal cells. Trends Biochem Sci 1986; 11:460-463.

102. Terasaki M. Recent progress on structural interactions of the endoplasmic reticulum. Cell Motil Cytoskel 1990; 15: 71-75.
103. Kreis TE. Role of microtubules in the organization of the Golgi apparatus. Cell Motil Cytoskel 1990; 15:67-70.
104. Vale RD. Intracellular transport using microtubule-based motors. Ann Rev Cell Biol 1987; 3:347-378.
105. McNiven MA, Porter KR. Organization of microtubules in centrosome-free cytoplasm. J Cell Biol 1988; 106:1593-1605.
106. Gibbons IR. Cilia and flagella of eukarotes. J Cell Biol 1981; 91:107s-124s.
107. Gundersen GG, Bulinski JC. Selective stabilization of microtubules toward the direction of cell migration. Proc Natl Acad Sci USA 1988; 85:5946-5950.
108. Gelfand VI, Bershadsky AD. Microtubule dynamics: mechanism, regulation, and function. Annu Rev Cell Biol 1991; 7:93-116.
109. Kirschner MW, Mitchison TJ. Beyond self assembly: from microtubules to morphogenesis. Cell 1986; 45:329-342.
110. Walker RA, O'Brien ET, Pryer NK, Sobeiro MF, Voter WA, Erickson HP, Salmon ED. Dynamic instability of individual, MAP-free microtubules analyzed by video light microscopy: rate constants and transition frequencies. J Cell Biol 1988; 107:1437-1448.
111. Carlier M, Pantaloni D. Kinetic analysis of guanosine 5'-triphosphate hydrolysis associated with tubulin polymerization. Biochemistry 1981; 20:1918-1924.
112. Simon JR, Parsons SF, Salmon ED. Buffer conditions and non-tubulin factors critically affect the microtubule dynamic instability of sea urchin egg tubulin. Cell Motil Cytoskel 1992; 21:1-14.
113. Avila J. Microtubule functions. Life Sci 1992; 50:327-334.
114. Edde B, Rossier J, Le Caer J, Desbruyeres E, Gros F, Denoulet P. Posttranslational glutamylation of α-tubulin. Science 1990; 247:83-85.
115. Boucher D, Larcher JC, Gros F, Denoulet P. Polyglutamylation of tubulin as a progressive regulator of in vitro interactions between the microtubule-associated protein tau and tubulin. Biochemistry 1994; 33:12471-12477.
116. Maccioni RB, Cambiazo V. Role of microtubule-associated proteins in the control of microtubule assembly. Physiol Rev 1995; 75:835-864.
117. Hirokawa N. Microtubule organization and dynamics dependent on microtubule-associated proteins. Curr Opin Cell Biol 1994; 6:74-81.
118. Mandelkow E, Mandelkow E-M. Microtubules and microtubule-associated proteins. Curr Opin Cell Biol 1995; 7:72-81.
119. Joshi HC. γ-Tubulin: the hub of cellular microtubule assemblies. BioEssays 1994; 15:637-643.
120. Zheng Y, Jung M, Oakley BR. γ-Tubulin is present in Drosophila melanogaster and Homo sapiens and is associated with the centrosome. Cell 1991; 65:817-823.
121. Horio T, Oakley BR. Human γ-tubulin functions in fission yeast. J Biol Chem 1994; 269:1465-1473.
122. Raff JW. Centrosomes and microtubules: wedded with a ring. Trends Cell Biol 1996; 6:248-251.
123. Schoenfeld TA, Obar RA. Diverse distribution and function of fibrous microtubule-associated proteins in the nervous system. Int Rev Cytol 1994; 151:67-137.
124. MacRae TH. Microtubule organization by cross-linking and bundling proteins. Biochim Biochim Acta 1992; 1160: 145-155.
125. Chen J, Kanai Y, Cowan NJ, Hirokawa N. Projection domains of MAP2 and tau determine spacings between microtubules in dendrites and axons. Nature 1992; 360:674-677.

126. Shea TB, Beermann ML, Nixon RA, Fischer I. Microtubule-associated protein tau is required for axonal neurite elaboration by neuroblastoma cells. J Neurosci Res 1992; 43:363-374.
127. Caceres A, Mautino J, Kosik K. Suppression of MAP2 in cultured cerebelar macroneurons inhibits minor neurite formation. Neuron 1992; 9:607-618.
128. Harada A, Oguchi K, Okabe S, Kuna J, Terada S, Ohshima T, Sato-Yoshitake R, Takei Y, Noda T, Hirokawa N. Altered microtubule organization in small-caliber axons of mice lacking tau protein. Nature 1994; 369:488-491.
129. Litman P, Barg J, Rindzooski L, Ginzburg I. Subcellular localization of tau mRNA in differentiating neuronal cell culture: implications for neuronal polarity. Neuron 1993; 10:627-638.
130. Kanai Y, Chen J, Hirokawa N. Microtubule bundling by tau proteins in vitro: analysis of functional domains. EMBO J 1992; 11:3953-3961.
131. Aniello FD, Couchie A, Bridoux A, Gripois D, Nunez J. The splicing of juvenile and adult tau mRNA variant is regulated by thyroid hormone. Proc Natl Acad Sci USA 1991; 88:4035-4038.
132. Doll T, Meichsner M, Riederer BM, Honegger P, Matus A. An isoform of microtubule-associated protein 2 (MAP2) containing four repeats of the tubulin binding motif. J Cell Sci 1993; 106:633-640.
133. Illenberger S, Drewes G, Trinczek B, Biernat J, Meyer HE, Olmsted JB, Mandelkow E-M, Mandelkow E. Phosphorylation of microtubule-associated proteins MAP2 and MAP4 by the protein kinase $p110^{mark}$. Phosphorylation sites and regulation of microtubule dynamics. J Biol Chem 1996; 271:10834-10843.
134. Patterson CL Jr, Flavin M. A brain phosphatase with specificity for microtubule-associated protein-2. J Biol Chem 1986; 261:7791-7796.
135. Vale RD. Severing of stable microtubules by a mitotically activated protein in Xenopus egg extracts. Cell 1991; 64: 827-839.
136. Caplow M. Microtubule dynamics. Curr Opin Cell Biol 1992; 4:58-65.
137. Harte PJ, Kankel DR. Genetic analysis of mutations at the Glued locus and interacting loci in Drosophila melanogaster. Genetics 1982; 101:477-501.
138. Gill SR, Schroer TA, Szilak I, Steuer ER, Sheetz MP, Cleveland DW. Dynactin, a conserved, ubiquitously expressed component of an activator of vesicle motility mediated by cytoplasmic dynein. J Cell Biol 1991; 115:1639-1650.
139. Rasenick MM, Caron MG, Dolphin AC, Kobilka BK, Schultz G. in Pharmacological Sciences: Perspectives for Research and Therapy in the Late 1990s. Cuello AC, Collier B, eds., Burkhauser Verlag, Basel, Switzerland, 1995; 91-103.
140. Rickard JE, Kreis TE. CLIPs for organelle-microtubule interactions. Trends Cell Biol 1996; 6:178-183.
141. Skoufias DA, Scholey JM. Cytoplasmic microtubule-based motor proteins. Curr Opin Cell Biol 1993; 5:95-104.
142. Thaler CD, Haimo LT. Microtubules and microtubule motors: mechanisms of regulation. Int Rev Cytol 1996; 164:269-327.
143. Barton NR, Goldstein LS. Going mobile: microtubule motors and chromosome segregation. Proc Natl Acad Sci USA 1996; 93:1735-1742.
144. Bloom GS. Motor proteins for cytoplasmic microtubules. Curr Opin Cell Biol 1992; 4:66-73.
145. Endow SA, Titus, MA. Genetic approaches to molecular motors. Annu Rev Cell Biol 1992; 8:29-66.
146. Vallee RB, Sheetz MP. Targeting of motor proteins. Science 1996; 271:1539-1544.
147. Moore JD, Endow SA. Kinesin proteins: a phylum of motors for microtubule-based motility. BioEssays 1996; 18:207-219.
148. Brady ST. A kinesin medley: biochemical and functional heterogeneity. Trends Cell Biol 1995; 5:159-164.

149. Selden SC, Pollard TD. Phosphorylation of microtubule-associated proteins regulates their interaction with actin filaments. J Biol Chem 1983; 258:7064-7071.

150. Olmsted JB. Microtubule-associated proteins. Annu Rev Cell Biol 1986; 2: 421-458.

151. Blose SH, Melttzer D, Feramisco J. 10 nm intermediate filaments induced to collapse in living cells microinjected with monoclonal and polyclonal antibodies against tubulin. J Cell 1983; 96:847-858.

152. Lazarides E. Intermediate filaments as mechanical integrators of cellular space. Nature 1980; 283:249-256.

153. Skalli O, Goldman RD. Recent insights into the assembly, dynamics, and function of intermediate filament networks. Cell Motil Cytoskel 1991; 19:67-69.

154. Steinert PM, Parry DAD. Intermediate filaments: conformity and diversity of expression and structure. Annu Rev Cell Biol 1985; 1:41-65.

155. Steinert PM, Roop DR. Molecular and cellular biology of intermediate filaments. Annu Rev Biochem 1988; 57:593-625.

156. Albers K, Fuchs E. The molecular biology of intermediate filament proteins. Int Rev Cytol 1992; 134:243-279.

157. Stewart M. Intermediate filament structure and assembly. Curr Opin Cell Biol 1993; 5:3-11.

158. Fuchs E, Weber K. Intermediate filaments: structure, dynamics, function, and disease. Annu Rev Biochem 1994; 63: 345-382.

159. Coulombe PA. The cellular and molecular biology of keratins: beginning a new era. Curr Opin Cell Biol 1993; 5:17-29.

160. Moll R, Franke WW, Schiller D, Geiger B, Krepler R. The catalog of human cytokeratins: patterns of expression in normal epithelia, tumors and cultured cells. Cell 1982; 31:11-24.

161. Quinlan RA, Franke WW. Heteropolymer filaments of vimentin and desmin in vascular smooth muscle tissue and cultured baby hamster kidney cells demonstrated by chemical crosslinking. Proc Natl Acad Sci USA 1982; 79:3452-3456.

162. Steinert PM, Idler WW, Cabral F, Gottesman MM, Goldman RD. In vitro assembly of homopolymer and copolymer filaments from intermediate filament subunits of muscle and fibroblastic cells. Proc Natl Acad Sci USA 1981; 78:3692-3696.

163. Monteiro MJ, Cleveland DW. Expression of NF-L and NF-M in fibroblasts reveals co-assembly of neurofilament and vimentin subunits. J Cell Biol 1989; 108:579-593.

164. Osborn M, Franke W, Weber K. Direct demonstration of the presence of two immunologically distinct intermediate-sized filament systems in the same cell by double immunofluorescence microscopy. Vimentin and cytokeratin fibers in cultured epithelial cells. Exp Cell Res 1980; 125:37-46.

165. Osborn MJ. Components of the cellular cytoskeleton: a new generation of markers of histogenetic origin? J Invest Dermatol 1983; 81:s104-107.

166. Portier MM, de Nechaud B, Gros F. Peripherin, a new member of the intermediate filament protein family. Dev Neurosci 1983; 6:335-344.

167. Liem RKH. Neuronal intermediate filaments. Curr Opin Cell Biol 1990; 2: 86-90.

168. Pachter JS, Liem RKH. α-Internexin, a 66-kD intermediate filament-binding protein from mammalian central nervous tissues. J Cell Biol 1985; 101:1316-1322.

169. Hisanaga S, Hirokawa N. Structure of the peripheral domains of neurofilaments revealed by low angle rotary shadowing. J Mol Biol 1988; 202:297-305.

170. Lowinger L, McKeon F. Mutations in the nuclear lamin proteins resulting in their aberrant assembly in the cytoplasm. EMBO J 1988; 7:2301-2309.

171. Lendahl U, Zimmerman LB, McKay RDG. CNS stem cells express a new class of intermediate filament protein. Cell 1990; 60:585-595.

172. Gounari F, Merdes A, Quinlan R, Hess J, FitzGerald PG, Ouzounis CA, Georgatos SD. Bovine filensin possesses primary and secondary structure similarity to intermediate filament proteins. J Cell Biol 1993; 121:847-853.

173. Klymkowsky MW. Intermediate filaments: new proteins, some answers, more questions. Curr Opin Cell Biol 1995; 7:46-54.

174. Heins S, Aebi U. Making heads and tails of intermediate filament assembly, dynamics and networks. Curr Opin Cell Biol 1994; 6:25-33.

175. Heins S, Wong PC, Muller S, Goldie K, Cleveland DW, Aebi U. The rod domain of NF-L determines neurofilament architecture, whereas the end domains specify filament assembly and network formation. J Cell Biol 1993; 123:1517-1523.

176. Foisner R. Dynamic organization of intermediate filaments and associated proteins during the cell cycle. BioEssays 1997; 19:297-305.

177. Inagaki M, Matsuoka Y, Tsujimura, K, Ando S, Tokui T, Takahashi T, Inagaki N. Dynamic property of intermediate filaments: regulation by phosphorylation. BioEssays 1996; 18:481-487.

178. Geisler N, Weber K. Phosphorylation of desmin in vitro inhibits formation of intermediate filaments: identification of three kinase A sites in the aminoterminal head domain. EMBO J 1988; 7:15-20.

179. Matsuoka Y, Nishizawa K, Yano T, Shibata M, Ando S, Takahashi T, Inagaki M. The different protein kinases act on a different time schedule as glial filament kinases during mitosis. EMBO J 1992; 11:2895-2902.

180. Hisanaga S, Kusubata M, Okumura E, Kishimoto T. Phosphorylation of neurofilament H subunit at the tail domain by CDC2 kinase dissociates the association to microtubules. J Biol Chem 1991; 266: 21798-217803.

181. Gutkind JS, Vitale-Cross J. The pathway linking small GTP-binding proteins of the Rho family to cytoskeletal components and novel signaling kinase cascades. Sem Cell Devel Biol 1996; 7: 683-690.

182. Omary MB, Baxter GT, Chou CF, Riopel CL, Lin WY, Strulovici B. PKC kinase associates with and phosphorylates cytokeratin 8 and 18. J Cell Biol 1991; 117: 583-593.

183. Chou C-F, Omary MB. Mitotic arrest-associated enhancement of O-linked glycosylation and phosphorylation of human keratins 8 and 18. J Biol Chem 1993; 268:4465-4472.

184. Ingber DE. Integrins as mechanochemical transducers. Curr Opin Cell Biol 1991; 3:841-848.

185. Eckert BS, Daley RA, Parysek LM. Assembly of keratin onto PtK1 cytoskeletons: evidence for an intermediate filament organizing center. J Cell Biol 1982; 92:575-578.

186. Georgatos SD, Weaver DC, Marchesi VT. Site specificity in vimentin-membrane interactions: intermediate filament subunits associate with the plasma membrane via their head domains. J Cell Biol 1985; 100:1962-1967.

187. Garrod DR. Desmosomes and hemidesmosomes. Curr Opin Cell Biol 1993; 5:30-40.

188. Troyanovsky SM, Eshkind LG, Troyanovsky RB, Leube RE, Franke WW. Contributions of cytoplasmic domains of desmosomal cadherins to desmosome assembly and intermediate filament anchorage. Cell 1993; 72: 561-574.

189. Foisner R, Wiche G. Intermediate filament-associated proteins. Curr Opin Cell Biol 1991; 3:75-81.

190. Wiche G. Plectin: general overview and appraisals of its potential role as a subunit protein of the cytomatrix. Crit Rev Biochem Mol Biol 1989; 24: 41-67.

191. Chou Y-H, Skalli O, Goldman RD. Intermediate filaments and cytoplasmic networking: new connections and more functions. Curr Opin Cell Biol 1997; 9:49-53.

192. Foisner R, Bohn W, Mannweiler K, Wiche G. Distribution and ultrastructure of plectin arrays in subclones of rat glioma C₆ cells differing in intermediate filament protein (vimentin) expression. J Struct Biol 1995; 115: 304-317.

193. Svitkina TM, Verkhovsky AB, Borisy GG. Plectin sidearms mediate interaction of intermediate filaments with microtubules and other components of the cytoskeleton. J Cell Biol 1996; 135: 991-1007.

194. Yang Y, Dowling J, Yu Q-C, Kouklis P, Cleveland DW, Fuchs E. An essential cytoskeletal linker protein connecting actin microfilaments to intermediate filaments. Cell 1996; 86: 655-665.

195. Leibowitz D, Kopan R, Fuchs E, Sample J, Kieff E. An Epstein-Barr virus transforming protein associates with vimentin in lymphocytes. Mol Cell Biol 1987; 7:2299-2308.

196. White E, Cipriani R. Specific disruption of intermediate filaments and the nuclear lamina by the 19-kDa product of the adenovirus E1B oncogene. Proc Natl Acad Sci USA 1989; 86:9886-9890.

197. Doorbar J, Ely S, Sterling J, McLean C, Crawford I. Specific interaction between HPV-16 E1-E4 and cytokeratins results in collapse of epithelial cell intermediate filament networks. Nature 1991; 352:824-827.

198. Klymkowsky MW, Maynell LAL, Nislow C. Cytokeratin phosphorylation, cytokeratin filament severing and the solubilization of the maternal mRNA *vg1*. J Cell Biol 1991; 114:787-797.

199. Venetianer A, Schiller DL, Magin T, Franke WW. Cessation of cytokeratin expression in a rat hepatoma cell line lacking differentiated functions. Nature 1983; 305:730-733.

200. Hedberg KK, Chen L-B. Absence of intermediate filaments in a human adrenal cortex carcinoma-derived cell line. Exp Cell Res 1986; 163:509-517.

201. Colucci-Guyon E. Portier M-M, Dunia I, Paulin D, Pournin S, Babinet C. Mice lacking vimentin develop and reproduce without an obvious phenotype. Cell 1994; 79:679-694.

202. Goldman RD, Khuon S, Chou YH, Opal P, Steinert PM. The function of intermediate filaments in cell shape and cytoskeletal integrity. J Cell Biol 1996; 134: 971-983.

203. Hendrix MJC, Seftor EA, Chu Y-W, Trevor KT, Seftor REB. Role of intermediate filaments in migration, invasion and metastasis. Cancer Metas Rev 1996; 15: 507-525.

204. Cleveland DW. Neuronal growth and death: order and disorder in the axoplasm. Cell 1996; 84:663-666.

205. Traub P, Shoeman RL. Intermediate filament proteins: cytoskeletal elements with gene-regulatory function? Int Rev Cytol 1994; 154:1-103.

206. Li H, Choudhary SK, Milner DJ, Munir MI, Kuisk IR, Capetanaki Y. Inhibition of desmin expression blocks myoblast fusion and interferes with the myogenic regulators myoD and myogenin. J Cell Biol 1994; 124:827-841.

207. Maniotis AJ, Chen CS, Ingber DE. Demonstration of mechanical connections between integrins, cytoskeletal filaments, and nucleoplasm that stabilize nuclear structure. Proc Natl Acad Sci USA 1997; 94:849-854.

CHAPTER 2

Signaling Components and Pathways

Introduction

The survival of a multicellular organism requires a carefully orchestrated net work of inter-cellular communications which regulate every aspect of growth, metabolic function, differentiation, development and programmed cell death (apoptosis). Cells of both unicellular and multicellular organisms respond to stimuli in their environments in intricately sensitive and selective manners. These responses are temporally ordered, varied and frequently complex, ranging in complexity from sensory signals such as photons of light, to proteins on adjacent cells. Analogously, cellular responses vary significantly from relatively simple modifications of metabolic pathways, to global structural rearrangements affecting the entire cell. In addition to the commonly recognized extracellular stimuli, cells may also respond to intracellular signals (inside-out signaling). Examples of inside-out signaling will be described in subsequent chapters. Chapter 2 will focus largely on the transduction of extracellular signals in the more classical outside-in mode and will describe in general terms the components and general molecular mechanisms for transduction of signals. The objective of this chapter is to acquaint the reader with the most basic aspects of this rapidly expanding field, the *concepts*, *classes of components* and *molecular mechanisms* important in the transduction of signals. Where appropriate, attention will be drawn to the signaling systems or components which have been shown to interact with the cytoskeleton or to regulate its organization. Of necessity, the descriptions will be cursory and review or overview articles will be cited prominently.

Signaling Components and Molecular Mechanisms

The behavior of cells within developing or adult tissues can be governed by signals emanating proximally, e.g., by neighboring cells, or from a distance, i.e., from distal organs or glands (1). Numerous peptide and steroid hormones act in an *endocrine* manner to influence the behavior of cells in distal tissues expressing specific receptors. These signals may influence the cellular growth state directly by eliciting changes in the cellular organization and metabolism, or indirectly by stimulating changes in transcriptional regulation. Cells can also influence the growth, metabolism or differentiative state of their neighbors within the same tissue by secreting cytokines, growth factors and hormones that act in a *paracrine* manner. This mechanism is particularly pertinent to signaling between cells of different types within the same organ or tissue, e.g., between mesenchymal and epithelial

Signaling and the Cytoskeleton, by Kermit L. Carraway, Coralie A. Carothers Carraway and Kermit L. Carraway III. © 1998 Springer-Verlag and R.G. Landes Company.

cells. Cells can act on their immediate neighbors through a number of mechanisms. *Gap junctions* allow direct exchange of small molecules and ions between contiguous cells. Extracellular components associated with one cell can elicit changes in an immediate neighbor cell by engaging transmembrane signaling receptors by a *juxtacrine* mechanism (2). Finally, some cells are capable of regulating their own growth through an *autocrine* mechanism whereby the cell secretes a factor that can augment or impede its own growth (3). This mechanism is commonly utilized by tumor cells to promote their growth or malignancy. A variant of this mechanism occurs without secretion of the factor by an *intracrine* or intracellular interaction between ligand and receptor within the cell (4). Finally, another variant of the autocrine response proposes that ligand and receptor in the same membrane can interact by an *intramembrane* mechanism either within the cell or at the cell surface (5). Both the intracrine and intramembrane mechanisms provide concentrative effects which can overcome weak ligand-receptor interactions.

Second Messengers

Target cells employ a variety of biochemical mechanisms to respond to extracellular stimuli. Specific *receptors* for external stimulating factors are present at the surface or in the cytoplasm of all cells. Upon binding of ligand, the receptor is conformationally altered, initiating a receptor-dependent series of events culminating in the appropriate cellular response. In the case of the *intracellular receptors*, the conformation change results in the activation of the receptor by releasing it from an inhibitory protein(s). The activated receptor, released from its spatial constraints, is targeted to the nucleus, where it modulates transcription. Signaling through *cell surface receptors* entails three stages: *reception, amplification* and *propagation* of signal. First, ligand binding to surface receptors elicits an alteration in receptor conformation, which either activates a cytoplasmic domain enzyme activity or recruits to the conformationally activated receptor, a cytoplasmic protein(s), which can transduce the signal. Second, a series of effector molecules triggered by engaged receptors serve to amplify and/or propagate the signal to appropriate cellular locations, sometimes by the generation of *second messenger molecules*. A growing list of such second messengers includes Ca^{2+}, *cyclic nucleotides*, certain *lipid metabolites* and *phosphorylated proteins*. These agents serve to propagate the signaling message by either directly or indirectly activating intracellular enzymes which are components of specific metabolic pathways or of signaling pathways. In some cases, such as the cyclic nucleotides, the known functions are restricted to the activation of very specific enzymes. At the other extreme, the Ca^{2+} ion has been directly implicated in a multiplicity of specific functions.

The intracellular concentration of *calcium ion*, which plays major roles in cellular signal transduction and metabolism, is very tightly regulated by the cell. In unstimulated or resting cells, free Ca^{2+} concentrations are $\leq 10^{-7}$ M, since most of the intracellular Ca^{2+} is organelle- or protein-bound. Free Ca^{2+} concentrations can be transiently increased, sometimes to $\approx 5 \times 10^{-6}$ M, by two general mechanisms. In the first, ligand binding to a gated Ca^{2+} channel permits transient entry of Ca^{2+} from the extracellular milieu. In the second, intracellular free Ca^{2+} concentrations can be transiently increased as a result of ligand binding to a receptor that is linked to a pathway which produces a second messenger responsible for the release of Ca^{2+} from intracellular stores (6). For example, ligands which bind some G-protein-linked receptors cause the activation of membrane-associated enzymes (described in the section on intracellular transducers) which hydrolyze membrane phospholipids to produce the second messenger lipid metabolites IP_3 and *diacylglycerol*. IP_3 can then bind to specific intracellular membrane receptors to trigger the transient release of Ca^{2+} (6, 7). Elevated levels of intracellular Ca^{2+} are then available for binding and activating numerous Ca^{2+}-dependent enzymes or binding proteins (see Table 1.5 for a partial listing). One of the important signaling enzymes modulated by elevated Ca^{2+} is protein kinase C (PKC), which requires both Ca^{2+} and diacylglycerol for activation. This signaling enzyme is discussed further in a later section.

Another important factor regulated by calcium is the *Ca^{2+}-activated neutral protease* or *calpain* (8), which constitutes a large family of proteases *implicated in cytoskeletal reorganization* (9). Calpain associates with the plasma membrane and is stimulated by Ca^{2+} ions, which cause the activation of the catalytic subunit. This protease, which has been implicated in important regulatory functions in many cell types, cleaves proteins at the membrane-microfilament interface in numerous cells, modulating their interaction with actin filaments (10). Calpain is proposed to participate in platelet activation, described further in chapter 4, and its hyperactivation has been implicated in diverse disease states, such as excessive platelet aggregation, cerebral and myocardial ischaemia, muscular dystrophy and cataracts (11). This protease may also be involved in the degradation of specific short-lived proteins which have PEST sequences, putative intramolecular signals for rapid proteolytic degradation (12, 13). The polypeptide chains of these proteins are rich in proline (P), glutamate (E), serine (S) and threonine (T). For example, several proteins which bind calmodulin, an important cytoplasmic protein in Ca^{2+}-regulated function (see below), contain PEST sequences and are susceptible to proteolysis by endogenous neutral proteases such as calpain I and calpain II (14).

Ca^{2+} is sequestered intracellularly in organelles, such as the endoplasmic reticulum, and by binding to specific proteins. The muscle protein troponin was the first Ca^{2+} sequestering protein to be identified. Non-muscle cells have a ubiquitous major calcium sequestering protein, calmodulin, which can constitute as much as 1% of the total cell protein mass. Ca^{2+}-dependent activation of numerous enzymes is mediated by interaction with Ca^{2+}/calmodulin (15). Calmodulin is a dumbbell-shaped protein having four Ca^{2+}-binding domains with high affinities for Ca^{2+}. Table 2.1 lists some important examples of enzymes and cytoskeletal proteins known to bind and become activated by Ca^{2+}/calmodulin. Non-muscle cells have other Ca^{2+}-binding proteins as well which likely mediate more specific cellular effects.

Lipid second messengers are produced as a consequence of the activation of different lipid metabolizing enzymes, as noted below. In other cases, ligand binding results in the modulation of enzymes which metabolize ATP or GTP to the second messengers *cyclic AMP (cAMP)* or *cyclic GMP (cGMP)*. The cyclic nucleotides are activators of specific types of kinases discussed in the section on intracellular transducers. cAMP is elicited as an early consequence of the activation of specific G-protein-linked receptors which lack intrinsic enzymatic activities. In contrast, cGMP production results from the direct stimulation by ligand of a membrane or cytoplasmic guanylate cyclase receptor, as described below. When activated, receptors which have intrinsic tyrosine kinase activity (following section) act primarily as recruiting agents for the assembly of multimeric *signal transduction complexes* or *particles* (16) which comprise signaling pathways. *In this case the phosphoproteins can be regarded as the second messengers*. The creation of phosphorylated tyrosine residues provides sites for binding specific proteins to initiate

Table 2.1. Examples of calmodulin-binding enzymes and cytoskeletal proteins

Enzymes	Cytoskeletal
phosphorylase kinase	spectrin
CaM kinases I-II	myosin I
calcineurin (phosphatase)	myosin V
EF-2 kinase	myosin light chain kinase
adenylate cyclase isoform	caldesmon
NAD kinase	MAP2
nitric oxide synthase	synapsin I
phosphodiesterase	tau
ATPase	

the assembly of signaling pathway complexes. The ultimate consequence of the activation of many pathways is the modulation of transcription. However, cellular responses to signals activating these signaling pathways are pleiotypic, and different signaling components must be activated which propagate the signal to the appropriate target pathways and organelle(s). Ligand activation of the receptor must therefore elicit coordinate activation of parallel pathways which can modulate a diversity of cellular activities, all of which are required for "priming" the cell to produce a certain type of behavior such as mitosis. Activation of some of these pathways is responsible for the *coordinate cytoskeletal rearrangements* required for the global structural remodeling which the cell undergoes during processes such as mitosis.

Domains Important
in Protein-Protein Interactions

Assembly of most signaling pathways is accomplished via a series of specific protein-protein interactions involving a number of recently characterized *binding domains* or *motifs*. Table 2.2. gives a listing of the currently known major classes of binding domains and their target motifs which provide a molecular basis for protein-protein interactions important in signaling. Some require posttranslational modification of a signaling protein as a consequence of an activation event. The prototypic example is the ligand-dependent tyrosine phosphorylation of a tyrosine kinase receptor such as the epidermal growth factor (EGF) receptor, or of a kinase-linked receptor such as the T cell receptor, both of which are described later in this chapter. The phosphorylated tyrosine residues serve as binding sites in the cytoplasmic domain of the receptor for specific proteins containing either *SH2* (*Src homology 2*) (17, 18, 19) or *PTB domains* (20) to begin the assembly of a signal transduction complex. SH2 domains comprise sequences of approximately 100 amino acid residues, initially described in c-Src (21), which bind specific protein sequences containing phosphotyrosine residues. Specificity for individual SH2 domains is determined primarily by the sequence of the short motif (≈6 amino acid residues) around the phosphotyrosine residue, particularly those residues immediately flanking the Tyr C-terminus (22). PTB domains bind phosphorylated tyrosine residues

Table 2.2. Binding domains and target motifs of signaling components.

Domain or motif	Binds to	Localization/Function
SH2	*Phosphorylated tyr* in specific motifs specified at C-terminus of motif	Assembly of signal transduction complexes by binding cytosolic and membrane proteins to activated receptors and other phosphorylated Y-containing signaling proteins
PTB	*Phosphorylated tyr* in specific motifs specified at N-terminus of motif (e.g., NPXpY)	same as for SH2?
14-3-3 in specific motifs	*Phosphorylated ser or thr* (Raf, RXRSXSXP)	Binds and activates signaling and other regulatory proteins (e.g. Raf1)
PH	*Phosphorylated ser or thr* in specific motifs; G-proteins/modulators, phosphoinositides	Found in a variety of signaling and cytoskeletal proteins, many of which are associated with plasma or organellar membranes; serve as signal-dependent membrane adaptors
SH3	Polypro (PXXP) in specific motifs	Linkage of adaptors to activated receptors and signaling pathway proteins; linkage of signaling proteins to the cytoskeleton
WWP/WW	Polyproline (PPP)ₓ in specific motifs usually containing tyr	Found in various structural, regulatory and signaling proteins; compete with SH3 domain-containing proteins?
PDZ	Extreme C-terminus of target proteins	Found in membrane-associated guanylate kinases located at different cell junction structures (MAGUK's)

in short motifs to which SH2 domains do not bind, possibly because they may require a specific conformation of the motif (23). PTB domain binding requires an essential asparagine residue (NPXpY) (23, 24, 25) found in numerous signal transduction proteins. PTB domains are longer than SH2 domains (≈160 amino acid residues) and structurally unrelated to them (26), but show structural similarities to the PH domain described below. This specificity of both SH2 and PTB domains for phosphorylated tyrosine residues in specific contexts creates the possibility for a large repertoire of intracellular complexes of activated receptors and other tyrosine-phosphorylated signaling proteins based on the specific phosphorylated sequences and the cytoplasmic proteins which bind to them. At present, the significance of the existence of two unrelated protein domains which bind very similar, but distinct phosphotyrosine-containing motifs is not understood.

The *PDZ/GLGF/DHR domain* also binds receptor proteins, in this case through a different mechanism. PDZ domains received their acronym from the first proteins in which they were found, PSD-95/SAP90 (channel clustering proteins), DlgA (*Drosophila* disc-large tumor suppressor protein) (27) and ZO-1 (a tight junction protein) (28). Each are found in numerous proteins located in regions of cell-cell contact, such as tight junctions, septate junctions and synaptic junctions (29). PDZ proteins contain motifs having the consensus residues GLGF (30) which bind to short (as few as 4 aa's) carboxy-terminal sequences of transmembrane proteins (31). *Many of these have sequence motifs for localization to the cytoskeleton* and have been *implicated in organization at the membrane-cytoskeleton interface* (32, 33). The oriented peptide library technique has identified specific 4 amino acid residue consensus motifs for binding to two families of PDZ-containing proteins (31) and will likely identify more (see chapter 8).

A critical step in the assembly of a signaling complex by a phosphorylated receptor is the binding of residues to the phosphorylated tyrosine *adaptors*, small proteins which have binding domains or motifs but no catalytic domains (17). Adaptors such as Grb2 of the RTK-Ras-MAPK pathway, both have SH2 and *SH3 (Src homology 3) domains*. The latter are domains of 55-60 amino acid residues which form a surface containing sites for binding specific residues in *polyproline motifs* (PXXP-containing motifs) of their target proteins (19). A partial listing of proteins containing SH3 domains is given in Table 2.3. Proteins containing SH2 domains cover a broad spectrum of signaling proteins, while SH3 domains are prominent features of adaptors and some *membrane skeletal proteins* (19). The adaptor Shc has both SH2 and SH3 domains and a PTB domain. In fact, *multiple binding domains are a common feature of many signaling proteins*. Examples of SH3 domain-containing skeletal proteins will be described in several chapters of this book. Like the SH2 domain, the SH3 domain has a related domain which recognizes a similar motif. The WW domain, which exhibits as a prominent feature, the presence of two conserved tryptophan (W) residues, also binds polyproline motifs (34). The WW domain found in numerous structural, regulatory and signaling molecules appears to have a preference for tyrosine-containing proline-rich sequences (35). As in the case of the SH2 and PTB domains, the significance of the similar, yet distinctive SH3 and WW domain binding motifs has yet to be elucidated.

The SH2-SH3 domain-containing adaptors link activated receptors to molecular *switches*, such as the GTPase Ras. Switches are key proteins in activating catalytic activities of the enzymes in signaling pathways. A major function for these adaptors and switches is *localization* to upstream activated receptors, or to downstream effectors or substrates as shown in Figure 2.1, for the early portion of the

Table 2.3. Selected proteins containing SH3 domains

	SH2	SH3	Other domains or activities
Non-receptor kinases and other enzymes			
Src family	1	1	tyrosine kinase
Abl family	1	1	tyrosine kinase, actin binding, DNA binding
CSK	1	1	tyrosine kinase
Tec	1	1	tyrosine kinase
PLC	2	1	phospholipid hydrolase
ISGF3	1	1	HLH[1], leucine zipper
Dlg	0	1	guanylate kinase
p47 and p67 phox	0	2	activate neutrophil cytochrome oxidase
GTPase-modulating proteins			
Ras GAP	2	1	Ras GTPase activator
CDC25, ste6	0	1	GNEF[2]
Cytoskeletal proteins			
Nonmuscle myosin	0	1	cytoskeleton-membrane
fodrin/spectrin	0	1	cytoskeleton-membrane
cortactin	0	1	at focal adhesions
p80/p85; HS1	0	1	at focal adhesions
ABP-1	0	1	yeast actin binding protein
BEM1	0	1	cytoskeleton/yeast
FUS1	0	1	cytoskeleton/yeast
Adaptor proteins			
PI3K p85	2	1	rhoGAP; links PI3K to activated receptor
GRB2 (Drk/Sem5)	1	2	human (*Drosophila*, yeast) adaption
Vav	1	2	dbl domain (GNEF?)
Crk	2	1	function unknown
Shc	1	2	adaptor
Nck	1	3	adaptor

[1]helix-loop-helix DNA binding domain
[2]guanine nucleotide exchange factor

EGF receptor-Ras-MAP kinase pathway for mitogenesis. (The end of the pathway is shown in Figure 2.7). There is no further transduction of the signal without this localization event (36). As expected for critical pathway components, switches are carefully modulated by the binding of a number of other proteins which can either turn on or turn off their activities, as described below. The binding of some proteins to GTPases is mediated through *pleckstrin homology (PH) domains*, regions of sequence similarity of ≈120 amino acid residues first identified in pleckstrin, the major substrate of PKC in platelets (37). Like SH3 domains, PH domains are also present in a wide variety of proteins involved in cell signaling or cytoskeletal

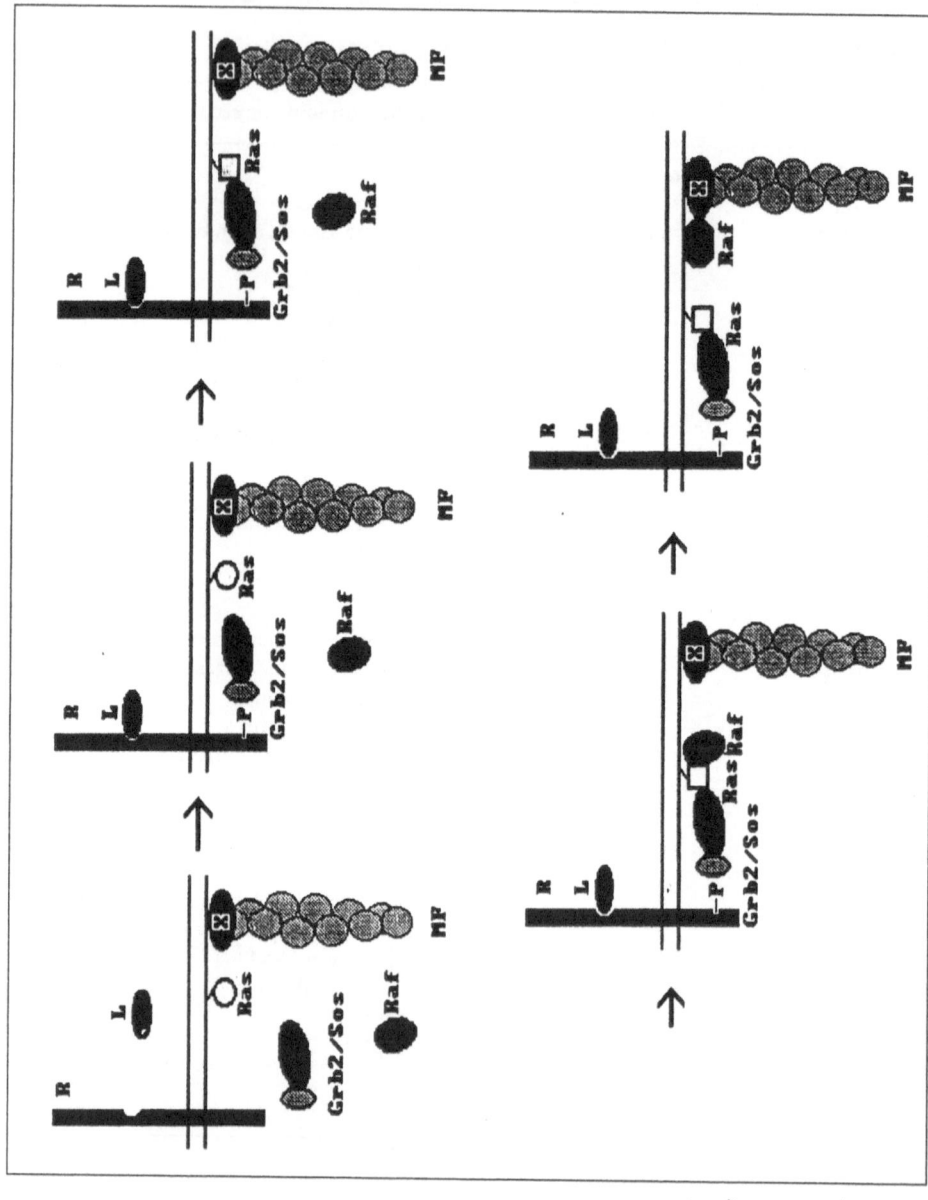

Fig. 2.1. Model for activation of Raf through the EGF receptor by recruitment of components to the membrane-cytoskeleton interface. Figure from ref. 36 (Carraway KL, Carraway CAC. BioEssays 1995; 17:171-175), courtesy of ICSU Press.

functions at the *membrane-cytoskeleton interface* (38). Recent studies suggest a role for PH domains in localizing these proteins at the membrane via interactions with specific phosphoinositides generated as a consequence of receptor activation (37). Structure determinations show that PTB/PI domains (23, 25) have the same topology as PH domains (39), although they share no significant sequence similarities. To complicate matters further, the PDZ domain shows an interesting resemblance. The relationships among these various domains and their binding proteins is not presently understood, and constitutes an important challenge for future studies on the assembly of signaling complexes (discussed further in chapter 8).

Another structural element which plays an adaptor role in signaling protein pathways and in cell cycle regulation is the 14-3-3 protein (40, 41, 42). First discovered as abundant acidic proteins in the brain, these proteins were later found to be ubiquitous proteins expressed widely from plants to yeast to man. The crystal structure of the 14-3-3 family of proteins shows that they form dimers having a negatively-charged groove down their length which might serve to bind common features of target proteins (41). The presence of conserved residues in the groove suggests that it plays a role in 14-3-3 adaptor function. Recently, 14-3-3 proteins have been shown to interact in a sequence-specific manner with conserved phosphorylated serine motifs (43). Thus, they could be an analog of SH2 and PTB domains for binding phosphorylated serine residues. A large number of proteins, including such important signaling elements as Raf, PKC, phosphatidylinositol 3-kinase (PI3K), Vav and cdc25 phosphatase have been shown to bind 14-3-3 proteins (41). Such diverse functional activities help to explain the widespread phenotypic effects of expression of 14-3-3 proteins, which have been implicated in proliferation and the cell cycle, transformation, differentiation and development (41).

Another class of proteins implicated in signal transduction are those which have *zinc finger domains.* Zinc fingers are a diverse set of small ($\approx$30 residues) protein motifs containing Cys and His. Many of these Cys-rich motifs are actually autonomously folded domains stabilized by the binding of Zn^{2+} (44), forming loop-turn-loop structures. There are several broad categories of zinc finger-containing proteins, some having catalytic functions, others having storage functions and a third binding to nucleic acids. The last class can be subdivided into those which bind double stranded DNA and those which bind single stranded nucleic acids such as retroviral nucleocapsid proteins (45). These domains have been most prominently identified with transcriptions factor; the first Zn^{2+} fingers identified in a transcription factor were found in a protein from *Xenopus* oocytes which was necessary for transcription of the 5S RNA gene (46). Transcription factors regulate cell growth, differentiation and development by binding to a specific site or set of sites on nucleic acids to regulate gene expression. Zn^{2+} fingers are frequently repeated in nucleic acid binding proteins, in some cases multiple times. DNA binding Zn^{2+} finger domain-containing proteins can be subdivided into families (47) which use structurally related motifs for recognition.

Recently, other Zn^{2+} finger proteins have been identified which appear to comprise a new family, the ZZ family, characterized by the presence of two or three C-X_2-C motifs and a conserved D-Y-D-L motif (32). A particularly noteworthy aspect of these proteins is that they appear in dystrophin and related proteins located at the membrane-microfilament interface. This family also contains motifs for binding to actin and to polyproline-containing proteins since they contain a WW domain as well. A related family, the TAZ (transcriptional adaptor) domain family, has triplicate repeats of a $C-X_4-C-X_8-C$ motif and appears along with ZZ domains in the transcriptional adaptor/co-activator proteins p300 and c-AMP response element binding protein (CREB) binding protein (CBP) and the yeast adaptor protein ADA2. Like the ZZ/TAZ family, the LIM domain is also a ubiquitous Cys-rich domain (48) found in transcription factors and in membrane skeletal proteins (49, 50). The Lim domain contains a dual zinc finger motif that may be repeated several times in clusters at the N- or C-terminus. Like the basic *leucine zipper* helix-loop-helix domain, which forms a dimerization motif that mediates the interaction of transcription factors, e.g., Fos and Jun to give the active AP-1 (51), the Lim domain is thought to be a dimerization motif. Interactions may occur in either a homodimeric or heterodimeric fashion. The Lim domain may or may not be associated with a homeodomain. In Lim homeodomain (Lim-HD) proteins, which are involved in the control of cell lineage and the regulation of differentiation, the Lim domains appear to function as negative regulatory domains (50). Lim only and Lim-HD proteins are localized differently in the cell, as are single Lim domain and multiple Lim domain proteins (52). Lim domains are found in the focal adhesion membrane skeletal proteins zyxin (53) and paxillin (54), and have been implicated in growth control and in adhesion-stimulated changes in gene expression (53).

A particularly important aspect of signaling is the localization of signaling components and complexes. Many signaling proteins are targeted to and intercalated into the plasma membranes and other specific sites via posttranslational modification with lipid moieties (55). Membrane association is required for the function of these proteins, and truncation of the sequences which delete these modification sites abolish both membrane association and function (55, 56). Examples of these include the (proto)oncogenes Src and Ras which are discussed in a later section of this chapter. Three general classes of these lipid modifications have been characterized: *glycosylphosphatidylinositol anchors* (GPI anchors or PIG-tails), which are often found in apical domains or in caveolae (57); acyl groups (myristate and palmitate) (55,56) and isoprenyl moieties (farnesyl and geranyl) (55,56). *Myristoyl* groups are added to N-terminal glycines as stable amides, *prenyl* groups to cysteines as stable thioethers and *palmitoyl* groups as labile thioesters. Myristoyl and farnesyl groups are insufficient for stable membrane association. Proteins with these modifications require contributions from peptide motifs or from additional acyl groups to form stable membrane linkages (56). Thus, modifications of the protein motifs or turnover of palmitate groups may contribute to protein localization.

All of these domains and motifs mediate protein-protein interactions among catalytic and/or *scaffolding proteins* which result, at least transiently, in the *assembly of signaling pathways*. These multimeric assemblies are not only important for the formation of pathways, but also for the *integration* of multiple signals through associations of components from different pathways. One example of this kind of *crosstalk* between pathways is the association of cytoskeletal components, frequently microfilaments with activated receptors (36,58). Another may be the integration of the signals initiated by cell-matrix interactions with those of growth factors (59). The complexity of such interacting pathways has been an impedance to understanding at the molecular level the nature of the coordinated modulation of cell responses to multiple extracellular signals. The important area of cross talk is envisioned as one of the final frontiers in the study of cell signaling (see chapter 8).

Ligands

A given cell in a multicellular organism is programmed to respond to a host of environmental signals, many in a temporally ordered manner. Over the course of the development of the organism, a cell can respond differently to a specific signal. For example, in early stages of the embryo, proliferation is a crucial activity and the cells respond maximally to growth factors. As development progresses, other signals such as cell-cell interactions, become more important and response to mitogenic factors becomes less significant. These temporal responses are carefully regulated since pathways initiated during mitogenesis frequently oppose those required during the establishment and maintenance of the differentiated state of the cell. Cessation of proliferation is an early requirement for the induction of polarity in the differentiating cell. Further, there are crucial signals which are required for a cell at any stage simply to survive. Deprivation of these signals requisite to basal level function, elicits the initiation of a programmed set of cellular pathways which culminate in *cell death* (50). Programmed cell death (*apoptosis*) is a key process for ridding the organism of excess cells in the normal pattern of cell differentiation in the developing organism (61). The roles of apoptosis in both the developing organism and in protecting the organism against damaged or altered cells will be further discussed in chapters 5 and 7.

The diversity of ligands to which cells respond is immense and includes a vast range of molecules and other stimuli. Transduction of signal, via these ligands, occurs through a relatively small number of classes of receptors which can be characterized most broadly by the chemical nature of the ligands they bind. *Hydrophobic ligands* such as sterols, retinoids, vitamin D and thyroxine readily diffuse through the phospholipid bilayer and bind to *intracellular receptors*. *Hydrophilic ligands*, as well as a diversity of *sensory signals*, bind to *cell surface receptors*. One known exception to this general rule is the eicosanoids, derivatives of arachidonic acid which are quite hydrophobic, yet transduce their signal via cell surface receptors. The cell surface class of receptors can be subdivided into a relatively small number of receptor families described further in the section below on receptors.

Hydrophobic Ligands

The lipid-soluble ligands fall into several major classes which are chemically different, but commonly have small hydrophobic ring structures. The steroid and steroid-like hormones have very hydrophobic complex fused ring structures, and vitamin D, the retinoids and thyroxine have less complex, though still hydrophobic, ring structures. Signaling via these ligands frequently requires transcription and protein synthesis; consequently, the cellular responses are slow. These hydrophobic ligands bind to cytosolic or nuclear receptors comprising a highly related superfamily having three domains, a highly variable N-terminal domain, a DNA binding domain and a C-terminal ligand binding domain, described below. The mechanism for signal transduction for this class of ligands is described in the following section. As modulators of gene expression, the steroid and steroid-like hormones display multiple and diverse effects in different tissues.

The eicosanoids are metabolites produced after phospholipase hydrolysis of phospholipids to yield arachidonic acid, a highly unsaturated fatty acid which is continuously synthesized, rapidly turned over and frequently involved in autocrine signaling. The eicosanoids comprise four major classes: prostaglandins, prostacyclins, leukotrienes and thromboxanes. The eicosanoids display a wide variety of activities, including smooth muscle contraction, platelet aggregation and participation in the inflammatory response. Though quite hydrophobic, they appear to act through cell surface receptors.

Table 2.4. Selected ligands for members of the EGF receptor family with examples of their sources or sites of action

Receptor	Ligand	Source/function
EGFR/ErbB-1	EGF	salivary gland, milk
	TGF-α	tumor cells; multiple tissues
	HB-EGF	smooth muscle
	amphiregulin	epithelial tissues
	betacellulin	tumor cells
	epiregulin	NIH3T3 cells
	VGF	vaccinia virus
ErbB-2/Neu/HER2	ASGP-2	ascites tumor cells
		epithelial tissues
ErbB-3	heregulin	heart/brain development
ErbB-4	heregulin	heart/brain development
	betacellulin	tumor cells
	HB-EGF	smooth muscle

Sensory Ligands

This class of ligands comprises a widely divergent group of receptor activators which initiate a cascade of events ultimately leading to sensory perception, i.e., vision, hearing, smell, taste and tactile perception. These stimuli include light, sound, a chemically diverse group of odorants, chemicals which activate taste receptors, and tactile stimuli which are perceived through mechanoreceptors. The largest group of these stimuli includes a broad diversity of chemicals which comprise odorants and activate odorant receptors. These receptors transmit signal via the activation of G-proteins, or large GTPases, a major class of intracellular transducers. The G-protein-coupled receptors and the mechanism for signal transduction through their transducers are discussed further in later sections of this chapter.

Hydrophilic Ligands

Small Compounds

Small hydrophilic ligands include inorganic compounds (NO), amino acid derivatives (epinephrine) and neuropeptides (corticotropin). All of these are derived from amino acids. NO is made by *nitric oxide synthase* from arginine. Its receptor is the intracellular form of guanylate cyclase described later in this chapter. Epinephrine and a number of other metabolic modulators are made from the amino acid tyrosine. Peptide hormones and neuropeptides are made from protein precursors which are produced by the protein translation apparatus, then cleaved into the individual peptides during processing in the organelles of the secretion pathway. Growth factors and cytokine peptides are often similarly made from membrane protein precursors which remain intact until they reach the cell surface where the factors are released by proteolytic cleavage.

Polypeptide Ligands

Many of the properties of soluble polypeptide ligands are exemplified by the EGF family which were the first growth factors discovered (62) and exhibit a broad range of biological functions (Table 2.4). Most of this family, as well as other peptide ligands, are synthesized as transmembrane precursors and released in soluble form during subsequent cellular processing (63). In some cases such as transforming growth factor-α (TGFα), the transmembrane form is active and can serve as an insoluble, juxtacrine ligand via intercellular interactions (2). At least 10 members of this family have been identified, most of which are redundant in their binding to the EGF receptor, but do not bind to the other members of the EGF receptor family. However, at least two members of the EGF receptor-binding group, betacellulin and heparin-binding growth factor (HB-EGF), are promiscuous and can bind and activate both the EGF receptor and ErbB-4 (64). The ligand for ErbB-3 and ErbB-4 is heregulin, also called neuregulin and Neu differentiation factor (NDF), and is similarly promiscuous (65), but does not bind either the EGF receptor or ErbB-2. As described later, *many of these ligands act by stimulating heterodimerization of the receptors preferentially to homodimerization. Ligand binding factors* may contribute to the activity of these ligands on cells. For example, HB-EGF, amphiregulin, heregulin and other family members bind heparin and specific glycosaminoglycans (66) which may modulate their activities. Interestingly, membrane HB-EGF forms a complex with a diphtheria toxin receptor protein which associates with α3β1 integrin at cell-cell contact sites (67), possibly providing a mechanism for focusing its site of action.

Insoluble Ligands

In addition to sensory and soluble activators, cells respond to ligands which are insoluble as a consequence of their association with a cell or with the extracellular matrix. The insoluble hydrophilic ligands fall into three major categories, the first of which comprises proteins that are *membrane forms of secreted proteins*. These are typified by the membrane-bound forms of growth factors. The second class is components of the *extracellular matrix* which bind to integral membrane receptors such as integrins. The third category includes membrane proteins which bind to membrane proteins on adjacent cells. The cadherins, membrane proteins which mediate cell-cell interactions at adherens junctions via homotypic interactions are a major example. In this case the membrane protein can be considered to be both a ligand and a receptor. In each of these cases interaction with the appropriate insoluble protein elicits a signaling response in the cells, frequently with *rapid reorganization of the actin cytoskeleton*. The components of the membrane-cytoskeleton assemblies induced by the binding of these ligands to their receptors will be described in the last section of this chapter.

Integral Membrane Growth Factors

A number of growth factors (cytokines) have been shown to function as membrane forms by a juxtacrine mechanism, including TGFα, HB-EGF, colony stimulating factor-1, kit ligand, tumor necrosis factor-α, ligands for the Eph receptor family and Boss (2). This mechanism may be particularly important during development where signals determining cell fate often need to be discretely localized (68). A good example is the development of the *Drosophila* eye (69), which is deficient in R7 photoreceptor cells in mutants of the ligand Boss (bride of sevenless) or its receptor Sevenless, a tyrosine kinase (70). Interestingly, Boss has seven transmembrane segments reminiscent of G-protein-linked receptors, raising the question of whether it may also have a receptor function. In mutants which fail to form the intercellular ligand-receptor complex, the target cell differentiates into a nonneuronal lens cell. Although the Boss-Sevenless interaction appears to result in a typical tyrosine kinase activation pathway, other juxtacrine ligands may act by different mechanisms. The receptor Notch and its analogs are activated by membrane ligands to specify cell fate during neural development (68). However, Notch is not an enzymic receptor. Instead, it appears to bind a transcription factor(s) and sequester it from the nucleus. Binding of the Notch ligand then releases the transcription factor to activate the target gene(s) in the nucleus.

One important question regarding the action of membrane ligands by juxtacrine mechanisms is whether the soluble form has a comparable activity (71). Although soluble and membrane forms of the kit ligand have similar receptor activation properties, animals expressing only the soluble form have developmental defects. These results suggest that the cellular localization and cell-cell interaction potential of the membrane form are important in its biological function. In the case of tumor necrosis factor (TNF), which induces apoptosis in some cell types, the soluble and membrane forms appear to elicit different cellular responses via different receptor isoforms (68). Soluble TNF activates the R1 receptor, while membrane TNF activates the R2 receptor. Since these receptors have different cytoplasmic domains, they mediate different cellular responses.

The possibility of reciprocal signaling arises for the juxtacrine mechanism, since the ligand and receptor are on different cells, and both are transmembrane proteins. Evidence for this phenomenon was recently described for a ligand for one of the Eph class of receptors, the largest known family of tyrosine kinase receptors (72). One of the functions of Eph family members is to mediate axonal pathfinding in the nervous system via interactions with cell surface ligands on other cells (73), as described further in chapters 3 and 4. If the ligand Lerk2 is expressed in NIH 3T3 cells, it becomes tyrosine-phosphorylated when the cells are exposed to the receptor Cek5, an Eph family member implicated in axonal growth during embryogenesis (74). Lerk2 was also found to be tyrosine-phosphorylated in the embryos and to suppress growth factor-induced proliferation when transfected into the 3T3 cells (73). These results suggest that both the ligand and receptor can participate in signaling pathways in their separate cells.

An alternative ligand-receptor interaction mechanism, an *intramembrane ligand-receptor interaction*, has been proposed for a putative ligand for the tyrosine kinase receptor ErbB-2 (5). Complexes of the ligand ASGP-2 (ascites sialoglycoprotein-2) and receptor can be immunoprecipitated from membranes of a highly metastatic rat mammary tumor. Similar complexes can be obtained when ASGP-2 and ErbB-2, but not the EGF receptor, are co-expressed in insect cells either in soluble or membrane forms. However, no complex is formed by mixing secreted soluble ligand and receptor or by adding soluble ligand to cells expressing cell surface receptor. These results suggest that the complex is formed inside the cells and that, in the case of the membrane species, it forms by an intramembrane mechanism (75). Since ASGP-2 can activate the ErbB-2 kinase under some conditions and may block the binding of other ligands, its expression in cells should act as a modulator of signaling through the EGF family of receptors in which ErbB-2 serves as a common subunit for heterodimerization (65).

Extracellular Matrix Protein Ligands

Another class of insoluble ligands consists of proteins residing in the extracellular matrix (ECM), a complex, insoluble network of secreted glycoproteins and proteoglycans elaborated by animal cells. The ECM serves as an adhesive lattice for organizing cells in tissues and as a reservoir for some growth factors and hormones (76) which regulate mitogenesis and differentiation. Important cellular phenomena, including cell adhesion, cell motility and gene expression, are mediated by the ECM and its components (77, 78). These functions are elicited, at least in part, via direct interactions between specific components of the matrix and their receptors on the cell surface. The most widely studied of receptors for ECM proteins is the family of *integrins*, different ones of which can interact with multiple matrix components, including collagens, laminins and fibronectin (Table 4.2) (79). These interactions trigger integrin aggregation, tyrosine phosphorylation and downstream signaling effects which are described in subsequent chapters. A notable consequence of the engagement of ECM receptors is the *assembly of actin at the activated receptor which leads to a global cytoskeletal reorganization*. One functional consequence of this membrane-microfilament interaction is anchorage of the cell to the matrix. Cell anchorage strongly affects the ability of cells to respond to mitogens and differentiation factors (80). The extracellular matrix can also play a second role as a modulator of growth factor activities. Specific matrix components may serve as passive participants to sequester growth factors in the vicinity of the cell until they can be released by a stimulus (71), as proposed for the regulation of heregulin by proteoglycan and proteolysis in skeletal muscle (81). Alternatively, the matrix component can serve as a co-activator, as in the case of cell surface proteoglycan, which is proposed to form a ternary complex with fibroblast growth factor (FGF) and its receptor to promote signaling (82, 83). As noted above for EGF family members, a number of polypeptide growth factors or cytokines are known to bind heparin and proteoglycans, suggesting that their activities can be modulated by extracellular matrix components (82, 84).

Receptors

Receptors can be most broadly characterized by their locale, either plasma membrane or intracellular. With the exception of the eicosanoids noted in the description of ligands, receptors for *hydrophobic ligands* are located intracellularly and those for *hydrophilic ligands* at the cell surface.

Intracellular Receptors

Small lipid-soluble hormones such as steroids, the retinoids, vitamin D and thyroxine diffuse freely through lipid membranes to activate cytoplasmic receptors which regulate specific members of a large superfamily of related *transcription factors*. These cytosolic or nuclear receptors comprise a highly related superfamily having three domains, a highly variable N-terminal domain (100-500 amino acid residues), a DNA binding domain showing a high degree of sequence similarity (42-94%) and a C-terminal binding domain having variable sequence similarity (15-57%). These intracellular receptors are anchored in a large complex with an inhibitor protein, the chaperone heat shock protein hsp90 (85), in the cytoplasm in

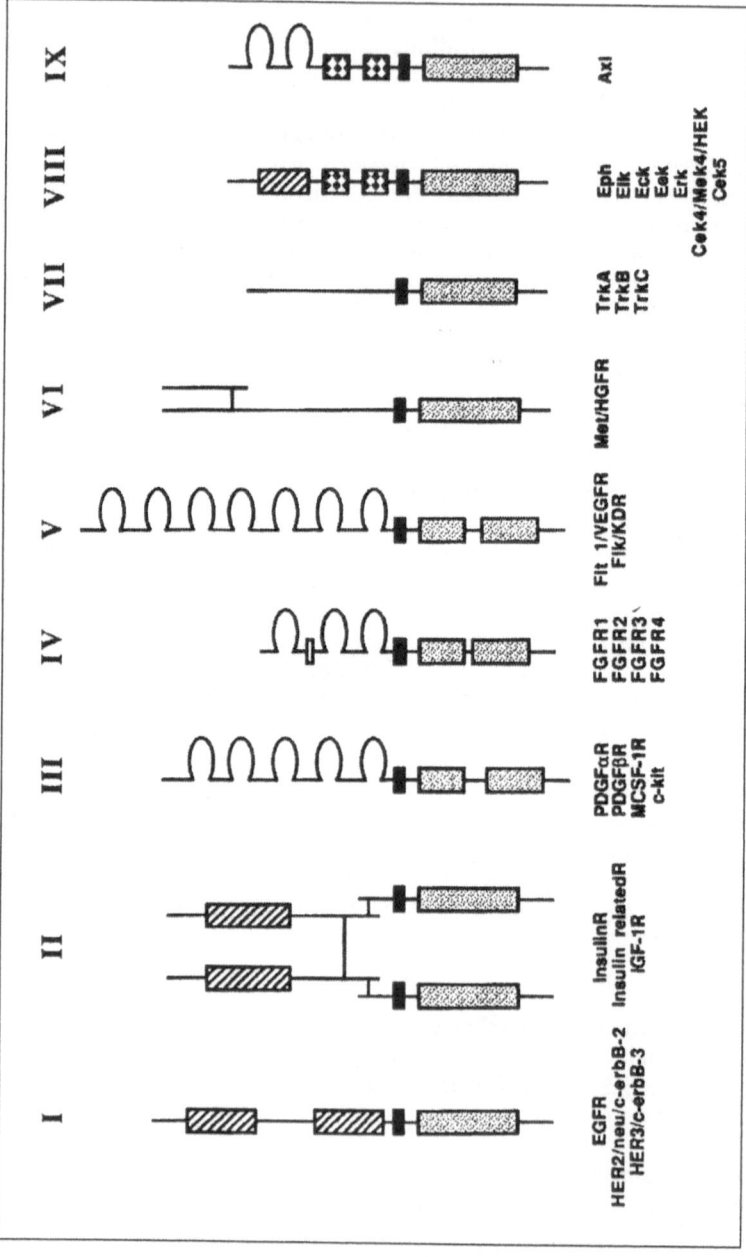

Fig. 2.2. Vertebrate receptor tyrosine kinases. Figure from ref. 88 (Fantl WJ, Johnson DE, Williams LT. Annu Rev Biochem 1993; 62: 453-481), with permission, from the Annual Review of Biochemistry, Volume 62, 1993, by Annual Reviews Inc.

resting cells. The mechanism for cytoplasmic anchorage is uncertain, but may involve *association with the cytoskeleton* (86). This intriguing possibility is discussed further in chapter 8. Deletion mutant analyses have shown that the hormone-binding domain inhibits transcriptional activation in the absence of ligand. Upon ligand binding, a conformational change results in the release of the inhibitor protein, and the ligand-receptor complex can translocate to the nucleus. The DNA-binding domain of the receptor then interacts with the appropriate response element, activating the target gene. Deletion of the hormone binding domain of the glucocorticoid receptor conferred hormone-independent, constitutive activation of the glucocorticoid response element (85). This observation suggests that the major function of the ligand is displacement of the inhibitory component(s), releasing the receptor from its cytoplasmic tether. The nature of the cytoplasmic anchoring mechanism is not understood, though instances of cytoplasmic complex formation and possible sequestration of other proteins that regulate transcriptional activators are being reported (87). An example is the proposal that a MAPKK analog (Pbs2p) in yeast acts as a scaffolding protein to block association with other similar pathways which use a MAPK cascade (see Fig. 2.7). *Thus, association with cytoskeletal components provides a plausible mechanism in unstimulated cells for the cytoplasmic sequestration in an inactive state of regulatory proteins which function in the nucleus.* Immunoelectron microscopy should provide insights into the molecular basis for these sequestration events.

Plasma Membrane Receptors

There are two general classes of cell surface or plasma membrane receptors for receiving extracellular signals: enzymic and nonenzymic. In the case of enzymic receptors, the extracellular signal can be transmitted directly via the receptor catalytic activity to the cell cytoplasm. These receptors are thus defined by their enzymic activities and are readily recognized and classified. The nonenzymic receptors are a more diverse group. Their signal must be passed to the cytoplasm via coupling to another protein(s), and can be further distingushed on the basis of their linkage either directly or indirectly to an enzyme. Each of these subclasses is activated by a different type of ligand and coupled to a different type of cytoplasmic protein. Receptors which have no intrinsic catalytic activity and are linked directly to kinases comprise the former group. The latter group is more heterogeneous and includes ion channel receptors and those receptors linked to enzymes via switches such as G-proteins.

Enzymic Receptors

The major enzymic receptors which have been identified are the *receptor tyrosine kinases* (RTKs), a Ser/Thr kinase, a class of receptor phosphatases and guanylate cyclase. The first class of these, the tyrosine kinases, is the most abundant and most frequently studied. At least fourteen separate classes of receptor tyrosine kinases have been described based on the structures of their extracellular domains (Fig. 2.2) (88). Several of these have been implicated in cellular changes involving the cytoskeleton (89). Except for the insulin receptor, which is a heterodimer of an integral membrane α subunit disulfide-linked to an extracellular peripheral membrane β subunit, these receptors have a single transmembrane domain and a cytoplasmic tyrosine kinase domain in common. They differ in sequence and structure most substantially in their extracellular regions which accounts for their ligand specificity. The extracellular domains are highly disulfide-linked and show a significant degree of structural complexity, having Ig, fibronectin and other structural motifs, whose functions in the receptors are not understood. Outside the kinase domains they also differ in their C-terminal regions which provide sites for tyrosine phosphorylation and for the transfer of signal to the cytoplasm. This signal is generated through receptor autophosphorylation. In most cases the phosphorylation is actually a transphosphorylation between receptor kinase subunits of a dimer or higher order multimer (65). This cross-phosphorylation creates an increased order of signal complexity and specificity when different members of the same receptor family associate to form heterodimers (65). The phosphorylated tyrosine residues can then serve as interaction sites for signaling-proteins with SH2 or PTB domains (Table 2.2) (20), such as the adaptor proteins, the first step in the formation of signal transduction complexes with the activated receptor at the membrane (90). Not all cytoplasmic signals are initiated by interactions of signaling proteins directly with the phosphorylated receptor. The insulin receptor kinase phosphorylates a cytoplasmic substrate IRS, which then acts as the docking site for formation of signaling complexes (91). In either case, the specificity of binding of signaling components and consequent determination of downstream signaling pathways is resident in the amino acid sequences adjacent to the phosphorylated tyrosines which bind the SH2 or PTB domains as noted above. For example, the PDGF β-receptor has nine defined tyrosine phosphorylation sites on its cytoplasmic domain for binding specific cytoplasmic signaling components (92).

Cell surface tyrosine kinase receptors have been implicated in several pathological conditions, including neoplasia (chapter 7); thus regulation of these activities is crucial. Several levels of regulation have been recognized for different receptors. For example, overexpression of the EGF receptor family member ErbB-2 in tumors can occur by either gene amplification or transcriptional activation (93). Overexpression may lead to activation in the absence of extracellular ligand (94), though the mechanism is unclear. Mutations resulting in deletion of the extracellular domain ligand binding site can cause constitutive activation of the receptor kinase, as found for v-Erb (95). A single point mutation in the extracellular region of the transmembrane domain of c-Neu, the rat form of ErbB2, also causes constitutive activation of that receptor kinase by a conformational activation (93). The kinase activities of receptors can be down-modulated by several mechanisms. One mechanism for inhibition is phosphorylation of specific serine or threonine residues by cytoplasmic kinases. An example is the inhibition of EGF receptor kinase activity by phosphorylation of a juxtamembrane Thr by protein kinase C (PKC) (96). Down-regulation of the EGF receptor, but apparently not other members of its family (5), occurs by internalization and degradation (97).

Another important aspect of the functioning of a signaling protein is its cellular locale. Information is limited for many receptors, but members of the EGF family have been located at both basal and apical surfaces of epithelial cells. The mechanisms for the regulation of the localizations are unclear, though *biosynthetic targeting and binding to the cytoskeleton or to components at cellular junctional complexes* undoubtedly contribute. The EGF receptor has been shown to directly bind to microfilaments (95), while ErbB-2 associates with microfilaments (58) or with β-catenin (98), a component of cadherin-mediated cell-cell junctions. Examples of the interactions of receptors with the cytoskeleton are further discussed at the end of this chapter and in chapters 3 and 4. Finally, receptor tyrosine kinases may also

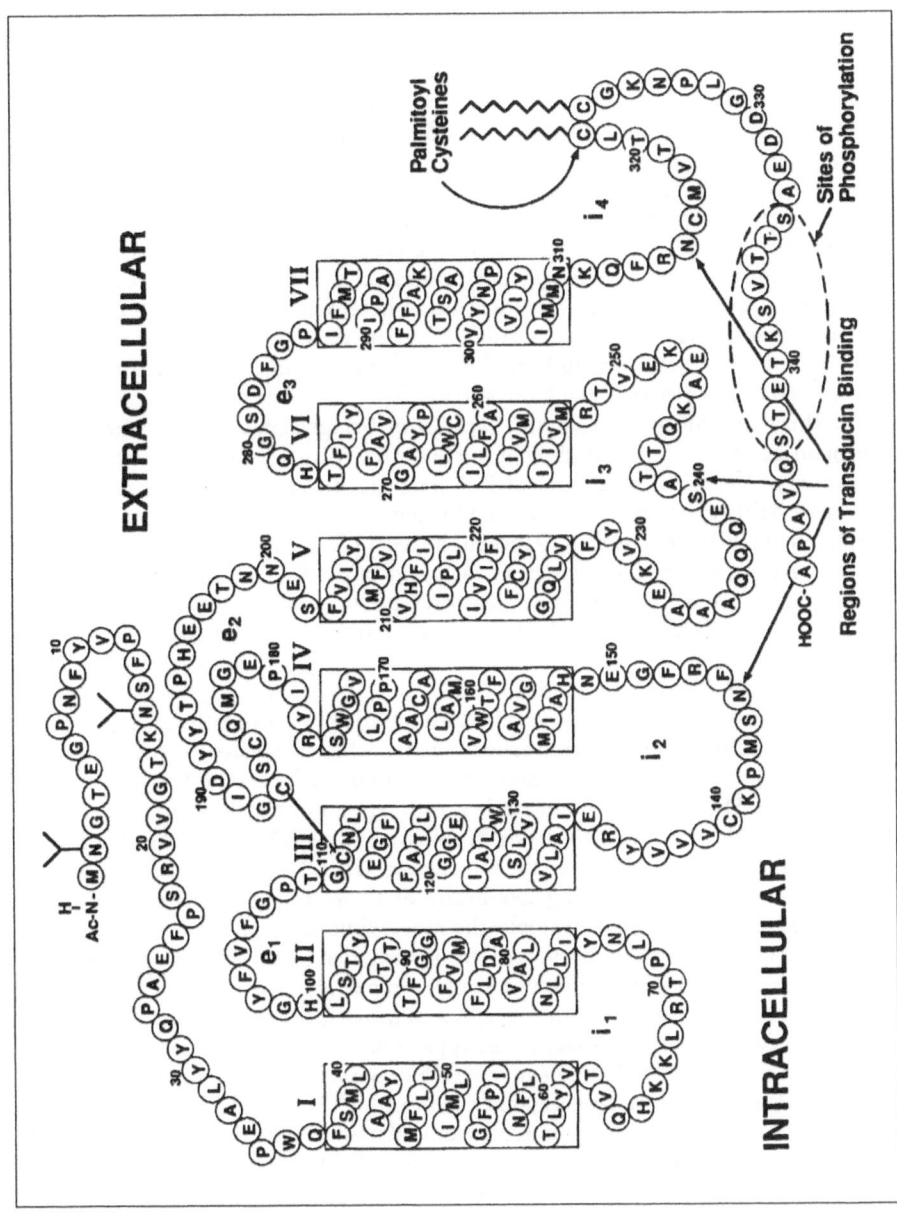

Fig. 2.3. Generalized model of G-protein-linked (serpentine) receptor. Figure from ref. 107 (Hargrave PA. Curr Opin Struct Biol 1991; 1:575-581), courtesy of Current Biology Ltd.

participate in other signaling pathways, such as those of the adhesion receptors or G-protein-linked receptors as a means of integrating multiple cellular inputs.

The TGF-β family of cytokines regulates cell proliferation, differentiation, recognition, death (apoptosis) and the cell cycle (see chapter 5) through interactions with a unique class of distantly related cell surface receptors having a *Ser/Thr kinase* in their cytoplasmic domains (99). TGF-β, a disulfide-linked dimer with a cystine knot structure, induces signaling by hetero-oligomerization of type I and type II transmembrane receptor Ser/Thr kinases to trigger cross-phosphorylation of the type I receptor. Although TGFβ is often regarded as a negative modulator of growth (100), the response is cell-specific and quite diverse. As in the case of Type I tyrosine kinase receptors, the induction of heteromeric receptor combinations by different ligands provides additional diversity of responses. Ligand binding to the receptors can also be modulated by other factors, including the membrane-anchored proteoglycans betaglycan and endoglin (101). For example, the presence of a third subunit of the receptor, Type III or betaglycan, a high Mr membrane protein, can enhance the binding of the low affinity isoform TGF-β2 to the type II "primary receptor". The actions of TGF-β on specific aspects of cell cycle regulation are being delineated, as described in chapter 5. Recent studies have also shed light on the intermediate steps between the receptor at the membrane and nuclear events regulating the cell cycle (102). Members of the Smad family of proteins are phosphorylated by the activated receptors, migrate into the nucleus and associate with DNA-binding proteins, which are involved in the regulation of transcription.

Ion Channel-Linked Receptors

Ion channel-linked receptors are best exemplified by the nicotinic acetylcholine receptor (AChR), a ligand-gated cation channel which regulates the passage of sodium and potassium ions in response to the binding of acetylcholine to receptors (chapter 3). The AChR is a pentameric $\alpha_2\beta\gamma\delta$ complex of transmembrane subunits which responds after cooperatively binding two acetylcholine molecules with a conformational change to open the ion channel (103). The *gated channel* is a common mechanism for regulating membrane ion fluxes, though many of these channels are not receptors. For example, the muscarinic acetylcholine receptors bind different neurotransmitters and are G-protein-linked (described below). Thus, the same ligand can activate different receptors in particular tissues and elicit different responses. The response to a neurotransmitter is then determined in specific target nerve and muscle tissues by the type of receptors in the cells and their coupling mechnisms. Ion channels may also be regulated by changes in membrane potentials as a result of opening other ion channels to provide a cascade mechanism. The nicotinic acetylcholine receptor and many other channels are *located in microfilament-associated junctional complexes*. The structure and formation of the neuromuscular junction, a well-studied example of channel signaling complexes, are described in chapter 3.

G-Protein-Linked Receptors

G-protein-linked receptors, or *serpentine receptors*, are recognized by their common structure of seven transmembrane helices of 22-24 hydrophobic residues. These receptors are similar in overall structure to rhodopsin, the prototype of the serpentine membrane proteins, which transmits signals resulting from activation of the receptor opsin by photons of light. Though similar in structure, this class of receptors binds a diverse group of ligands and stimulates a multitude of cellular

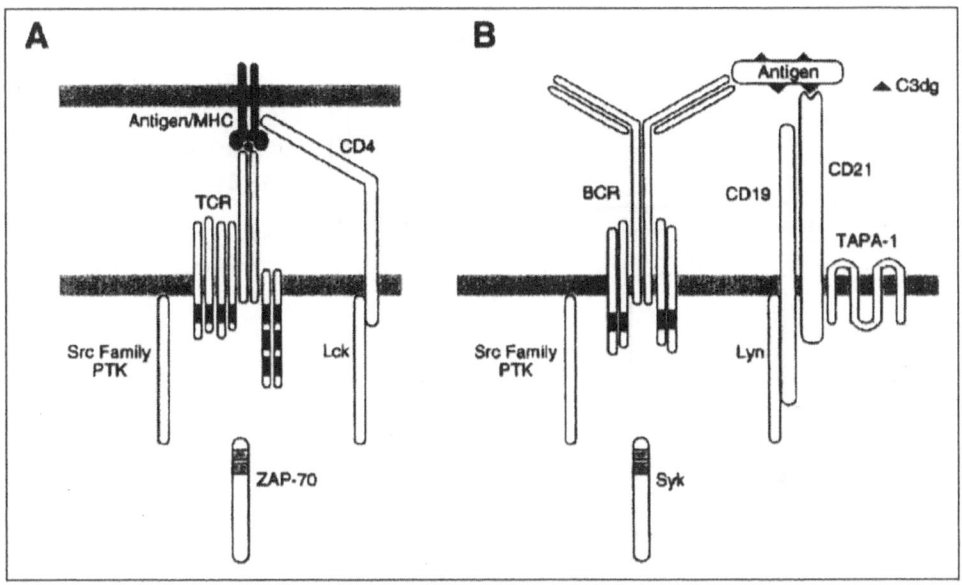

Fig. 2.4. Model for T cell receptor complex interaction with tyrosine kinases. Figure from ref. 11c (Weiss A. Cell 1993; 73: 209-212), courtesy of Cell Press.

functions. More than 100 different members of this receptor family have been cloned and sequenced, demonstrating their considerable structural homologies, despite their diverse ligands and functional activities (104). In addition to the seven trans-membrane helices, other important features of the general structure include an extracellular N-terminal region whose glycosylation may contribute to stability; three extracellular loops, two of which are disulfide-linked; three cytoplasmic loops, the third of which is larger and more variable; and a C-terminal cytoplasmic do-main, which may be palmitoylated to provide an additional binding site to the plasma membrane (Fig. 2.3) (105, 106, 107). For most of these receptors, the ligands bind to hydrophobic pockets formed by the transmembrane helices, all of which except helix I have been implicated in ligand binding for different types of recep-tors. The cytoplasmic C-terminal domain and the loop between α-helices 5 and 6 provide the binding site for G-proteins, heterotrimeric GTPases which associate with activated receptors. The G-proteins (discussed in more detail under Intracel-lular Transducers) serve as molecular switches, transducing signal from the acti-vated receptor to cytoplasmic effectors. Transduction occurs through interaction of the α or βγ subunits of the activated G-protein with membrane-associated chan-nels or with enzymes such as adenylate cyclase and phospholipase C, which pro-duce second messengers (described above). Signals from the G-protein-linked re-ceptors are turned off by three mechanisms: desensitization (loss of coupling to G-protein), internalization and down-regulation (degradation). Desensitization oc-curs through phosphorylation of receptor cytoplasmic residues by Ser/Thr kinases, including specific G-protein receptor kinases, which phosphorylate only agonist-occupied receptors (108), PKA and PKC (107). Phosphorylated receptors are bound by arrestins, proteins which contribute to desensitization, presumably by disrupt-ing the interaction between the receptor and G-protein (108).

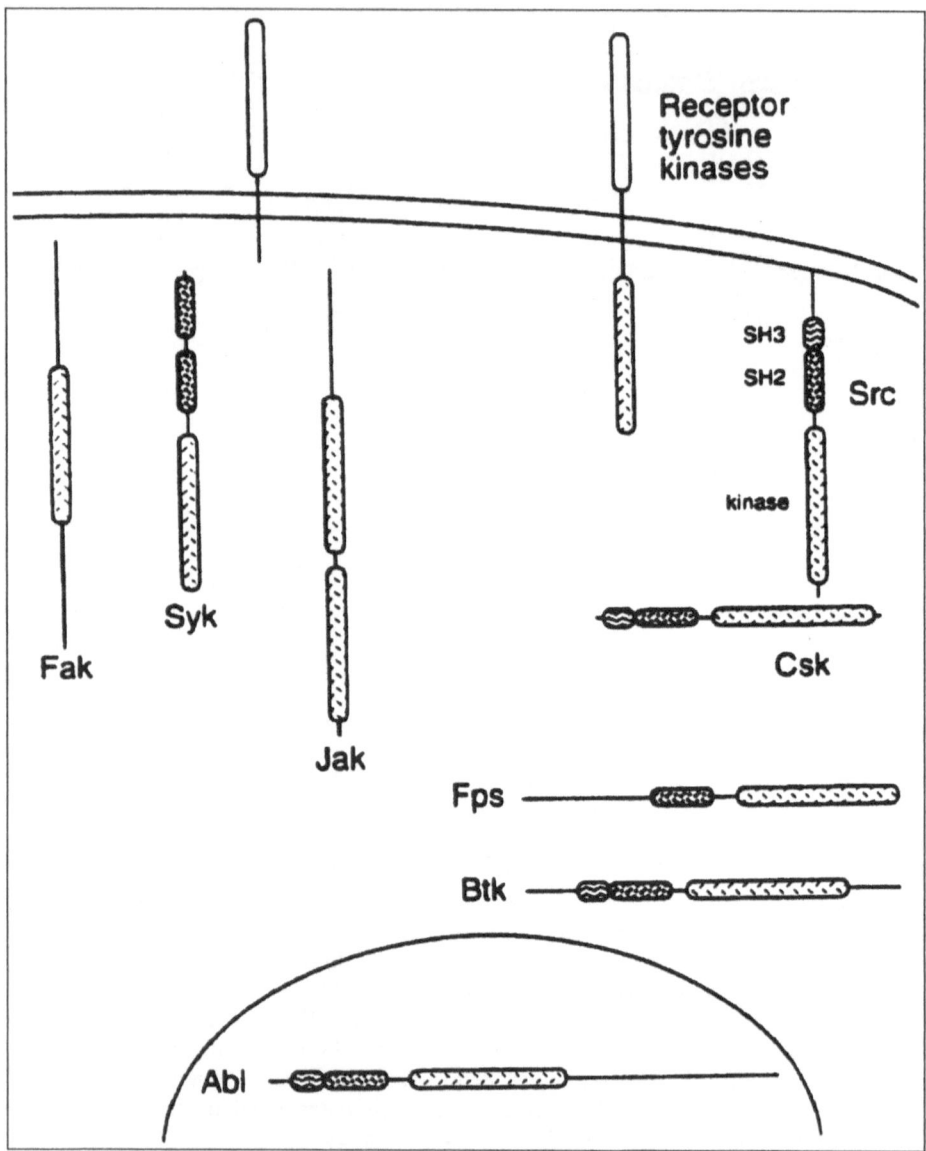

Fig. 2.5. Domain structures of major nonreceptor tyrosine kinases. Figure from ref. 122, (Courtneidge SA. Sem Cancer Biol 1994; 5: 239-246), courtesy of Academic Press.

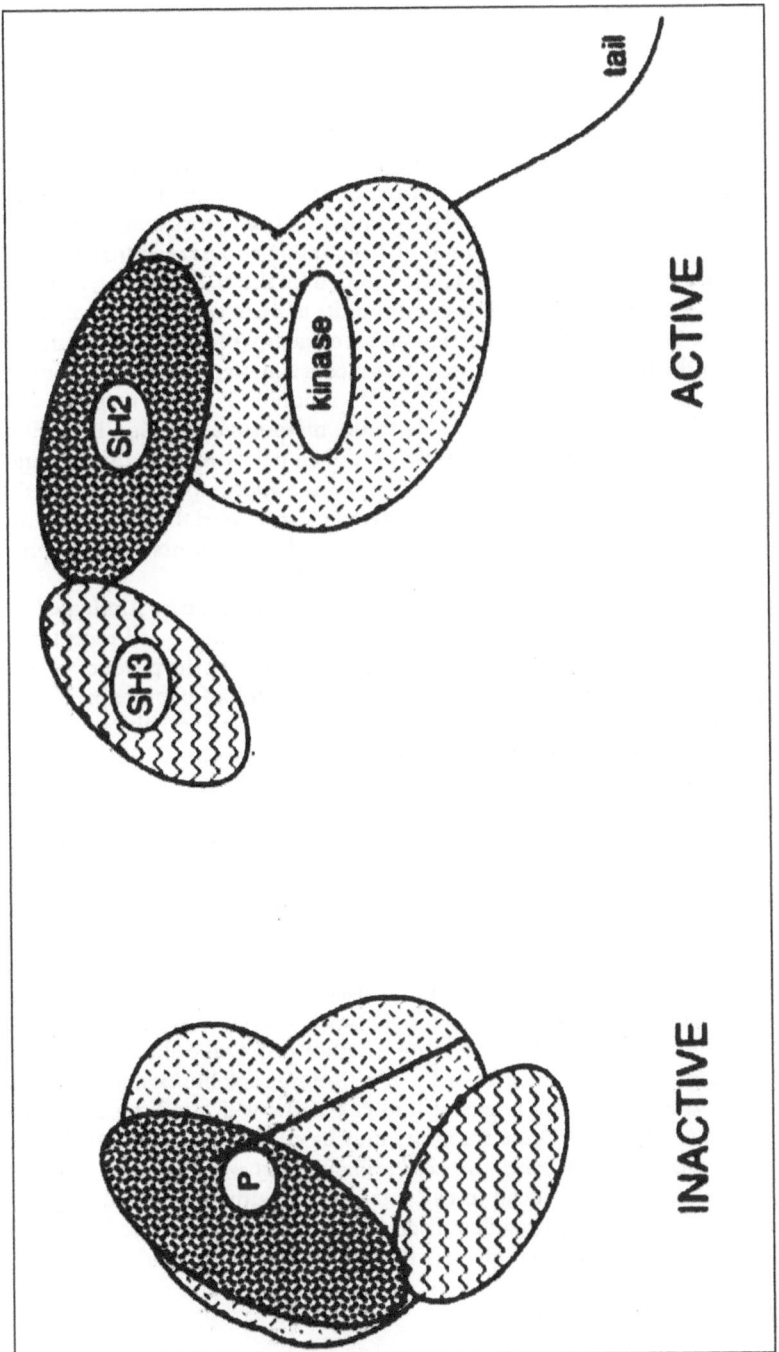

Fig. 2.6. Model for regulation of Src kinase activity. Figure from ref. 122 (Courtneidge SA. Sem Cancer Biol 1994; 5: 239-246), courtesy of Academic Press.

Kinase-Linked Receptors

Receptors which lack intrinsic kinase activity can nonetheless transduce signals via tyrosine phosphorylation as the first or a very early step. *Kinase-linked receptors* are defined by their association with cytoplasmic kinases (following section), specifically nonreceptor tyrosine kinases. Their ligands may be either soluble, such as the cytokines, or associated with other cells. *Cytokine receptors* generally consist of two of more subunits in a membrane complex and transmit their signals through nonreceptor tyrosine kinases (109). Individual receptors may share a receptor subunit, accounting for the functional redundancy observed for different cytokines. The signaling mechanism is exemplified by the IL-2 receptor, consisting of a complex of three transmembrane subunits associated with the polypeptide ligand. The Src family kinase Lck is also associated with the complex and is activated upon the binding ligand, possibly as a consequence of the oligomerization of the receptor and phosphorylation of its subunits. The IL-2 receptor can recruit and activate other tyrosine kinases to the membrane complex, including Syk and members of the JAK kinase family, described below. As with the tyrosine kinase receptors association and cross-phosphorylation appear to be the most likely mechanisms involved. The phosphotyrosine residues produced then provide sites for the recruitment of other signaling molecules, such as adaptors, small G-proteins or lipid kinases, into these complexes.

Many of the known examples of enzyme-linked receptors can be stimulated by interaction with cell-associated ligands, as exemplified by the T cell receptor, a multimeric complex of transmembrane proteins composed of the products of six genes (Fig. 2.4) (110). Signaling occurs with the recognition of peptide presented on the cell surface of an apposing antigen-presenting cell and recruitment of a T cell coreceptor CD4 molecule complexed with Lck, a Src tyrosine family member. The transmembrane phosphatase CD45 is necessary for dephosphorylating and activating the Src family member(s) of the complex to initiate tyrosine phosphorylation, as described below for Src. Phosphorylation of tyrosine residues of specific "activation motifs" is required for recruitment of SH2-containing signaling components for downstream signaling (111), which includes activation of the phosphoinositide pathway, increase of cytoplasmic calcium and stimulation of transcription of specific genes (112).

Adhesion receptors are located at cell-cell or cell-matrix interaction sites and are listed as a separate class because their ligands are insoluble as a result of their residence in the extracellular matrix or on another cell. Cell adhesion proteins comprise four major classes: integrins, cadherins, immunoglobulins and selectins. The primary examples are the integrins and cadherins which will be discussed further in chapters 3 and 4. These receptors may be considered a subclass of the kinase-linked receptors. *Integrins and cadherins interact with microfilaments in stable structures* which are crucial to cell interactions with matrix and adjacent like cells, respectively. A more detailed discussion of associations of cell adhesion receptors with the cytoskeleton is given at the end of this chapter and in chapter 3.

Intracellular Transducers

Nonreceptor Tyrosine Kinases

Nonreceptor tyrosine kinases are the first signaling elements in the transduction of signals from many of the nonenzymic receptors, including cytokine receptors, lymphoid cell receptors and possibly, adhesion-linked receptors. Although several classes of non-receptor tyrosine kinases have been identified, as indicated in Figure 2.5, the study of these enzymes has been dominated by work on the *Src family kinases*, largely because of the position of Src as the first tyrosine kinase discovered, its role as an oncogene product (113) and the importance of other Src family members in numerous signaling events and cellular behaviors (114) including lymphoid cell activation (115, 116). Src not only serves as the prototype of the Src class of kinases, but also for the structural domains SH2 and SH3, as described in the introduction to this chapter (19). These SH2 and SH3 domains permit Src family kinases to form stable complexes with other signaling molecules including substrates (117). N-terminal myristoylation of Src and some other members of its family provides a site through which they may associate with membranes, and plasma membrane localization of Src by myristoylation is involved in both c-Src and v-Src cell function (118). *Oncogenic forms of Src have been shown to be associated with the cytoskeleton*, though the mechanism of association is still unclear (89). Src kinase activity is regulated in part by an intramolecular interaction between its SH2 domain and a phosphorylated C-terminal tyrosine, thus providing a facile mechanism for control (Fig. 2.6). Because of its folded structure, Src can be activated not only by dephosphorylation of the C-terminal tyrosine, but also by agents which bind the SH2 and particularly, the SH3 domains (113). The oncogenic form v-Src is truncated at its C-terminus and lacks the autoinhibitory C-terminal phosphorylation site. As a consequence, v-Src is constitutively active. Cellular inactivation of c-Src is attributed to the *C-terminal Src tyrosine kinase CSK*. Multiple tyrosine phosphatases have been implicated in activation of Src family members depending on cell context. An example is the involvement of CD45 in T cell activation described above. Interestingly, Src knockouts have relatively mild phenotypic changes, suggesting significant redundancies in functions among family members (113).

A structurally related kinase *Abl* is an important oncogene for lymphoid malignancies (119) and has been implicated in transcriptional regulation and cell cycle functions (120). The N-terminal half of Abl is similar in structure to the Src family, but Abl also contains nuclear translocation, F-actin-binding and DNA-binding domains in its C-terminal half (121). These multiple domains suggest that Abl's nuclear activities may be regulated by translocation from the nucleus to the cytoplasmic cytoskeleton, as well as by mechanisms similar to those of the Src family. Abl has several tyrosine phosphorylation sites, some of them autophosphorylation sites, but the significance of each of these in the activity or binding of Abl to SH2 domain- or PTB-containing proteins is not understood. Another kinase involved in lymphoid cell signaling is *Syk*, which is recruited to receptor complexes early in lymphocyte activation (115, 116) and also plays a role in platelet signaling (see chapter 4).

Recent studies have demonstrated the importance of a number of other non-receptor tyrosine kinases, some of which are illustrated with their cellular locations in Figure 2.5 (122). In addition to Abl, these include FAK, SYK, JAK, CSK, FPS, and BTK. The *focal adhesion kinase FAK*, a key component in signaling through integrins, is another nonreceptor tyrosine kinase with structural and signaling implications (chapters 4 and 5). Like Src and Abl, FAK is autophosphorylated, though this event is probably more important in creating a site for binding and activating Src family kinases than for FAK activation (123). Although FAK does not contain

Table 2.5. PKC binding proteins

Binding Protein	Substrate	Location	Function/Properties
Vinculin and talin	yes	Focal contacts	Substratum adhesion
Annexins I and II	yes	Vesicles,	Vesicle trafficking,
MARCKS		plasma	secretion
MARCKS-related protein		membrane	
Kinesin light chain			
Desmoyokin/ AHNAK	N.D.	Desmosomes/ nucleus	Cell-cell contact and communication
α-Adducin			
γ-Adducin (Clone 35H)	yes	Cell polarity	Stabilizes actin-spectrin interactions
Actin	yes	Cytoskeleton	PKC ε-selective
PKB/akt/Rac	yes		Cell signaling
PICK1	yes	N.D.	Unknown
Fcε receptor	yes	Plasma membrane	Immune function, PKC δ-selective
AKAP79	yes	Post-synaptic densities	Synaptic transmission
RACK	no	N.D.	Caveolae, others?
HIV protein Nef	N.D.		Viral function

Abstracted from materials provided by Dr. Susan Jaken.

SH2 and SH3 domains, it has a number of other sites for forming complexes with signaling or cytoskeletal components, including a specific C-terminal site for association with focal adhesions, proline-rich motifs which can bind SH3 domains and several phosphorylatable tyrosine residues, whose phosphorylation creates binding sites for signaling components such as Grb2 (124). Thus, one of the primary functions of FAK may be to recruit Src family kinases and other components to adhesion signaling complexes as part of the downstream signaling pathways. A curious observation is that FAK recruitment to integrin-mediated adhesion complexes is an early event in fibroblast adhesion, but a late event in platelet activation. The cellular roles of FAK are discussed more fully in chapters 4 and 5.

Cytokine signaling involves *Janus kinase (JAK)* as well as Src family members. JAKs provide a more direct pathway for regulating gene expression because of their ability to phosphorylate and regulate activity of the *STAT family of transcription factors* (125). As noted above, JAK family kinases are activated as a consequence of ligation of plasma membrane cytokine receptors. JAK phosphorylation of the receptors permits recruitment of STATS, which have SH2 domains. Subsequent JAK tyrosine phosphorylation of the STATs stimulates their homo- or heterodimerization, translocation to the nucleus and involvement in transcriptional activation (126). The specificity of STAT activation results from the STAT SH2-receptor phosphotyrosine interaction. The JAK-STAT pathway is not limited to

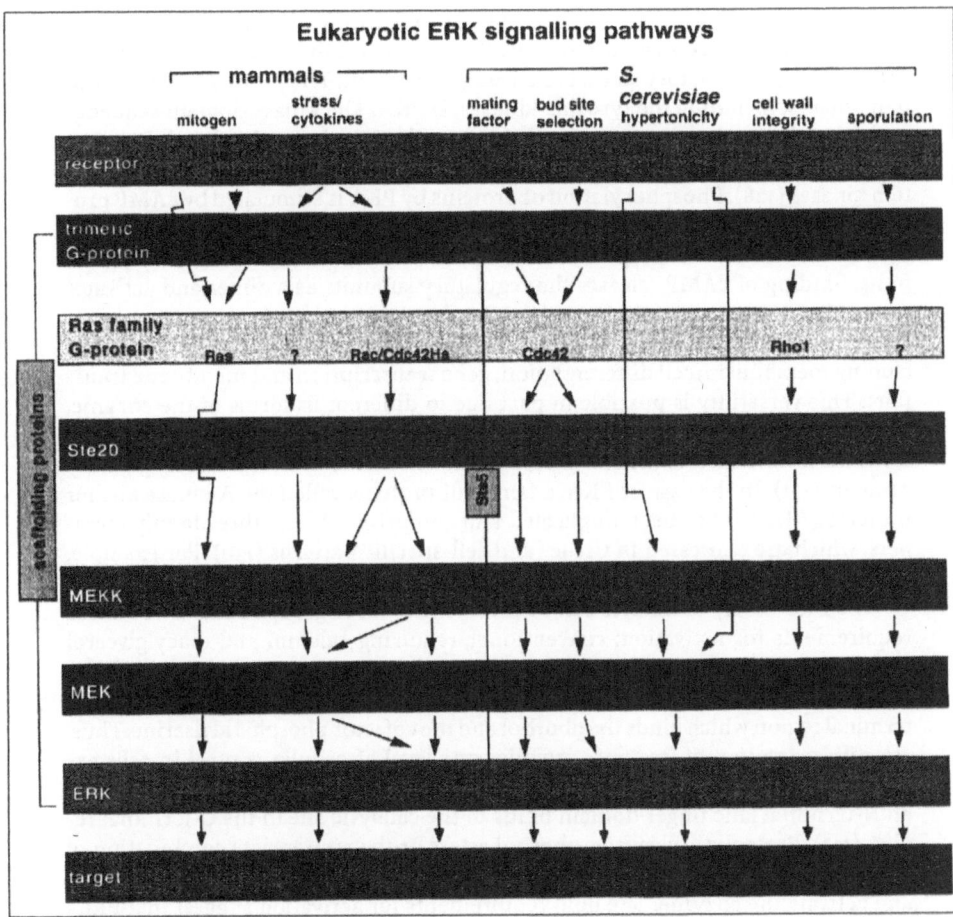

Fig. 2.7. Eukaryotic MAPK signaling pathways. Figure from ref. 138 (Kyriakis JM, Avruch J. BioEssays 1996; 18: 567-577), courtesy of ICSU Press.

cytokine activation. JAKs and STATs can be activated by other tyrosine kinases such as the EGF receptor. STAT activity can be modulated by serine phosphorylation, though the details of the mechanisms and kinases involved remain unclear (126).

Serine/Threonine Kinases

Intracellular serine/threonine kinases are particularly important in the transmission of signals from the membrane to cytoplasmic components and the nucleus, and for the integration of signaling pathways. Although the TGFβ receptor, described above, is the only known instance of a receptor serine/threonine kinase, there is quite a diversity of intracellular kinases of this type. Among the most important families are *PKA, PKC, calmodulin-dependent kinase* (CamK), components of the *MAP kinase* or *ERK* cascade (*MEKK, MEK, MAPK*) and *cyclin-dependent*

kinases (cdk) (127), which are discussed more extensively in chapter 5. The serine/
threonine kinases are structurally more heterogeneous than the tyrosine kinases,
some having a regulatory as well as a catalytic subunit. The prototype for all protein kinases, including the tyrosine kinases, is PKA. The kinase domain sequence
and three-dimensional structure of PKA were the first determined and thus provided the homology and structure comparisons for the recognition of other protein kinases (128). Phosphorylation of proteins by PKA is stimulated by cAMP produced via activation of G-protein-linked receptors, described below. Unactivated
PKA is a tetramer composed of two catalytic subunits and two regulatory subunits. Binding of cAMP releases the regulatory subunits as a dimer and activates
the catalytic monomers, which catalyze the phosphorylation of substrate proteins.
PKAs are important in the regulation of a number of physiological processes, including metabolism, cell differentiation, gene transcription and membrane transport. This versatility is possible in part due to different isoforms of the enzyme,
three different genes in mammals (129, 130). However, another important contributor is the *localization of the enzymes*, a persistent theme in the story of regulatory
proteins (131). In the case of PKA, a family of proteins called the A kinase anchor
proteins (AKAPs) has been implicated and comprises at least three family members, which are expressed in tissue- and cell-specific patterns (130). For example,
AKAP150 in forebrain neurons is located with microtubules of proximal dendrites.

The PKC family of kinases can be divided into three groups based on cofactor
requirements for activation: conventional, requiring calcium and diacylglycerol
(DAG); novel, requiring only DAG; and atypical, requiring neither (132). All classes
can be activated by the tumor promoter phorbol ester and contain a regulatory N-
terminal region which binds the phorbol and the cofactor phosphatidylserine. Thus,
selective activation of isozymes provides one level of cellular control in cells expressing PKCs. Regulation is imposed via an autoinhibitory mechanism in which
an N-terminal zinc finger domain binds to the catalytic site in the C-terminal region (133). Proteolytic cleavage, phorbol esters or the appropriate combination of
lipid cofactors and calcium can reverse that interaction to free the catalytic site
and activate the enzyme. The lipid requirements for activation suggest that PKCs
may exert most of their effects at membranes, and translocation to membrane
fractions can be observed (131). The activation by phorbol ester tumor promoters
implies a role in cell proliferation, but direct analyses of PKC effects on cell growth
have given mixed results (133). These observations may be a consequence of the
role of specific localization of the PKC isozymes in different cellular contexts. Stimulation of cells with hormones or phorbol esters results in the translocation of most
of the PKCs to new cellular sites, including the cytoskeletal elements, nucleus and
plasma membrane, with site specificity observed for different isozymes (134). A
number of different PKC binding proteins have been identified (Table 2.5), including some proposed to localize the enzyme, such as PKC substrate/binding protein,
receptors for activated C kinase (RACKs) and proteins that interact with C kinase
(PICKs) (135). Since substrate specificity of PKC isozymes is relatively low, these
results suggest that *localization and complex formation* are the key elements in
controlling their actions. Recent studies indicate that PKCs are important integrators of signaling pathways (136), including the JAK/STAT, G-protein and receptor
tyrosine kinase pathways, actions which may be facilitated by binding of the PKCs
to multimeric complexes with components of these other pathways.

Calcium is an important factor in the regulation of eukaryotic cell processes
such as contractility (chapter 4), cell division (chapter 5) and exocytosis (chapter

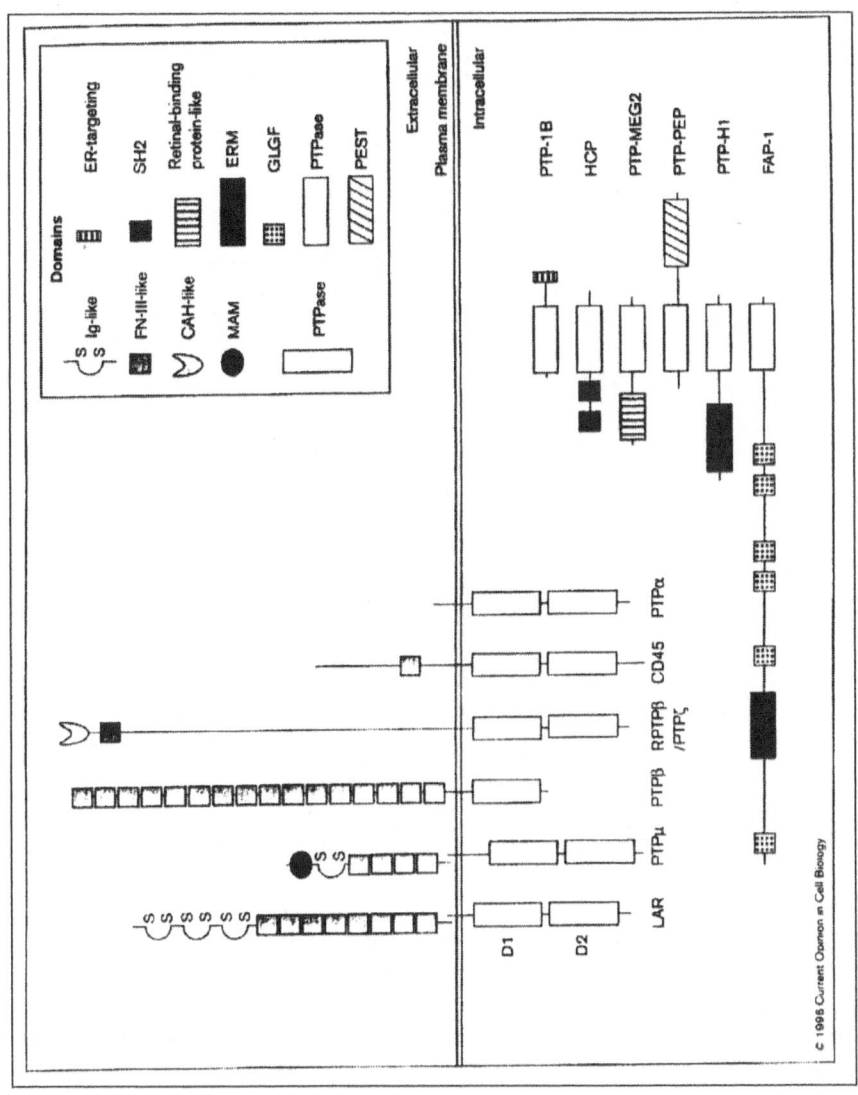

Fig. 2.8. Domain structures of major tyrosine phosphatases. Figure from ref.150 (Streuli M. Curr Opin Cell Biol 1996; 8: 182–188), courtesy of Current Biology Ltd.

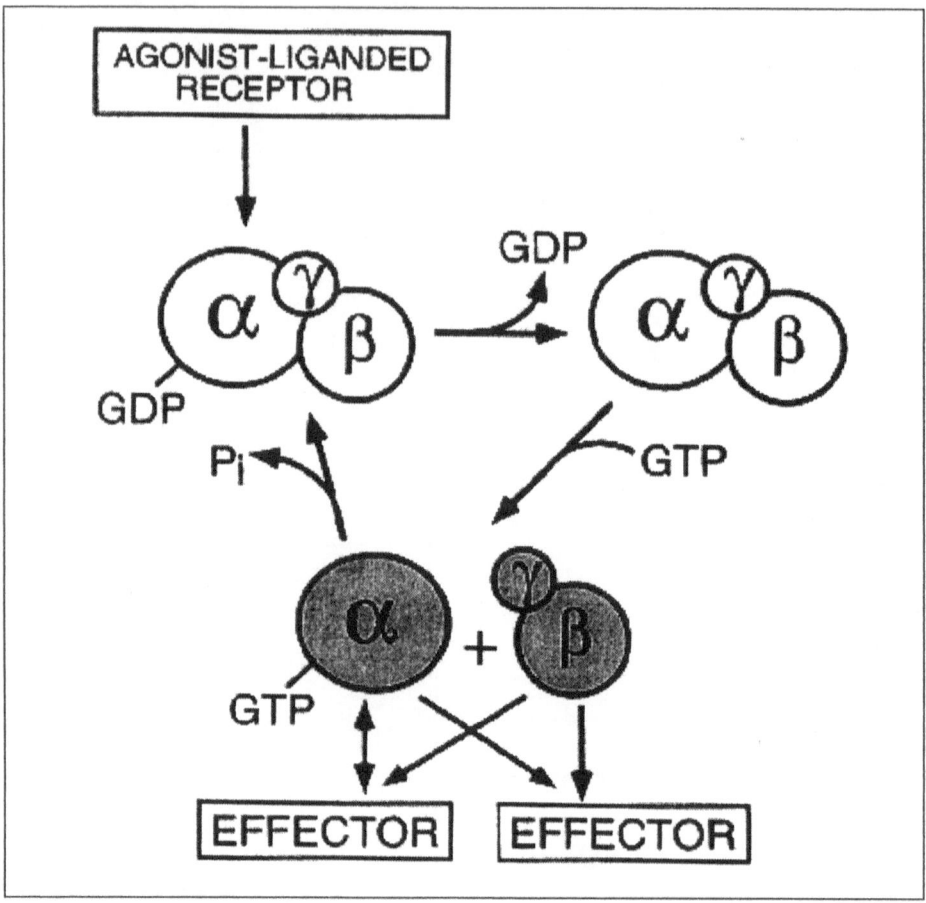

Fig. 2.9. Regulatory cycle of heterotrimeric G-proteins. Figure from ref. 158 (Neer EJ. Cell 1995; 80: 249-257), courtesy of Cell Press.

6). In addition to its effects on PKCs, calcium acts through the enzyme modulator calmodulin to stimulate multiple cellular functions. These effects result in part from actions on two classes of protein *CamKs* (Table 2.1): multifunctional and dedicated (137). Six different isozymes have been characterized, including the dedicated isozymes myosin light chain kinase, phosphorylase kinase and EF-2 kinase which are involved in regulation of contractility, glycogen metabolism and protein synthesis, respectively (14). How the nonspecific CamKs are regulated and contribute to calcium-mediated signaling in cellular processes is still unclear.

The *MAP kinase pathway* is comprised of a set of 3-4 enzymes involved in multiple functions, including the mitogenic pathway and stress pathways in mammals and several pathways in budding yeast, *Drosophila* and *C. elegans* (138). There are at least six separate pathways in mammals and four in yeast (139), including the eight shown in Figure 2.7, which are coupled to different extracellular signals and functions. This multiplicity undoubtedly plays a significant role in the diversity of responses available for different cell types and for the same cell under different conditions. In mammals the targets of the cascade include transcription factors

Table 2.6. Classes of G subunits

Class	Members	Modifying Toxin	Some Functions
α_s	α_s, α_{olf}	Cholera	Stimulate adenylyl cyclase, regulate Ca²⁺ channels
α_i	α_{i-1}, α_{i-2}, α_{i-3}, α_o, α_{t-1}, α_{t-2}, α_{gust}, α_z	Pertussis (except α_z)	Inhibit adenylyl cyclase, regulate K⁺ and Ca²⁺ channels, activate cGMP phosphodiesterase
α_q	α_q, α_{11}, α_{14}, α_{15}, α_{16},	–	Activate PLC
α_{12}	α_{12}, α_{13}	–	Regulate Na⁺/K⁺ exchange[b]

Modified from ref. 158 (Neer EJ. Cell 1995; 80: 249-257), courtesy of Cell Press.

regulating gene expression (140) which are phosphorylated by the terminal kinase of the cascade, called *mitogen-activated kinase (MAPK)* or *extracellular signal-regulated kinase (ERK)*. Specific family members have their own individual names. For example, the MAPK which phosphorylates the transcription factor Jun in mammals is called Jnk. ERKs (MAPKs) are activated by the *dual function* serine/threonine and tyrosine kinases, *mitogen-activated ERK-activating kinase (MEK)* (141). The activators for MEKs are called *MEK kinases (MEKK)*. In the case of the mammalian mitogenic pathway the MEKK is the protooncogene *Raf (MAP kinase kinase kinase)*. The activation of Raf is the least well understood aspect of this pathway, probably because it can be activated by multiple mechanisms (142). A critical step is the recruitment of Raf to the plasma membrane by activated Ras, the small G-protein switch (Fig. 2.1) (89) and to the membrante—microfilament interface (36). PKC and other factors can also activate Raf under some conditions, but which of these is involved in the normal pathway is unclear (143). Interestingly, Raf is inactivated by phosphorylation with PKA (144). In addition to the pathways shown in Figure 2.7, ERKs may contribute to the regulation of gene expression through other mechanisms, including the JAK-STAT pathway and by phosphorylating *ribosome specific kinases (Rsks)* which can then phosphorylate transcription factors as well as components of the protein synthesis system (145).

Phosphatases

Protein phosphatases regulate protein kinase signaling by catalyzing the removal of phosphate groups from proteins. Interestingly, phosphatases can act as either inhibitors or co-activators of kinases. The early opinion that phosphatases serve solely to reverse kinase effects was quickly modified when it was found that dephosphorylation of Src Tyr527, the autoinhibitory phosphorylation site, activates c-Src (113). Like protein kinases, there are two major classes of protein phosphatases, categorized on the basis of their specificities, serine/threonine-specific and tyrosine-specific. A third class, exemplified by the cdc25 phosphatases important in the cell cycle (chapter 5) and MAPK phosphatases, can act on both types of phosphorylated residues as a dual function phosphatase, though only in specific sequence contexts (146). Unlike the kinases, the two major classes appear to be unrelated in

Table 2.7. Effectors regulated by G-protein subunits

| Effector | Regulator | | Type of Gα/Gβγ |
	Gα	Gβγ	interaction
K⁺ channel ($I_{K.ACh}$)	Yes	Yes	Independent
K⁺ channel ($I_{K.ATP}$)	Yes	No	
PLA₂	?	Yes	
Pheromone response	No	Yes	
Adenylyl cyclase I	Yes	Yes	Antagonistic
Adenylyl cyclase II(IV)	Yes	Yes	Synergistic
Adenylyl cyclase III	Yes	No	
PLCβ1-3	Yes	Yes	Independent
Ca²⁺ channels (L and N)	Yes	Unknown	
cGMP PDE	Yes	No	
Muscarinic receptor kinase	No	Yes	
βARK	No	Yes	

Modified from ref. 163 (Clapham DE, Neer EJ. Nature 1993; 365: 403-406), courtesy of Nature.

sequence (147), suggesting that they arose by convergent evolution. As hydrolases, their mechanism of action appears simpler than that of the kinases. Like the tyrosine kinases the *protein tyrosine phosphatases* (*PTPs*) can be further divided into two classes based on structure and localization, membrane and nonmembrane (Fig. 2.8). The membrane tyrosine phosphatases are sometimes called receptor phosphatases because of their extracellular regions, which contain Ig, fibronectin and cadherin-like domains (148) that can serve as binding sites for other proteins. In fact, transfection studies have shown that the phosphatase extracellular domains can promote intercellular adhesion (149). Although it is unclear whether extracellular interactions modulate phosphatase activity as required for a receptor function, one model for phosphatase regulation suggests that dimerization of membrane tyrosine phosphatases may cause their inhibition (146). In addition, the extracellular domains may provide the information for localization of phosphatases at sites of adhesion to other cells or extracellular components as part of larger complexes at these sites.

Perhaps because of their relatively low specificities, *protein interactions and localization appear to be an important aspect of phosphatase behavior*. In addition to their catalytic domains, many *nonmembrane tyrosine phosphatases* have binding domains whose primary function appears to be to regulate interactions with other proteins and to determine cellular localization (150). These domains may be critical to determining functions of the PTPs. For example, two different PTPs with SH2 domains have been described. SH-PTP1 is primarily expressed in hematopoietic cells and blocks proliferative signaling (150). In contrast, SH-PTP2 is more widely expressed and is a positive transducer of mitogenic signaling. Interestingly, most membrane PTPs, but not nonmembrane PTPs, contain a second intracellular

Table 2.8. Small G-proteins

Subclass	Approximate number Mammals	Yeast	Representative proteins	Cellular function
Ras and Ras-like	6	3	Ha-, Ki-, N-ras (mammals) R-ras, Rap proteins (mammals) Ras1, Ras2 (*S. cerevisiae*)	Multiple
Ypt1/Sec4	8	2	Rab3 (mammals) Ypt1 (*S. cerevisiae*) Sec4 (*S. cerevisiae*)	Localized in synaptic vesicles Transport, ER to Golgi Vesicle transport to plasma membrane
Rho	3	3	Rho C (mammals) CDC42 (*S. cerevisae*)	Exoenzyme C3 of *C. botulinum* ADP-ribosylates Rho protein and disrupts actin microfilaments Inactivating mutation disrupts cytoskeleton and cell polarity
ARF-like	1	3	ADP-ribosylation factor (ARF)	Required for modification of α_s, catalyzed by cholera toxin

The subclasses of small GTPases are largely defined by similarities of primary structure in regions known to participate in formation of the guanine nucleotide binding pocket. Modified from ref. 155 (Bourne HR, Sanders DA, McCormick F. The GTPase superfamily: a conserved switch for diverse cell functions. Nature 1990; 348: 125-132), courtesy of Nature.

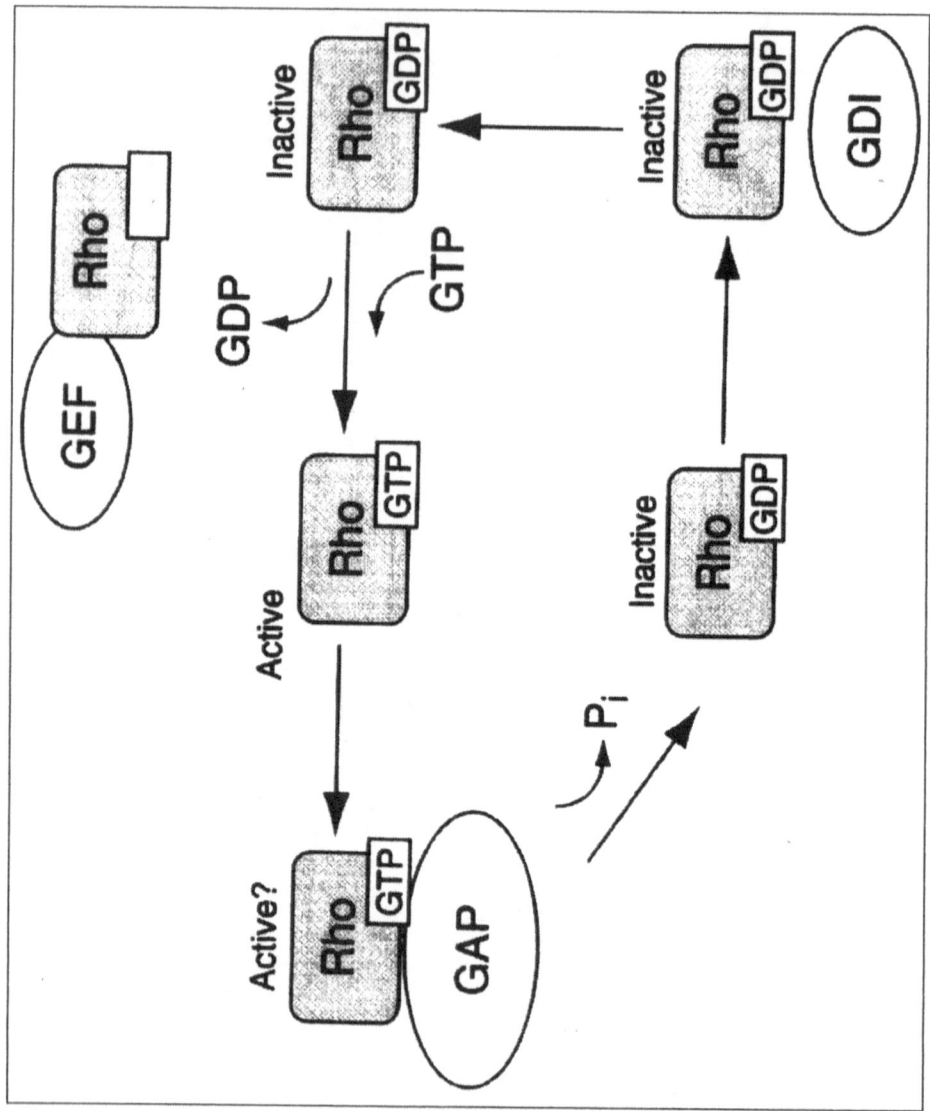

Fig. 2.10. GTPase cycle for Rho. Figure from ref. 211 (Machesky LM, Hall A. Trends Cell Biol.1996; 6: 304-310), courtesy of Current Biology Ltd.

domain related to the catalytic domain, but with little catalytic activity, suggesting an alternative role in binding cytoplasmic proteins (147).

Protein serine/threonine phosphatases (PPs) were originally discovered during studies of hormone regulation of glycogen metabolism, but have since been found to be involved in multiple, diverse aspects of cellular regulation (151). Four major forms of catalytic subunits (PP1, PP2A, PP2B, PP2C) have been described, plus a number of more minor members. Functional diversity among the different forms is generated by the interactions of the major forms with targeting or regulatory subunits which control their localization to cellular organelles, including cytoskeletal elements and their activities (152, 153). Both PP1 and PP2A have been implicated in cell cycle regulation, the former in mitosis and the latter in progression from G2 to M (see chapter 5) (152). PP1 activity can be regulated by phosphorylation of its regulatory subunit, while PP2A is regulated by phosphorylation of its catalytic subunits (152). Natural and synthetic inhibitors have been useful in defining the roles of these phosphatases (154). For example, the shellfish toxin okadaic acid is an inhibitor of PP1 and PP2A and a potent tumor promoter, implicating these phosphatases in important proliferation events. The immunosuppressant cyclosporin A forms complexes with cyclophilin which inhibit PP2B (calcineurin) and block IL-2 expression and T cell activation.

G-Proteins

There are two major classes of GTPases involved in signaling: heterotrimeric receptor-coupled large G-proteins and small G-proteins (155-157). Their primary similarity is that they transduce information by a switch mechanism in which the switch is on when GTP is bound and off when GDP is bound. The switch is regulated by the hydrolysis of GTP. Agents which increase this catalytic rate turn the switch off; those which facilitate exchange of GDP for GTP turn the switch on. The heterotrimeric G-proteins act specifically to couple signals from membrane serpentine receptors to a variety of enzymes and ion channels. In contrast, the small G-proteins regulate a multitude of cell functions, including proliferation, differentiation, gene expression, cytoskeleton assembly and membrane trafficking.

The role of heterotrimeric G-proteins as coupling factors is illustrated by the G-protein cycle (Fig. 2.9) (158). In unstimulated cells, G-protein is present as a GDP-associated heterotrimer in the plasma membrane which can bind to its complementary serpentine receptor. Activation of the receptor by ligand binding causes a conformational change in the receptor which is transmitted to G-protein and initiates an exchange of the GDP for GTP on the α subunit. GTP binding releases α from a complex of the βγ subunits, which allows both α and βγ to interact with downstream effectors to transmit the signal. This activated state is maintained until hydrolysis of the GTP by the α subunits permits reassociation into the heterotrimer. Specificity of the signal transmission is resident in both different types of receptors and different isoforms of the G-proteins. The four classes of G-proteins, defined by their α subunits, are listed in Table 2.6, along with some of their functions and toxins which are used experimentally to differentiate between the classes. Mammals have more than 20 α subunits, five β subunits and at least six γ subunits, but not all combinations are found. Membrane association is facilitated by myristoylation of some α subunits and prenylation of γ subunits (159).

A key aspect of the G-protein system is its ability to transmit and regulate opposing signals to modulate cell and tissue behavior (158). In the heart β-adrenergic stimulation increases contraction rate, while the muscarinic cholinergic response decreases the rate. The former is transmitted through the Gs class of G-proteins; the latter is transmitted through Gi and Gq classes. Since there are more types of receptors than G-proteins, there is also promiscuity in the receptor-G-protein interactions. In vitro studies in reconstituted vesicle systems also suggest a surprising lack of specificity of some receptors, which can act with both stimulatory and inhibitory G-proteins. Additional specificity may be achieved by variations in the expression levels of the G-proteins, by localization into specialized membrane regions, such as caveolae, focal adhesions or cytoskeletal complexes (160, 161) and by specific interactions of either α or βγ subunits with downstream effectors (162). Although many effectors are activated by either α or βγ, others are modulated by both. The effects may be either independent, synergistic or antagonistic (Table 2.7) (163). In some systems βγ also serves to facilitate phosphorylation of receptors as part of a feed-back loop for desensitization.

The Ras superfamily of small (20-35 kDa) G-proteins comprises a group of more than 50 members which can be divided into five primary families: Ras, Rho, Rab, Ran and ARF (Table 2.8) (164). Ras is noteworthy for its role in cell proliferation, Rho for cytoskeleton organization, Rab and ARF for membrane trafficking and Ran for nuclear import. However, recent demonstrations of interactions between pathways modulated by the different families suggest that these assignments may be too simplistic (165). All small G-proteins act through a GTPase cycle similar to that described for the heterodimeric G-proteins (166) (Fig. 2.10, illustrated for Rho). GDP-containing G-proteins in unstimulated cells undergo a GTP for GDP exchange catalyzed by *guanine nucleotide exchange factors (GEF)* (167). GTP-activated G-protein binds its effector to promote its activation. The G-protein is returned to its ground state by a slow GTP hydrolysis, which can be stimulated by *GTPase-activating proteins (GAPs)*. A third class of regulatory protein inhibits guanine nucleotide exchange (168). Multiple GAPs and GEFs are being identified for different G-proteins. Moreover, many of these proteins contain multiple binding domains (166, 169), suggesting their involvement in the formation of multimeric protein complexes as part of the signal transduction mechanism.

Mechanistically, the role of Ras in coupling receptor tyrosine kinase activation to the MAPK cascade is the best understood example of small G-protein function (Fig. 2.1) (36). Ras, which is prenylated, is a plasma membrane protein, as is the receptor kinase. Phosphorylation of the receptor recruits a complex of an SH2-containing adaptor Grb2 and a GEF Sos to the plasma membrane, where the GEF activates Ras. Activated Ras then recruits Raf to the plasma membrane, allowing it to interact with an unknown activator(s) and to associate with the cytoskeleton (36). This recruitment of effectors into multicomponent complexes has been proposed to be a general model for members of the Ras superfamily (170). Interestingly, Ras appears also to be activated through the heterotrimeric G-protein βγ dimer in a mechanism that links the large and small G-proteins at the membrane (157). A second concept for these proteins is that they are linked to multiple downstream effectors. For Ras these include Raf, PI3K, Rin1 and a Ral GTPase GEF. Interactions among the different small G-proteins or their complexes appear to be important for some cellular functions. For example, Rac and Rho are required for cell transformation by Ras (165). Recently, superoxide has been implicated in the pathway linking Ras through Rac to cell proliferation (171).

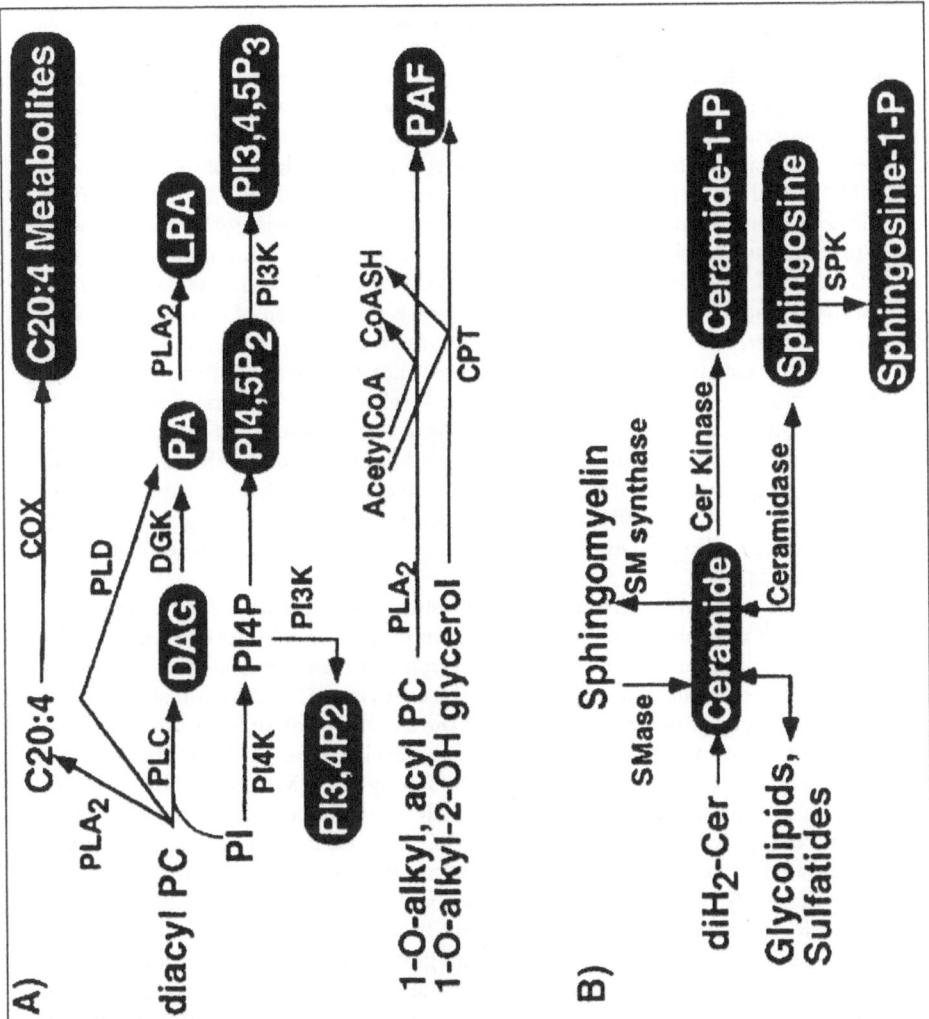

Fig. 2.11. Glycerolipid and sphingolipid pathways and second messengers. Figure from ref. 181 (Ghosh S, Strum JC, Bell RM. FASEB J 1997; 11: 45-50), courtesy of Federation of American Societies for Experimental Biology.

Regulators of Cyclic Nucleotide Levels

The *cyclic nucleotide cAMP* was the first intracellular transducer recognized and provided the original definition for the term second messenger (172). Studies on the enzymes involved in its synthesis (*adenylyl cyclase*) and degradation (*cAMP phosphodiesterase*) showed the importance of the regulation of second messenger levels in specific metabolic responses. Analysis of the regulation of the synthesis of cAMP gave the first example of membrane signal transduction, as binding of the hormone epinephrine to its serpentine receptor led to the receptor interaction with G-proteins, as described above, to trigger their dissociation, interaction with and activation of adenylyl cyclase and production of cAMP (172). The role of phosphodiesterase is similar to that of phosphatase, destruction of a second messenger. Moreover, cAMP phosphodiesterases contain regulatory domains involved in their cellular localization, as do phosphatases. Thus, the roles of both of these classes of enzymes may be to spatially restrict the levels of their substrates as a mechanism for the regulation of signaling (173). As described above, the effects of cAMP often result from its activation of PKA. We will not try to review the vast literature on the regulation of cAMP and the effects of cAMP and PKA on metabolism here. Instead, we will concentrate on their effects on regulating and integrating signaling pathways leading to changes in cell behavior through regulation of gene transcription (174).

Increases in cAMP levels cause opposite effects in different types of cells: growth in rat-1 or 3T3 fibroblasts and differentiation into neuronal cells of PC12 cells derived from rat adrenal medullary tissue (175). The transcription factor Elk-1 has been implicated in the differentiation of PC12 cells via cAMP activation of the MAPK cascade (176). Specifically, PKA activates the small G-protein Rap1, which is a selective activator of kinase isoform B-Raf and an inhibitor of Raf-1. Thus, the expression levels of these members of the Raf family of the MAPK cascade appear to regulate specific transcription factors which determine cell fate. This effect may depend on sustained activation of the ERKs because factors such as EGF which cause transient activation in PC12 cells do not induce differentiation (175), possibly because EGF acts via Ras, instead of Rap1 as a small G-protein switch. How these effects are integrated in a cell is one of the key puzzles in signal transduction, but based on other examples of signal integration, it is tempting to propose that the organization of multimeric complexes of signaling components of different pathways may be important. In other signaling pathways cAMP inhibits activation of transcription factors. In the Ras-mediated pathway, PKA can phosphorylate and decrease the activity of Raf-1, the first component of this MAPK cascade (174). Similarly, PKA can phosphorylate some isoforms of phospholipase C to block their hydrolysis of phospholipids and inositol lipid turnover, an important component of some signaling pathways, as indicated below.

One of the activities directly regulated by cyclic nucleotides is the *gated ion channel*. For example, discrimination of odors is achieved by specific olfactory receptors coupled to G-proteins which stimulate adenylyl cyclase to synthesize cAMP and open *cAMP-gated ion channels* (177). *cGMP-gated ion channels* are involved in the visual system. cGMP is synthesized by guanylate cyclase in retinal rod cells in the dark and maintains the channels in an open state. Absorption of light by the G-protein-linked receptor rhodopsin activates the G-protein transducin to stimulate phosphodiesterase, which cleaves cGMP to GMP, resulting in the closure of the channels and hyperpolarization of the cell membrane (178). Thus, cGMP

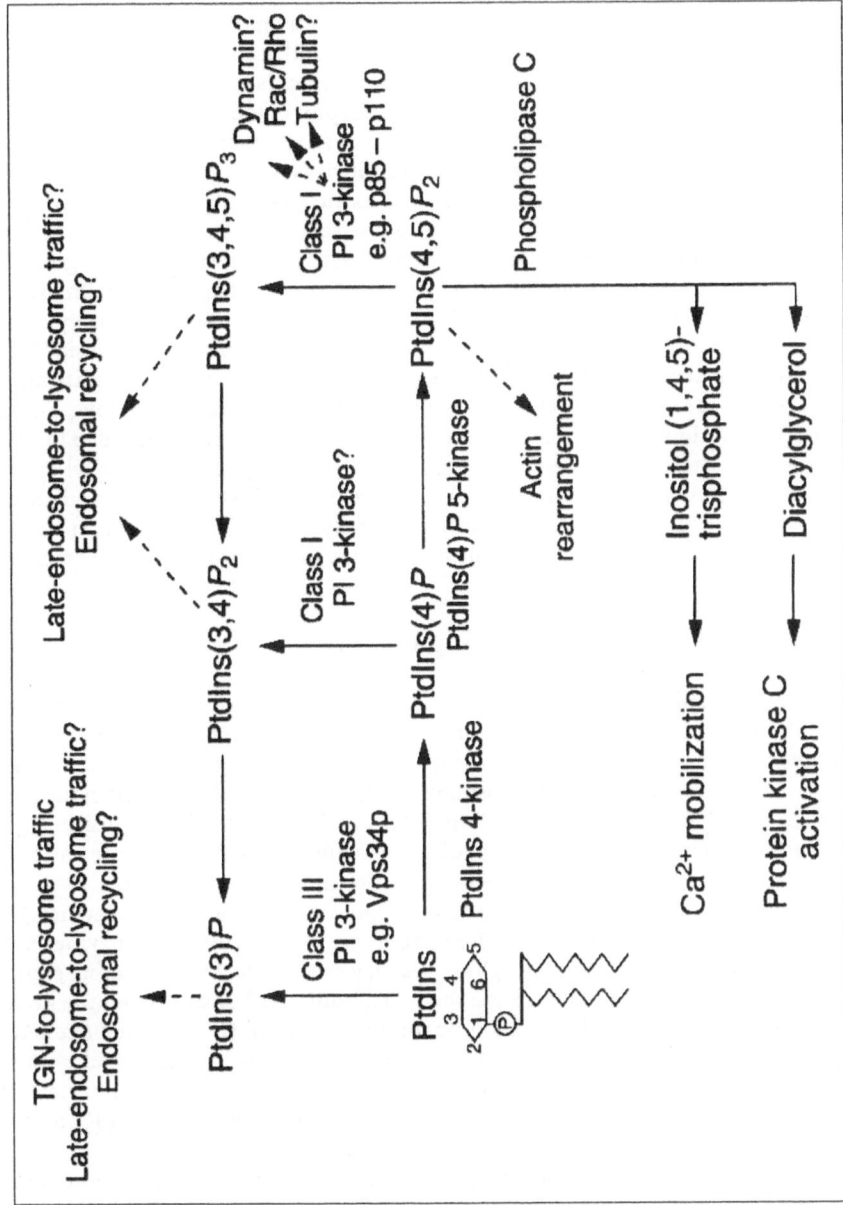

Fig. 2.12. Synthesis of phosphoinositides and their roles in cellular functions. Figure from ref. 194 (Shepherd PR, Reaves BJ, Davidson HW. Trends Cell Biol 1996; 6: 92-97), courtesy of Current Biology Ltd.

levels are acutely regulated by phosphodiesterase for amplification and a rapid response. In other systems cGMP is regulated by synthesis via the guanylyl cyclase receptors. Like tyrosine kinases, this enzymic receptor is found in two forms, membrane and cytoplasmic (179). Unlike the nonreceptor cytoplasmic tyrosine kinases, the cytoplasmic guanylyl cyclases are themselves receptors for small molecules, specifically nitric oxide and drugs such as nitrovasodilators. The ligands for the membrane receptors are a separate class of extracellular peptides, the natriuretic peptides involved in the control of blood pressure. The membrane guanylyl cyclase receptors have extracellular, transmembrane and intracellular regions. In addition to the catalytic domain, the intracellular region has a protein kinase homology domain, though this domain has little or no kinase activity. The cytoplasmic guanylyl cyclase receptors have a heme-containing domain for binding NO. In spite of their structural differences both types of receptors appear to require multimerization for activation. The analogy to the protein kinases is obvious, but the structural rationale is unclear unless two or more subunits are required to form an active site. Studies in sea urchin sperm showed that membrane guanylyl cyclase receptors are normally highly phosphorylated, but lose the phosphate and are desensitized upon ligand binding (180). Mammalian guanylyl cyclase receptors are desensitized by a similar process or by receptor internalization and degradation (179).

Lipids and Lipid-Metabolizing Enzymes

Glycerophospholipids

Products of both major classes of membrane lipids, glycerolipids and sphingolipids can act as second messengers for important signaling pathways (181). Moreover, as indicated in chapter 1, phosphoinositides have been implicated in the regulation of the actin cytoskeleton. Thus, a knowledge of the role of lipids in signal transduction is critical to understanding the regulation of cell structure and function. Figure 2.11 shows a summary of the major lipid species implicated as second messengers and their derivation. The key aspect in each case is the enzyme(s) involved in the production of the lipid products. In particular, the enzymes catalyzing hydrolysis of glycerophospholipids play critical roles in generating signals. *Phospholipases C* (PLCs) hydrolyze the phosphodiester bond between the glyceride backbone and the phosphate to form diacylglycerol (DAG), while *phospholipases D (PLDs)* hydrolyze between the phosphate and hydrophilic head group to form phosphatidic acid (Fig. 2.11). *Phospholipases A (PLAs)* hydrolyze the ester linkages between the fatty acids and glycerol to generate lysophospholipids and fatty acids.

The first indication of a role for phospholipids in signaling was the observation of the rapid turnover of inositol phospholipids in response to extracellular agents (6). Subsequent studies showed that activation of either G-protein-linked receptors or tyrosine kinase receptors could trigger these events through the activation of PLC isoforms. In the former case, the β isoform is involved, while in the latter case, the γ isoform is required (182). PLC-β isozymes are activated by interaction with either Gα or βγ, depending on the cell involved (183). In contrast, PLC-γ isoforms which contain SH2 domains, are activated by tyrosine phosphorylation after binding to phosphorylated tyrosine residues on the receptor cytoplasmic domain (184). In either case, the products of the hydrolysis of PIP_2 are DAG and IP_3, which are critical to the stimulation of PKC as described earlier. Hydrolysis also lowers the level of PIP_2 which is known to *associate with actin binding proteins*, such as gelsolin or profilin, and increase the level of DAG which can act as a *membrane site for actin filament nucleation* (185). Changes in PIP_2 levels may also affect the localization and functions of proteins with PH domains and the activation of some PLDs, for which PIP_2 serves as a cofactor (186).

PLDs have also been implicated in mitogenesis and cytoskeleton reorganizations, though the mechanism(s) for such effects remains unclear (187). One possibility is the production of lysophosphatidic acid, an effector of cell adhesion which acts through a heterotrimeric G-protein-linked receptor (see chapter 4) (188, 189) via sequential actions of PLD and PLA2 (Fig. 2.11). Alternatively or additionally, phosphatidic acid produced by PLD could be converted to DAG by a phosphatase. A number of mechanisms for the activation of PLD have been suggested, implicating such diverse components as tyrosine kinases, PLC, PKC and both small and heterotrimeric G-proteins (187). Src activation of PLD involves a cascade of small G-proteins including Ras and Ral, the latter of which directly binds the PLD (182).

PLA2s are notable for their ability to produce lysophospholipids and to release arachidonic acid from inositol phospholipids (190). The arachidonic acid can then be further metabolized to eicosanoids, which have potent regulatory effects, particularly in inflammatory responses. Two different forms of PLA2 have been described, secretory and cytoplasmic, which differ in structure, substrate specificity and regulation. Signaling components implicated in PLA2 activation include the βγ subunits of heterotrimeric G-proteins, PKC and tyrosine kinases, though the details of the activation mechanisms remain unclear.

Turnover of phosphoinositides requires the synthetic abilities of lipid kinases shown in Figure 2.11. Thus, it is of interest to understand how regulation of these enzymes contributes to lipid turnover and signaling. One additional component of these processes is the phophatidylinositol transfer protein, which binds the inositides and presents them to the appropriate enzymes (191). A possibly important example for the cytoskeleton is the regulation of phosphatidylinositol 4-phosphate 5-kinase by the small GTP-binding protein Rho (192), which has been implicated in microfilament organization. However, much more attention in recent years has been devoted to the phosphatidylinositol 3-kinases because of their implication in cell proliferation (193). Interestingly, the 3-phosphorylated inositols are resistant to PLC hydrolysis, indicating that they are not contributing to stimulated PI turnover. PI3K was first discovered in complexes with Src from transformed cells. More recent studies have identified three classes of the enzymes involved in the synthesis of the different 3-substituted inositides (Fig. 2.12) (194). Potential roles and mechanisms for the class I PI3Ks were clarified with the discovery that the enzyme is a heterodimer, consisting of catalytic (110 kDa) and regulatory (85 kDa) subunits. The presence of SH2 and SH3 domains in the latter permits its interaction with other signaling components in complexes at the membrane or other sites. For example, PI3K can be recruited to phosphorylated tyrosines on a number of receptor tyrosine kinases, both activating it and bringing it into juxtaposition with potential substrate molecules in the membrane (195). This mechanism is only one of several for activating PI3K which responds to signals through nonreceptor tyrosine kinases and G-protein-linked receptors as well as tyrosine kinase receptors (196, 197).

In spite of the large effort in studying PI3Ks, the mechanisms for their participation in cellular functions remain uncertain. One particularly important question is whether the enzyme itself plays a role in signaling or whether its product 3-phosphorylated inositides serve as second messengers in signaling or both. One suggested role is as an activator of atypical PKC isoforms (198) which are not regulated by DAG and calcium. PI3K has been shown to serve as a Ras effector (199). PI3K also binds Rac and may play a role through that mechanism in its regulation of microfilament organization in some cells (157). PI3Ks have been implicated in a number of other cellular functions (196). However, many of these studies used inhibitors whose specificity, particularly between PI3K isozymes, is not definitive (194). Moreover, PI3K has a protein kinase activity which is susceptible to the commonly used inhibitors and whose contributions to signaling is unclear. The best established role for PI3K is in intracellular membrane trafficking, though the mechanism remains unclear, particularly whether the effects are due to the enzyme itself or the products of the enzyme. Interactions of PI3K via its SH3 domain with dynamin increase the dynamin GTPase activity necessary for endocytosis (see chapter 6). However, results with PI3K inhibitors indicate that the PI3K activity is necessary for its effects on trafficking, observations that are most compatible with a role for the inositides, though the nature of that role remains unclear (194).

Sphingolipids

A second major lipid signaling pathway involves the sphingolipids (Fig. 2.11). Specifically, the sphingolipid precursor ceramide has been implicated in cellular functions as diverse as proliferation, differentiation, growth arrest and apoptosis (200). As in the case of the hydrolysis of glycerolipids by PLC to generate DAG, the sphingolipid signal is generated predominantly by the hydrolysis of the membrane phospholipid sphingomyelin by sphingomyelinase, which removes a choline phosphate group to generate ceramide (201). Sphingomyelinase is activated by a number of extracellular agents, including vitamin D derivatives, tumor necrosis factor α, γ-interferon, complement, dexamethasone, interleukin-1 and the apoptosis factor Fas ligand. Kinetic responses to these agents are complex and varied, ranging from seconds to hours and from less than 1-fold to 15-fold, thus confusing analyses of the coupling mechanism for sphingomyelinase activation (202). Ceramide has been shown to modulate at least three different target signaling components: a Ser/Thr protein kinase which can phosphorylate Raf-1, an isoform of PKC and a heterotrimeric member of the PP2A phosphatase family (200).

At least three different sphingomyelinases appear to be involved in responses to different effectors and in different subcellular compartments, including plasma membrane caveolae, endosomes and lysosomes (200). Two topologically and functionally distinct pathways for sphingomyelinase activation have been proposed in reponse to signaling through the TNF-α receptor, one involving a neutral sphingomyelinase linked to cell survival mediated through Raf and the MAPK cascade, and the other involving an acidic sphingomyelinase linked to cell death and mediated through MEKK and the JNK pathway (200). Each of these requires a second signal, survival through a PKC isoform and the NF-κB transcription factor, also possibly involving the acidic sphingomyelinase, and cell death via the ICE-like cysteine proteases (caspases) (200). These two pathways also appear to be compartmentalized differently. How cell survival is promoted is uncertain, but ceramides can induce cell cycle arrest at the G0/G1 boundary (see chapter 5) by mediating dephosphorylation of the retinoblastoma protein Rb.

Other sphingolipids may also have roles in signaling. Sphingosine is an inhibitor of some PKC isoforms, and both sphingosine and sphingosine 1-phosphate have been implicated in other cellular functions (182). However, defining the mechanisms of these actions requires delineation of the specific components involved.

Transcription Factors

The ultimate effect of many signaling pathways is a change in gene expression induced by modification of *transcription factors*. Transcription factors are multidomain proteins with DNA-binding, activation, nuclear localization and ligand binding domains (203). Most transcription factors also have dimerization domains, which are involved in regulation. There are several classes of DNA binding domains in transcription factors, including *helix-turn-helix* (HTH), *zinc finger* and *pleated sheet* motifs (47). *Homeodomains*, which play a critical role in orchestrating development, are a special class of transcription factors which contain HTH domains. Dimerization plays an important role in regulating transcription by increasing affinities for repeated sequences (homodimers) and expanding the repertoire of sequence combinations which can be recognized (heterodimers). Both *leucine zipper* and *helix-loop-helix* motifs mediate dimerization and DNA binding of transcription factors. Two types of transcription factors are necessary for activating specific gene expression. *General transcription factors* assemble into complexes at TATA sequences on the gene and promote the association of RNA polymerase. Phosphorylation of the polymerase by a kinase in the complex allows the initiation of RNA synthesis. Specific activation of transcription occurs when *specific transcription factors* bind both DNA and the complex of general transcription factors. The latter binding occurs on the activation domains which are rich in aspartic, glutamic, glutamine or proline residues.

Transcription factors may act as both activators and repressors of transcription, as proposed for Myc (204). Myc is a helix-loop leucine zipper type of transcriptions factor which is regulated by its heterodimerization with Max, a transcription factor subunit which also interacts with other factors such as Mad. Myc-Max acts as a modest transcriptional activator, while Mad-Max is a transcriptional repressor which may antagonize Myc-Max functions (204). Understanding the roles of specific transcription factors requires identification of their target genes. This is difficult for Myc because its activation can contribute to cell proliferation which stimulates expession of many genes. Nevertheless, several genes which should contribute to proliferation have been identified as specific Myc targets, including ornithine decarboxylase, involved in polyamine biosynthesis necessary for S phase, elF-2α, the rate-limiting enzyme in translation, cdc25A, a dual function phosphatase which activates cyclin-dependent kinases (see chapter 5) and p53, the tumor repressor (see chapters 5 and 7).

A number of other transcription factors play important roles in the signaling pathways mentioned above or later in this volume (203). The *AP-1/Fos/Jun* family form active heterodimers and less active homodimers via leucine zipper motifs and are targets for phorbol ester tumor promoters and the glucocorticoid receptor. *CREB* is a cAMP- and calcium-responsive transcription factor which can cross dimerize with members of the Jun family through its leucine zipper motifs. Such cross interactions of course amplify the diversity of responses which can be achieved by a given number of proteins. A superfamily of *steroid hormone receptors*, presumably derived from a common ancestral gene, have two highly conserved zinc finger domains and recognize DNA promoter/enhancer sequences called *hormone response elements.* Other transcription factors regulate tissue specific gene expression (*MyoD* in muscle) or during early development (*Oct* factors), providing the program necessary for the cellular changes which drive the organization of complex organisms.

Although the role of transcription factors in the nucleus is becoming clearer, the mechanisms for coupling cell surface and cytoplasmic signaling events to the nuclear functions remain uncertain. Three general strategies for the passage of information from the cell surface into the nucleus have been proposed: activation and translocation of cytoplasmic kinases such as MAPK, to the nucleus to modify nuclear transcription factors; direct activation of cytoplasmic transcription factors such as the STATs by phosphorylation before translocation to the nucleus; and release of transcription factors such as NF-κB from cytoplasmic sequestering proteins by phosphorylation (205). The key step in each case is phosphorylation by a kinase activated by a pathway derived from an external signal. Biochemical aspects of these pathways have largely been delineated. Specificity of the phosphorylations appears to be determined by interactions between the kinases and their substrates (206). What remains uncertain is the mechanism of translocation of components from the plasma membrane into the nucleus. Thus, a significant aspect of regulating gene expression is the nuclear import of kinases or transcription factors from the cytoplasm (207). The presence of nuclear localization signals is necessary, but not sufficient for transport through the nuclear membrane, because the signal sequence may be masked by nearby phosphorylated residues or by associated proteins. Phosphorylation by cytoplasmic kinases or dephosphorylation by phosphatases may therefore be necessary for nuclear import. Both PKA and PKC have been implicated in these processes.

The question of how kinases or transcriptional factors move to the nucleus is largely unanswered. The problem can be stated simply by looking at the expression of an "immediate-early" response gene such as *c-fos* (205). Its transcriptional induction occurs within minutes after stimulation with growth factors and requires phosphorylation of the transcription factor Elk-1 in the nucleus by MAPK (140). MAPK is activated by phosphorylation at the plasma membrane and translocated to the nucleus within 5-10 min. The consistency and organization of the cytoskeleton in the cytoplasm suggest that the translocation is not possible by protein diffusion. Alternatively, one must postulate either a transient disruption of the cytoskeleton to create a path or a cytoskeleton-associated transport mechanism to move the proteins. Possible mechanisms for translocation of transcription factors and other signaling proteins to the nucleus after an activation event are discussed in chapter 8.

Signaling and the Cytoskeleton

Since the 1970s, a plethora of studies have suggested the importance of cytoskeletal structures in cellular function. Early morphological and biochemical studies established the plasticity of microfilaments and microtubules during cellular processes. As described in chapter 1, intermediate filaments are much less dynamic, and a role largely in the maintenance of morphology was proposed for these structures. In this premolecular biology era, a significant number of concepts were established which paved the way for later investigation of molecular mechanisms involved in normal cell function. Roles for microtubules in cell division and in cilia and flagella were established, and investigations of molecular motors were initiated. In addition to the more obvious role of actin in myosin-based contraction of muscle, roles for microfilaments in motility and morphogenesis in non-muscle cells were proposed. The concept of physical associations between cytoskeletal structures and of cytoskeleton-associated proteins was developed, and studies on several important actin- and tubulin-associated proteins made major conceptual contributions to the understanding of the role of cytoskeletal organization in cell function. Studies comparing normal cell architecture and that of neoplastically transformed cells provided insights into the correlations between cytoskeletal organization and cell function alterations which afforded insights into the roles of specific cytoskeletal structures in normal cell processes.

Another important concept arising from these early studies is that of *membrane-associated actin*, also referred to as *cortical actin* or *membrane skeletal actin*. Not identifiable in the microscope or biochemically in the manner in which F-actin and actin bundles are, membrane-associated actin nonetheless was perceived as an important element in the transduction of signal. Studies beginning with the erythrocyte established that there are assemblies of proteins at the membrane whose function includes association with actin oligomers for the purpose of structural and organizational stability. This concept, developed in the largely metabolically unresponsive erythrocyte, was expanded with studies on structures resulting from activation events in more environmentally responsive cells. Studies on assemblies containing integrins in the focal adhesions of fibroblasts and in the integrin-containing membrane-microfilament interaction sites in stimulated platelets (described further in chapter 4) extended the scope of the significance of such complexes. The initial concept of *membrane-cytoskeletal interaction sites* as important *structural* elements (208, 209) was expanded to include transduction of extracellular signals to the cytoskeleton as a major function for these assemblies. The paradigm that *the assembly of large complexes associated with microfilaments is a requisite step in the transduction of signals* is a common one, evolving from studies in a multitude of biological systems. These vary in complexity from single cell organisms to oocytes to developing and adult tissues in organisms ranging from insects and worms to man. Observations on alterations in these complexes in tumor cells have served to substantiate their importance in the normally functioning cell. Such assemblies for signaling have been partially characterized for several of the receptor types described in previous sections of this chapter and will be discussed in more detail in subsequent chapters. Some of the better characterized examples include the focal adhesion, the activated platelet membrane-microfilament interaction site, the cadherin-mediated adherens junction, and microvilli of a constitutively activated mammary tumor cell. One noteworthy observation is that specialized cytoplasmic protein assemblies frequently mediate actin association with different receptor types and in different cell systems. The commonalities and differences in these complexes will be examined in greater detail in chapters 3, 4 and 7.

Another significant paradigm evolving from studies on a number of fronts is the importance of the *role of protein localization* in cellular regulation (36). This concept may also apply to mRNA localization as discussed in subsequent chapters. As an example, the separate observations that membrane localization is required for c-Src function (55) and that activated c-Src and v-Src are localized to microfilaments (210) implicated sites of membrane association with microfilaments as key elements in the transduction of signal through Src. A critical element in the initial targeting of Src to the membrane is its association via its posttranslational modification with a myristyl group (55). A similar requirement for localization to the membrane was found for Ras, and its mechanism of membrane association is another lipid modification, prenylation (55). In each case the lipid modification was required for membrane localization, demonstrating the *critical nature of such lipid posttranslational modifications* in protein function. Mechanisms for the assembly of actin at membranes are of great interest as well, since initial targeting to the membrane is a requisite first step in assembly of microfilament-associated signaling complexes. The role of profilin and phosphoinositides and of other actin binding membrane skeletal proteins in membrane targeting of actin was discussed in chapter 1. A subset of these membrane skeletal proteins contain SH3 domains, implicating these proteins in possible adaptor mechanisms for assembly of microfilaments with the membrane. Other mechanisms, such as direct binding of receptors to actin, have been observed, although this is an area which has not been widely explored. Known examples of direct receptor binding to actin will be discussed in chapters 3 and 7.

Summary

Cells respond to a wide array of influences impinging on them from their extracellular environment. These include such diverse signals as photons of light, sensory stimuli, physical forces such as stretch, steroid hormones and related molecules, growth factors and cytokines, and a host of metabolism-modulating factors of various types. Another class of important signals includes proteins associated with the extracellular matrix or adjacent cells which usually stimulate assembly of complexes containing cytoskeletal structures. Cellular responses to incoming signals are mediated by receptors which may be located intracellularly or at the cell surface, depending on the nature of the ligand. Cell surface receptors of two general types, enzymic or nonenzymic, transduce ligand-generated signals by the assembly of signaling complexes comprised of several classes of intracellular transducers. These include kinases, GTPases (G-proteins) and their modulators, cyclic nucleotide-regulating enzymes and specific classes of lipid-metabolizing enzymes. These complexes are assembled via specific protein-protein interactions mediated by several known classes of binding domains or motifs. Different cellular pathways can be activated after a signaling event by second messengers elicited in response to the signal, including Ca^{2+}, cyclic nucleotides and a number of products of membrane lipid degradation.

Binding of ligand to enzymic receptors activates their enzyme activities and usually stimulates a protein association mechanism which is involved in the transmission of the signal to the cytoplasm. In the case of receptor tyrosine kinases, the most prevalent type, the signal involves an autophosphorylation; in this case the second messenger can be considered to be the phosphorylation event itself. There are three primary classes of nonenzymic receptors, ion channel-linked, G-protein-linked and tyrosine kinase-linked. Ligand binding triggers channel modulation for the first of these and associations with G-proteins and tyrosine kinases, respectively, for the last two. Tyrosine kinase-associated nonenzymic receptors assemble signaling complexes in a similar fashion to the receptor tyrosine kinases. G-protein association with receptor causes dissociation into its α and βγ subunits, which can then activate ion channels or membrane or cytoplasmic enzymes. The consequence of all enzyme activations is the production of second messenger molecules, including phosphorylated proteins which can transmit signal to cytoplasmic sites or to nuclear functional sites via transcription factors. The second messengers then regulate such diverse functions as metabolic pathways, gene expression and cytoskeletal rearrangements. Both the cytoskeleton and individual membrane sites contribute to the last effect and play a major role in determining how cells respond to multiple signal inputs.

References

1. Alberts B, Bray D, Lewis J, Raff M, Roberts K, Watson JD, eds. Molecular Biology of the Cell, 3rd ed., Garland Publishing Co., New York, Chap. 18.
2. Massague J, Pandiella A. Membrane-activated growth factors. Annu Rev Biochem 1993; 62:515-541.
3. Sporn MB, Todaro GJ. Autocrine secretion and malignant transformation of cells. New Engl J Med 1980; 303:878-880.
4. Bejcek BE, Li DY, Deuel TF. Transformation by v-*sis* occurs by an internal autoactivation mechanism. Science 1989; 245:1496-1499.
5. Carraway KL, Carraway CAC, Carraway KL III. Roles of ErbB-3 and ErbB-4 in the physiology and pathology of the mammary gland. J Mammary Gland Biol Neoplasia 1997; 2:187-198.
6. Berridge MJ, Irvine RF. Inositol phosphates and cell signaling. Nature 1989; 341:197-205.
7. Nishizuka Y. The role of protein kinase C in cell surface signal transduction and tumour promotion. Nature 1984; 308: 693-697.
8. Suzuki K, Sorimachi H, Yoshizawa T, Kinbara K, Ishiura S. Calpain: novel family members, activation, and physiologic function. Biol Chem Hoppe Seyler 1995; 376:523-529.
9. Kawasaki H, Kawashima S. Regulation of the calpain-calpastatin system by membranes. Mol Membr Biol 1996; 13: 217-224.
10. Fox JE. Transmembrane signaling across the platelet integrin glycoprotein IIb-IIIa. Ann NY Acad Sci 1994; 714:75-87.
11. Wang KK, Yuen PW. Calpain inhibition: an overview of its therapeutic potential. Trends Pharmacol Sci 1994; 15:412-419.
12. Rogers S, Wells R, Rechsteiner M. Amino acid sequences common to rapidly degraded proteins: the PEST hypothesis. Science 1986; 234:364-368.
13. Rechsteiner M, Rogers SW. PEST sequences and regulation by proteolysis. Trends Biochem Sci 1996; 21:267-271.
14. Barnes JA, Gomes AV. PEST sequences in calmodulin-binding proteins. Mol Cell Biochem 1995; 149-150:17-27.
15. Nairn AC, Picciotto MR. Calcium/calmodulin-dependent protein kinases. Sem Cancer Biol 1994; 5:295-303.
16. Ullrich A, Schlessinger J. Signal transduction by receptors with tyrosine kinase activity. Cell 1990; 61:203-212.
17. Mayer BJ, Baltimore D. Signalling through SH2 and SH3 domains. Trends Cell Biol 1993; 3:8-13.
18. Pawson T, Schlessinger J. SH2 and SH3 domains. Curr Biol 1993; 3:434-442.
19. Cohen GB, Ren R, Baltimore D. Modular binding domains in signal transduction proteins. Cell 1995; 80:237-248.
20. van der Geer P, Pawson T. The PTB domain: a new protein module implicated in signal transduction. Trends Biochem Sci 1995; 20:277-280.
21. Koch CA, Anderson D, Moran MF, Ellis C, Pawson T. SH2 and SH3 domains: elements that control interactions of cytoplasmic signaling proteins. Science 1991; 252:668-674.
22. Songyang Z, Cantley LC. Recognition and specificity in protein tyrosine kinase-mediated signalling. Trends Biochem Sci 1995; 20:470-475.

23. Kavanaugh WM, Turck CW, Williams LT. PTB domain binding to signaling proteins through a sequence motif containing phosphotyrosine. Science 1995; 268:1177-1179.

24. Cowburn D. Adaptors and integrators. Structure 1996; 4:1005-1008.

25. Bork P, Margolis B. A phosphotyrosine interaction domain. Cell 1995; 80:693-694.

26. Eck MJ. A new flavor in phosphotyrosine recognition. Structure. 1995; 3:421-424.

27. Woods DF, Bryant PJ. The discs-large tumor suppressor gene of Drosophila encodes a guanylate kinase homolog localized at septate junctions. Cell 1991; 66: 451-464.

28. Itoh M, Nagafuchi A, Yonemura S, Kitani-Yasuda T, Tsukita S, Tsukita S. The 220-kD protein colocalizing with cadherins in non-epithelial cells is identical to ZO-1, a tight junction-associated protein in epithelial cells: cDNA cloning and immunoelectron microscopy. J Cell Biol 1993; 121:491-502.

29. Doyle DA, Lee A, Lewis J, Kim E, Sheng M, MacKinnon R. Crystal structures of a complexed and peptide-free membrane protein-binding domain: molecular basis of peptide recognition by PDZ. Cell 1996; 85:1067-76.

30. Saras J, Heldin CH. PDZ domains bind carboxy-terminal sequences of target proteins. Trends Biochem Sci 1996; 21: 455-458.

31. Songyang Z, Fanning AS, Fu C, Xu J, Marfatia SM, Chishti AH, Crompton A, Chan AC, Anderson JM, Cantley LC. Recognition of unique carboxyl-terminal motifs by distinct PDZ domains. Science 1997; 275:73-77.

32. Ponting CP, Blake DJ, Davies KE, Kendrick-Jones J, Winder SJ. ZZ and TAZ: new putative zinc fingers in dystrophin and other proteins. Trends Biochem Sci 1996; 21:11-13.

33. Marfatia SM, Cabral JH, Lin L, Hough C, Bryant PJ, Stolz L, Chishti AH. Modular organization of the PDZ domains in the human discs-large protein suggests a mechanism for coupling PDZ domain-binding proteins to ATP and the membrane cytoskeleton. J Cell Biol 1996; 135: 753-766.

34. Sudol M, Chen HI, Bougeret C, Einbond A, Bork P. Characterization of a novel protein-binding module—the WW domain. FEBS Lett 1995; 369:67-71.

35. Staub O, Rotin D. WW domains. Structure 1996; 4:495-499.

36. Carraway KL, Carraway CAC. Signaling, mitogenesis and the cytoskeleton: Where the action is. BioEssays 1995; 17:171-175.

37. Shaw G. The pleckstrin homology domain: an intriguing multifunctional protein module. BioEssays 1996; 18:35-46.

38. Gibson TJ, Hyvonen M,. Musacchio A, Saraste M, Birney E. PH domain: the first anniversary. Trends Biochem Sci 1994; 19:349-353.

39. Lemmon MA, Ferguson KM, Schlessinger J. PH domains: diverse sequences with a common fold recruit signaling molecules to the cell surface. Cell 1996; 85:621-624.

40. Aitken A. 14-3-3 proteins on the MAP. Trends Biochem Sci 1995; 20:95-97.

41. Aitken A. 14-3-3 and its possible role in co-ordinating multiple signalling pathways. Trends Cell Biol 1996; 6:341-347.

42. Morrison D. 14-3-3: modulators of signaling proteins? Science 1994; 266:56-57.

43. Muslin AJ, Tanner JW, Allen PM, Shaw AS. Interaction of 14-3-3 with signaling proteins is mediated by the recognition of phosphoserine. Cell 1996; 84:889-897.

44. Klug A. Protein motifs 5. Zinc fingers. FASEB J 1995; 9:597-604.

45. Coleman JE. Zinc proteins: enzymes, storage proteins, transcription factors, and replication proteins. Annu Rev Biochem 1992; 61:897-946.

46. Hanas JS, Hazuda DJ, Bogenhagen DF, Wu FY-H, Wu C-W. Xenopus transcription factor A requires zinc for binding to the 5S RNA gene. J Biol Chem 1993; 258:14120-14125.

47. Pabo CO, Sauer RT. Transcription factors: structural families and principles of DNA recognition. Annu Rev Biochem 1992; 61:1053-1095.

48. Sanchez-Garcia I, Rabbitts TH. The LIM domain: a new structural motif found in zinc-finger-like proteins. Trends Genet 1994; 10:315-320.

49. Crawford AW, Pino JD, Beckerle MC. Biochemical and molecular characterization of the chicken cysteine-rich protein, a developmentally regulated LIM-domain protein that is associated with the actin cytoskeleton. J Cell Biol 1994; 124: 117-127.

50. Dawid IB, Toyama R, Taira M. LIM domain proteins. Compt Rend 1995; 318: 295-306.

51. Busch H. The final common pathway of cancer. Cancer Res 1990; 50:4830-4838.

52. Arber S, Caroni P. Specificity of single LIM motifs in targeting and LIM/LIM interactions in situ. Genes Devel 1996; 10:289-300.

53. Sadler I, Crawford AW, Michelsen JW, Beckerle MC. Zyxin and cCRP: two interactive LIM domain proteins associated with the cytoskeleton. J Cell Biol 1992; 119:1573-1587.

54. Turner CE, Miller JT. Primary sequence of paxillin contains putative SH2 and SH3 domain binding motifs and multiple LIM domains: identification of a vinculin and pp125Fak-binding region. J Cell Sci 1994; 107:1583-1591.

55. Resh MD. Regulation of cellular signalling by fatty acid acylation and prenylation of signal transduction proteins. Cell Signal 1996; 8:403-412.

56. Bhatnagar RS, Gordon JI. Understanding covalent modifications of proteins by lipids: where cell biology and biophysics meet. Trends Cell Biol 1997; 7:14-20.

57. Udenfriend S, Kodukula K. How glycosylphosphatidylinositol-anchored membrane proteins are made. Annu Rev Biochem 1995; 64:563-591.

58. Carraway CAC, Carvajal ME, Li Y, Carraway KL. Association of p185[neu] with microfilaments via a large glycoprotein complex in mammary carcinoma microvilli. Evidence for a microfilament-associated signal transduction particle. J Biol Chem 1993; 268:5582-5587.

59. Clark EA, Brugge JS. Integrins and signal transduction pathways: the road taken. Science 1995; 268:233-239.

60. Raff MC. Social controls on cell survival and cell death. Nature 1992; 356:397-400.

61. Nagata S. Apoptosis by death factor. Cell 1997; 88:355-365.

62. Carpenter G, Cohen S. Epidermal growth factor. Ann Rev Biochem 1979; 48: 193-216.

63. Mroczkowski B, Reich M, Chen K, Bell GI, Cohen S. Recombinant human epidermal growth factor precursor is a glycosylated membrane protein with biological activity. Mol Cell Biol 1989; 9: 2772-2779.

64. Riese DJ II, Bermingham Y, van Raaij TM, Buckley S, Plowman GD, Stern DF. Betacellulin activates the epidermal growth factor receptor and erbB-4, and induces cellular response patterns distinct from those stimulated by epidermal growth factor or neuregulin-beta. Oncogene 1996; 12: 345-353.

65. Carraway KL III, Cantley LC. A neu acquaintance for ErbB3 and ErbB4: a role for receptor heterodimerization in growth signaling. Cell 1994; 78:5-8.

66. Johnson GR, Wong L. Heparan sulfate is essential to amphiregulin-induced mitogenic signaling by the epidermal growth factor receptor. J Biol Chem 1994; 269: 27149-27154.

67. Nakamura K, Iwamoto R, Mekada E. Membrane-anchored heparin-binding EGF-like growth factor (HB-EGF) and diphtheria toxin receptor-associated protein (DRAP27)/CD9 form a complex with integrin $\alpha 3\beta 1$ at cell-cell contact sites. J Cell Biol 1995; 129:1691-1705.

68. Fagotto F, Gumbiner BM. Cell contact-dependent signaling. Devel Biol 1996; 180:445-454.

69. Yamamoto D. Signaling mechanisms in induction of the R7 photoreceptor in the developing *Drosophila* retina. BioEssays 1994; 16:237-244.
70. Kramer H. Patrilocal cell-cell interactions: sevenless captures its bride. Trends Cell Biol 1993; 3:103-105.
71. Flaumenhaft R, Rifkin DB. The extracellular regulation of growth factor action. Mol Biol Cell 1992; 3:1057-1065.
72. Tuzi NL, Gullick WJ. *eph*, the largest known family of putative growth factor receptors. Br J Cancer 1994; 69:417-421.
73. Bruckner K, Pasquale EB, Klein R. Tyrosine phosphorylation of transmembrane ligands for Eph receptors. Science 1997; 275:1640-1643.
74. Holland SJ, Gale NW, Mbamalu G, Yancopoulos GD, Henkemeyer M, Pawson T. Bidirectional signaling through the eph-family receptor Nuk and its transmembrane ligands. Nature 1996; 383:722-725.
75. Carraway KL, Fregien N, Carraway KL III, Carraway CAC. Tumor sialomucin complexes as tumor antigens and modulators of cellular interactions and proliferation. J Cell Sci 1992; 103:299-307.
76. Taipale J, Keski-Oja J. Growth factors in the extracellular matrix. FASEB J 1997; 11:51-59.
77. Juliano RL, Haskill S. Signal transduction from the extracellular matrix. J Cell Biol 1993; 120:577-585.
78. Lafrenie RM, Yamada KM. Integrin-dependent signal transduction. J Cell Biochem 61:543-53, 1996
79. Hynes RO. Integrins: versatility, modulation and signaling in cell adhesion. Cell 1992; 69:11-25.
80. Juliano R. Cooperation between soluble factors and integrin-mediated cell anchorage in the control of cell growth and differentiation. BioEssays 1996; 18: 911-917.
81. Loeb JA, Fischbach GD. ARIA can be released from extracellular matrix through cleavage of a heparin-binding domain. J Cell Biol 1995; 130:127-135.
82. Schlessinger J, Lax I, Lemmon M. Regulation of growth factor activation by proteoglycans: what is the role of the low affinity receptors? Cell 1995; 83:357-360.
83. Vlodavsky I, Miao HQ, Medalion B, Danagher P, Ron D. Involvement of heparan sulfate and related molecules in sequestration and growth promoting activity of fibroblast growth factor. Cancer Metastasis Rev 1996; 15:177-86.
84. Schubert D. Collaborative interactions between growth factors and the extracellular matrix. Trends Cell Biol 1992; 2: 63-66.
85. Wahli W, Martinez E. Superfamily of steroid nuclear receptors: positive and negative regulators of gene expression. FASEB J 1991; 5:2243-2249.
86. Pratt WB, Sanchez ER, Bresnick EH, Meshinchi S, Scherrer LC, Dalman FC, Welsh MJ. Interaction of the glucocorticoid receptor with the Mr 90,000 heat shock protein: an evolving model of ligand-mediated receptor transformation and translocation. Cancer Res 1989; 49: 2222s-2229s.
87. Posas F, Saito H. Osmotic activation of the HOG MAPK pathway via Ste11p MAPKKK: scaffold role of Pbs2p MAPKK. Science 1997; 276:1702-1705.
88. Fantl WJ, Johnson DE, Williams LT. Signalling by receptor tyrosine kinases. Annu Rev Biochem 1993; 62:453-481.
89. Carraway CAC, Carraway KL. In: Hesketh HE, Pryme IF, eds. *Treatise on the Cytoskeleton*, Greenwich, CT: JAI Press, 1996:207-238.
90. Panayotou G, Waterfield MD. The assembly of signalling complexes by receptor tyrosine kinases. BioEssays 1993; 15: 171-177.
91. White MF, Kahn CR. The insulin signaling system. J Biol Chem 1994; 269:1-4.
92. Claesson-Welch L. Platelet-derived growth factor receptor signals. J Biol Chem 1994; 269:32023-32026.

93. Hynes NE, Stern DF. The biology of *erbB-2/neu/HER-2* and its role in cancer. Biochim Biophys Acta 1994; 1198: 165-184.

94. Weiss FU, Daub H, Ullrich A. Novel mechanisms of RTK signal generation. Curr Opin Genet Devel 1997; 7:80-86.

95. Boonstra J, Rijken P, Humbel B, Cremers F, Verkleij A, van Bergen en Henegouwen P. The epidermal growth factor. Cell Biol Int 1995; 19:413-430.

96. Yarden Y, Ullrich A. Growth factor receptor tyrosine kinases. Annu Rev Biochem 1988; 57:443-478.

97. Sorkin A, Waters CM. Endocytosis of growth factor receptors. BioEssays 1993; 15:375-382.

98. Kanai Y, Ochiai A, Shibata T, Oyama T, Ushijima S, Akimoto S, Hirohashi S. c-erbB-2 gene product directly associates with β-catenin and plakoglobin. Biochem Biophys Res Commun 1995; 208:1067-1072.

99. Massague J. TGF signaling: receptors, transducers, and Mad proteins. Cell 1996; 85:947-950.

100. Massague J, Polyak K. Mammalian antiproliferative signals and their targets. Curr Opin Genet Devel 1995; 5:91-96.

101. Massague J, Weis-Garcia F. Serine/threonine kinase receptors: mediators of transforming growth factor beta family signals. Cancer Surv 1996; 27:41-64.

102. Massague J, Hata A, Liu F. TGF-β signalling through the Smad pathway. Trends Cell Biol 1997; 7:187-192.

103. Karlin A. Structure of nicotinic acetylcholine receptors. Curr Opin Neurobiol 1993; 3:299-309.

104. Strader CD, Fong TM, Tota MR, Underwood D, Dixon RAF. Structure and function of G-protein-coupled receptors. Annu Rev Biochem 1994; 63: 101-132.

105. van Biessen T, Luttrell LM, Hawes BE, Lefkowitz RJ. Mitogenic signaling via G-protein-coupled receptors. Endoc Rev 1996; 17:698-714.

106. Collins S, Lohse MJ, O'Dowd B, Caron MG, Lefkowitz RJ. Structure of G-protein-coupled receptors: the beta 2-adrenergic receptor as a model. Vitamins Hormones 1991; 46:1-39.

107. Hargrave PA. Seven-helix receptors. Curr Opin Struct Biol 1991; 1:575-581.

108. Bohm SK, Grady EF, Bunnett NW. Regulatory mechanisms that modulate signalling by G-protein-coupled receptors. Biochem J 1997; 322:1-18.

109. Taniguchi T. Cytokine signaling through nonreceptor protein tyrosine kinases. Science 1995; 268:251-255.

110. Weiss A. T cell antigen receptor signal transduction: a tale of tails and cytoplasmic protein-tyrosine kinases. Cell 1993; 73:209-212.

111. Defranco AL. Transmembrane signaling by antigen receptors of B and T lymphocytes. Curr Opin Cell Biol 1995; 7: 163-175.

112. Weiss A, Littman DR. Signal transduction by lymphocyte antigen receptors. Cell 1994; 76:263-274.

113. Brown MT, Cooper JA. Regulation, substrates and functions of src. Biochim Biophys Acta 1996; 1287:121-149.

114. Erpel T, Courtneidge SA. Src family protein tyrosine kinases and cellular signal transduction pathways. Curr Opin Cell Biol 1995; 7:176-182.

115. Perlmutter RM, Levin SD, Appleby MW, Anderson SJ, Alberola-Ila J. Regulation of lymphocyte function by protein phosphorylation. Annu Rev Immunol 1993; 11:451-499.

116. Chan AC, Desai DM, Weiss A. The role of protein tyrosine kinases and protein phosphatases in T cell antigen receptor signal transduction. Annu Rev Immunol 1994; 12:555-592.

117. Superti-Fulga G, Courtneidge SA. Structure-function relationships in Src family and related protein tyrosine kinases. BioEssays 1995; 17:321-330.

118. Cross FR, Garber EA, Pellman D, Hanafusa H. A short sequence in the p60src N terminus is required for p6osrc myristylation and membrane association and for cell transformation. Mol Cell Biol 1984; 4:1834-42.

119. Chung S-W, Wong PMC. The biology of Abl during hemopoietic stem cell differentiation and development. Oncogene 1995; 10:1261-1268.

120. Pendergast AM. Nuclear tyrosine kinases: from Abl to WEE1. Curr Opin Cell Biol 1996; 8:174-181.

121. Wang JYJ. Abl tyrosine kinase in signal transduction and cell-cycle regulation. Curr Opin Genetics Devel 1993; 3:35-43.

122. Courtneidge SA. Protein tyrosine kinases, with emphasis on the Src family. Sem Cancer Biol 1994; 5:239-246.

123. Schaller MD, Parsons JT. Focal adhesion kinase and associated proteins. Curr Opin Cell Biol 1994; 6:705-710.

124. Hanks SK, Polte TR. Signaling through focal adhesion kinase. BioEssays 1997; 19: 137-145.

125. Darnell JE Jr, Kerr IM, Stark GM. JAK-STAT pathways and transcriptional activation in response to IFNs and other extracellular signaling proteins. Science 1994; 264:1415-1421.

126. Briscoe J, Kohlhuber F, Muller M. JAKs and STATs branch out. Trends Cell Biol 1996; 6:336-340.

127. Hanks SK, Quinn AM, Hunter T. The protein kinase family: conserved features and deduced phylogeny of the catalytic domains. Science 1988; 241:42-52.

128. Taylor SS, Bubis J, Toner-Webb J, Sarawat LD, First EA, Buechler JA, Knighton DR, Sowadski J. cAMP-dependent protein kinase: prototype for a family of enzymes. FASEB J 1988; 2: 2677-2685.

129. Beebe SJ. The cAMP-dependent protein kinases and cAMP signal transduction. Sem Cancer Biol 1994; 5:285-294.

130. Rubin CS. A kinase anchor proteins and the intracellular targeting of signals carried by cyclic AMP. Biochim Biophys Acta 1994; 1224:467-479.

131. Mochly-Rosen D. Localization of protein kinases by anchoring proteins; a theme in signal transduction. Science 1995; 268: 247-251.

132. Jaken S. Protein kinase C isozymes and substrates. Curr Opin Cell Biol 1996; 8: 168-173.

133. Stabel S. Protein kinase C-an enzyme and its relatives. Sem Cancer Biol 1994; 5: 277-284.

134. Inagaki N, Ito M, Nakano T, Inagaki M. Spatiotemporal distribution of protein kinase and phosphatase activities. Trends Biochem Sci 1994; 19:448-452.

135. Faux MC, Scott JD. More on target with protein phosphorylation: conferring specificity by location. Trends Biochem Sci 1996; 21:312-315.

136. Nishizuka Y. Protein kinase C and lipid signaling for sustained cellular responses. FASEB J 1995; 9:484-496.

137. Cohen P. Signal integration at the level of protein kinases, protein phosphatases and their substrates. Trends Biochem Sci 1992; 17:408-413.

138. Kyriakis JM, Avruch J. Protein kinase cascades activated by stress and inflammatory cytokines. BioEssays 1996; 18: 567-577.

139. Robinson MJ, Cobb MH. Mitogen-activated protein kinase pathways. Curr Opin Cell Biol 1997; 9:180-186.

140. Davis RJ. Transcriptional regulation by MAP kinases. Molec Reprod Devel 1995; 42:459-467.

141. Seger R, Krebs EG. The MAPK signaling cascade. FASEB J 1995; 9:726-735.

142. Morrison DK, Cutler RE Jr. The complexity of Raf-1 regulation. Curr Opin Cell Biol 1997; 9:174-179.

143. Cobb MH, Goldsmith EJ. How MAP kinases are regulated. J Biol Chem 1995; 270:14843-14846.

144. Magnuson NS, Beck T, Vahidi H, Hahn H, Smola U, Rapp UR. The Raf-1 serine/threonine protein kinase. Sem Cancer Biol 1994; 5:247-253.

145. Ferrari S, Thomas G. S6 phosphorylation and the $p70^{s6k}/p85^{s6k}$. Crit Rev Biochem Molec Biol 1994; 29:385-413.

146. Neel BG, Tonks NK. Protein tyrosine phosphatases in signal transduction. Curr Opin Cell Biol 1997; 9:193-204.

147. Denu JM, Stuckey JA, Saper MA, Dixon JE. Form and function in protein dephosphorylation. Cell 1996; 87:361-364.

148. Brady-Kalnay S, Tonks NK. Receptor protein tyrosine phosphatases, cell adhesion and signal transduction. Adv Prot Phosphatases. 1996; 8:227-257.

149. Brady-Kalnay S, Tonks NK. Protein tyrosine phosphatases as adhesion receptors. Curr Opin Cell Biol 1995; 7: 650-657.

150. Streuli M. Protein tyrosine phosphatases in signaling. Curr Opin Cell Biol 1996; 8:182-188.

151. Shenolikar S. Protein serine/threonine phosphatases-new avenues for cell regulation. Annu Rev Cell Biol 1994; 10: 55-86.

152. Wera S, Hemmings BA. Serine/threonine protein phosphatases. Biochem J 1995; 311:17-29.

153. Barford D. Molecular mechanisms of the protein serine/threonine phosphatases. Trends Biochem Sci 1996; 21:407-412.

154. MacKintosh C, MacKintosh RW. Inhibitors of protein kinases and phosphatases. Trends Biochem Sci 1994; 19:444-448.

155. Bourne HR, Sanders DA, McCormick F. The GTPase superfamily: a conserved switch for diverse cell functions. Nature 1990; 348:125-132.

156. Bourne HR, Sanders DA, McCormick F. The GTPase superfamily: conserved structure and molecular mechanism. Nature 1991; 349:117-127.

157. Bokoch GM. Interplay between Ras-related and heterotrimeric GTP binding proteins: lifestyles of the BIG and little. FASEB J 1996; 10:1290-1295.

158. Neer EJ. Heterotrimeric G-proteins: organizers of transmembrane signals. Cell 1995; 80:249-257.

159. Casey PJ. Lipid modification of G-proteins. Curr Opin Cell Biol 1994; 6: 219-225.

160. Neubig RR. Membrane organization in G-protein mechanisms. FASEB J 1994; 8: 939-946.

161. Rodbell M. The role of GTP-binding proteins in signal transduction: from the sublimely simple to the conceptually complex. Curr Top Cell Regulat 1992; 32: 1-47.

162. Neer EJ. G-proteins: critical control points for transmembrane signals. Prot Sci 1994; 3:3-14.

163. Clapham DE, Neer EJ. New roles for G-protein βγ-dimers in transmembrane signaling. Nature 1993; 365:403-406.

164. Hall A. Ras-related proteins. Curr Opin Cell Biol 1993; 5:265-268.

165. Symons M. Rho family GTPases: the cytoskeleton and beyond. Trends Biochem Sci 1996; 21:178-181.

166. Boguski MS, McCormick F. Proteins regulating Ras and its relatives. Nature 1993; 366:643-654.

167. Feig LA. Guanine-nucleotide exchange factors: a family of positive regulators of Ras and related GTPases. Curr Opin Cell Biol 1994; 6:204-211.

168. Khosravi-Far R, Der CJ. The Ras signal transduction pathway. Cancer Metas Rev 1994; 13:67-89.

169. Cerione RA, Zheng Y. The Dbl family of oncogenes. Curr Opin Cell Biol 1996; 8:216-222.

170. Macara IG, Lounsbury KM, Richards SA, Mckiernan C, Bar-Sagi D. The Ras superfamily of GTPases. FASEB J 1996; 10: 625-630.

171. Irani K, Xia Y, Zweier JL, Sollott SJ, Der CJ, Fearon ER, Sundaresan M, Finkel T, Goldschmidt-Clermont PJ. Mitogenic signaling mediated by oxidants in Ras-transformed fibroblasts. Science 1997; 275:1649-1651.

172. Krebs EG. Nobel lecture: protein phosphorylation and cellular regulation I. Biosci Rep 1993; 13:127-142.

173. Degerman E, Belfrage P, Manganiello VC. Structure, localization, and regulation of cGMP-inhibited phosphodiesterase (PDE3). J Biol Chem 1997; 272:6823-6826.

174. Graves LM, Lawrence JC Jr. Insulin, growth factors, and cAMP. Trends Endocrinol Metab 1996; 7:43-50.

175. Marshall CJ. Specificity of receptor tyrosine kinase signaling: transient versus sustained extracellular signal-regulated kinase activation. Cell 1995; 80:179-185.

176. Vossler MR, Yao H, York RD, Pan M-G, Rim CS, Stork PJS. cAMP activates MAP kinase through a B-Raf- and Rap1-dependent pathway. Cell 1997; 89:73-82.

177. Buck LV. The olfactory multigene family. Curr Opin Neurobiol 1992; 2: 282-288.

178. Kaupp UB, Koch KW. Role of cGMP and Ca^{2+} in vertebrate photoreceptor excitation and adaptation. Annu Rev Physiol 1992; 54:153-176.

179. Garbers DL, Lowe DG. Guanylyl cyclase receptors. J Biol Chem 1994; 269:30741-30744.

180. Garbers DL. Molecular basis of fertilization. Annu Rev Biochem 1989; 58: 719-742.

181. Ghosh S, Strum JC, Bell RM. Lipid biochemistry: functions of glycerolipids and sphingolipids in cellular signaling. FASEB J 1997; 11:45-50.

182. Spiegel S, Foster D, Kolesnick R. Signal transduction through lipid second messengers. Curr Opin Cell Biol 1996; 8: 159-167.

183. Katan M. The control of inositol lipid hydrolysis. Cancer Surv 1996; 27: 199-211.

184. Wahl M, Carpenter G. Selective phospholipase C activation. BioEssays 1991; 13:107-113.

185. Shariff A, Luna EJ. Diacylglycerol-stimulated formation of actin nucleation sites at plasma membranes. Science 1992; 256: 245-247.

186. Lee SB, Rhee SG. Significance of PIP_2 hydrolysis and regulation of phospholipase C isozymes. Curr Opin Cell Biol 1995; 7:183-189.

187. Boarder MR. A role for phospholipase D in control of mitogenesis. Trends Pharm Sci 1994; 15:57-62.

188. Moolenar WH. Lysophosphatidic acid, a multifunctional phospholipid messenger. J Biol Chem 1995; 270:12949-12952.

189. Moolenar WH. Lysophosphatidic acid signaling. Curr Opin Cell Biol 1995; 7: 203-210.

190. Glaser KB, Mobilio D, Chang JY, Senko N. Phospholipase A2 enzymes: regulation and inhibition. Trends Pharm Sci 1993; 14:92-98.

191. Liscovitch M, Cantley LC. Signal transduction and membrane traffic: the PITP/phosphoinositide connection. Cell 1995; 81:659-662.

192. Chong LD, Traynor-Kaplan A, Bokoch GM, Schwartz MA. The small GTP-binding protein rho regulates a phosphatidylinositol 4-phosphate 5-kinase in mammalian cells. Cell 1994; 79:507-513.

193. Kapeller R, Cantley LC. Phosphatidylinositol 3-kinase. BioEssays 1994; 16: 565-576.

194. Shepherd PR, Reaves BJ, Davidson HW. Phosphoinositide 3-kinases and membrane traffic. Trends Cell Biol 1996; 6: 92-97.

195. Varticovski L, Harrison-Findik D, Keeler ML, Susa M. Role of PI 3-kinase in mitogenesis. Biochim Biophys Acta 1994; 1226:1-11.
196. Fry MJ. Structure, regulation and function of phosphoinositide 3-kinases. Biochim Biophys Acta 1994; 1226:237-268.
197. Divecha N, Irvine RF. Phospholipid signaling. Cell 1995; 80:269-278.
198. Malarkey K, Belham CM, Paul A, Graham A, McLees A, Scott PH, Plevin R. The regulation of tyrosine kinase signaling pathways by growth factor G-protein-coupled receptors. Biochem J 1995; 309:361-375.
199. Rodriguez-Viciana P, Marte M, Warne PH, Downward J. Phosphatidylinositol 3-kinase: one of the effectors of Ras. Trans R Soc Lond B Biol Sci 1996; 35: 225-231.
200. Testi R. Sphingomyelin breakdown and cell fate. Trends Biochem Sci 1996; 21: 468-471.
201. Hannun YA. The sphingomyelin cycle and the second messenger function of ceramide. J Biol Chem 1994; 269: 3125-3128.
202. Hannun YA. Functions of ceramide in coordinating cellular responses to stress. Science 1996; 274:1855-1859.
203. Ruddon RW. Cancer Biology, Chap. 5, New York: Oxford University Press, 1995.
204. Grandori C, Eisenman RN. Myc target genes. Trends Biochem Sci 1997; 22: 177-181.
205. Edwards DR. Cell signalling and the control of gene transcription. Trends Pharm Sci 1994; 15:239-244.
206. Karin M. Signal transduction from the cell surface to the nucleus through the phosphorylation of transcription factors. Curr Opin Cell Biol 1994; 6:415-424.
207. Vandromme M, Gauthier-Rouviere C, Lamb N, Fernandez A. Regulation of transcription factor localization: fine-tuning of gene expression. Trends Biochem Sci 1996; 21:59-64.
208. Carraway KL, Carraway CAC. Membrane-cytoskeleton interactions in animal cells. Biochim Biophys Acta 1989; 988: 147-171.
209. Luna EJ, Hitt A. Cytoskeleton-plasma membrane interactions. Science 1992; 258:955-964.
210. Krueger JG, Garber EA, Goldberg AR. Subcellular localization of pp60^{src} in RSV-transformed cells. Curr Topics Microbiol Immunol 1983; 107:51-124.
211. Machesky LM, Hall A. rho: a connection between membrane receptor signalling and the cytoskeleton. Trends Cell Biol 1996; 6: 304-310.

CHAPTER 3

Cell Polarity and Morphology

Introduction

Cell responses to stimuli in their environments are sensitive, selective and tem porally or- dered. One of the most frequent responses is an induction of polarity, or asymmetric morphology, in one or more regions of the cell. Establishment and maintenance of cell polarity are requisite to the differentiation and development of organisms and to many cellular functions (1), including cell motility, localized membrane growth, vectorial transport across cell layers and activation of immune response mechanisms. The ultimate morphology of a cell is determined by a progression of events through three temporally ordered stages:
  1) determination of the site(s) of origin of the alteration in morphology;
  2) establishment of polarity; and
  3) maintenance (or reversal) of the polarized state.

The association of membrane components with cytoskeletal structures plays major roles in both the establishment of polarized morphology and its maintenance (2, 3). The triggers of morphogenesis are many and varied, covering a broad diversity of ligands and receptors. The mechanisms for the maintenance of polarity appear more universal and usually involve the cytoskeleton. Organization of cells into tissues, organs and ultimately, into organisms also involve similar morphogenetic processes, many of which employ the same cellular components and mechanisms used in the establishment and maintenance of polarity in single cells. The focus of this chapter is a consideration of the basic components and mechanisms involved in the generation of polarity in individual cells. The analogous processes in tissues and whole organisms are discussed further in chapter 7.

Morphogenetic alterations can be subtle changes in cell architecture or global changes in morphology which encompass all of the cytoskeletal systems. A typical eukaryotic cell responds to a signal by the development of polarity in *membrane domains*, the *cytoskeleton* and *organelles*. In some cases, all of the cytoskeletal structures are involved in the morphological changes, and the result is a global morphological rearrangement such as that occurring during cell division. In other cases, specific cytoskeletal structures, most often microfilaments, are selectively involved, especially in the earliest stages. The development of polarity by a cell is sometimes a transient phenomenon, occurring as a response to a temporal stimulus. Unlike the permanent shape change occurring upon platelet activation (chapter 4) or the more permanent but somewhat reversible changes resulting from cell-cell contact, some cellular processes are normally completely reversible upon removal of the stimulus. This type of reversible morphological change occurs even when a cell undergoes the massive shape changes associated with cell division. During mitosis the cytoskeleton displays immense plasticity, with breakdown of microfilament

Signaling and the Cytoskeleton, by Kermit L. Carraway, Coralie A. Carothers Carraway and Kermit L. Carraway III. © 1998 Springer-Verlag and R.G. Landes Company.

assemblies, intermediate filaments, nuclear lamins, and microtubules. The disassembly phase is followed by the reorganization of microtubules into the specialized structures required for chromatin rearrangement and cell division (discussed in chapter 5). After mitosis, the daughter cells completely reassemble the cytoskeletal structures necessary for initiating morphogenesis of the resting cell. For example, fibroblasts return to their polarized states after undergoing cell division, (chapter 5), although the morphology of the daughter cells is not identical to that of the parental cell. Less extensive changes occur where local deformation of the cell is involved, such as during the budding process in yeast after stimulation by a chemoattractant. In that case, only the portion of the cell forming a bud is grossly deformed.

Both the signal initiating polarization and the designation of the site of origin of morphogenesis can be either externally or internally (extrinsically or intrinsically) imposed, or both. These modes of signaling, known as outside-in and inside-out, respectively, are further discussed in chapter 4. In the induction of epithelial cell polarization and tracking of a neuron, extracellular signals interacting with cell surface proteins are the key modulators of morphogenesis. In other cases, such as proliferation of budding yeast, the site of origin which determines both budding and ultimately spindle formation is designated by internal cues remaining in the maternal cell after the previous cell division as described below. Three sequential stages of signal propagation are required: marking the site and decoding the cue; reinforcement of the cue; and propagation of the cue (4). At each stage feedback mechanisms serve to coordinate and reinforce the temporal and spatial ordering imposed by the signal. Figure 3.1. shows a comparison of this complex modulation for budding yeast and for epithelial cells. The similarity in the mechanisms for the induction and maintenance of polarity in these phylogenetically distant cell types underscores their likely general nature (4).

Membrane Receptors, the Cytoskeleton and Signaling

A still-evolving paradigm describing morphogenesis entails *the association of membrane components with cytoskeletal structures* as a major feature. Membrane-cytoskeleton interactions are a major mechanism for both the *establishment of polarized morphology* and for its *maintenance* (5), especially in those cells or cellular structures which undergo a more prolonged polarization. The molecular mechanisms involved at each stage of remodeling for each of the cytoskeletal structures has been an ongoing subject of interest for over two decades. Although the story has evolved slowly, it has been clear since the 1970s that an important aspect of morphology and morphogenesis lay in the interactions of cytoskeletal assemblies with the membrane. *Assembly of a signaling complex at the site of ligand-receptor interaction* and *association of the complex with the cytoskeleton* imposes constraints on both motility and plasticity. These constraints dictate the subsequent events which proceed temporally as a consequence of receptor engagement. Abnormalities in membrane-cytoskeletal interactions are correlated with morphological changes observed in pathogenesis, particularly with neoplastic transformation (6).

Early work on the erythrocyte, which served for several years as the primary model system for the study of membrane-cytoskeleton interactions, identified and characterized a number of membrane skeletal proteins which were postulated to play general roles in these interactions (Fig. 1.2) (7). While some of these, such as

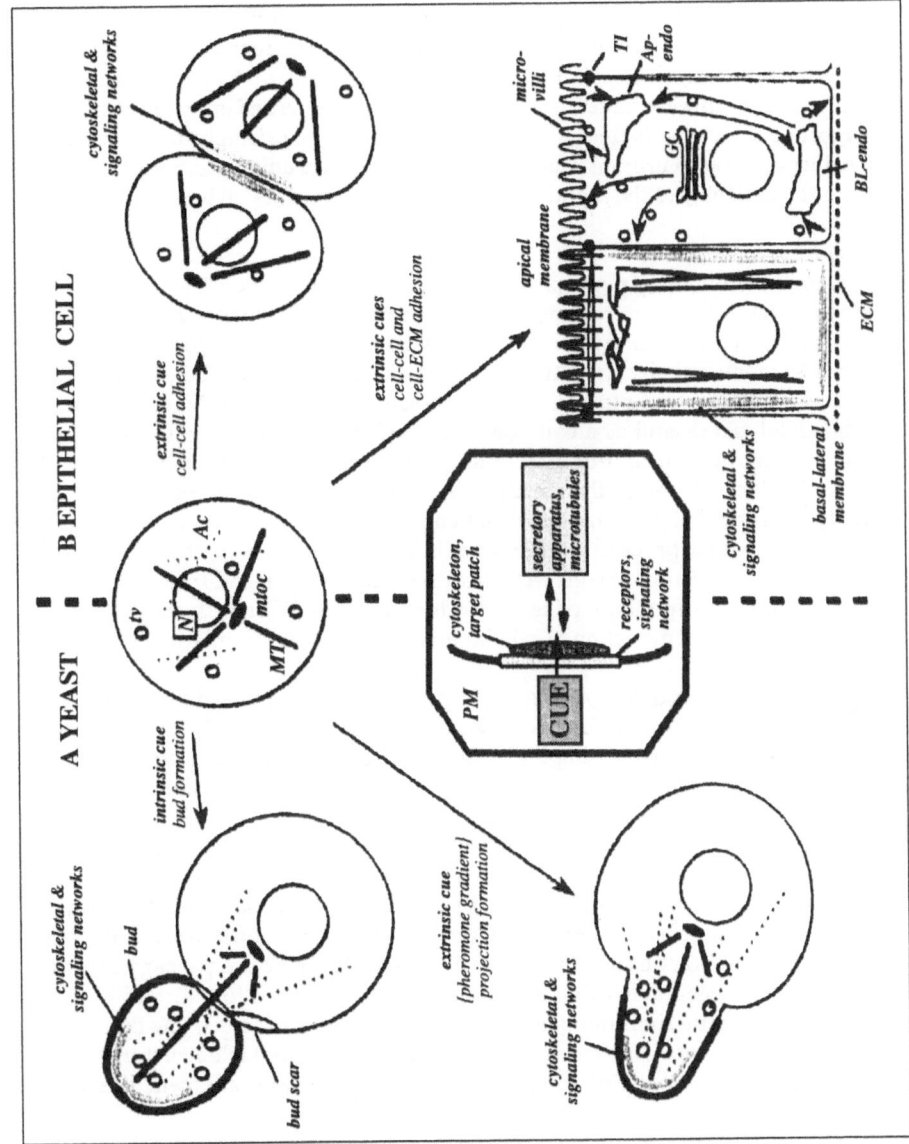

Fig. 3.1. Comparison of polarization of yeast and epithelial cells. Figure from ref. 4 (Drubin DG, Nelson WJ. Cell 1996; 84: 335-344), courtesy of Cell Press.

ankyrin and spectrins (fodrins), are ubiquitous, more complex cells express specialized assemblies of cytoplasmic proteins which provide the primary mechanisms for association of specific classes of receptors with the cytoskeleton. For example, cadherins are linked to cytoskeletal structures via a family of specialized membrane skeletal proteins called catenins (8). Not surprisingly, more complex structures such as neuromuscular junctions tend to have unique components critical for the assembly of the junction and for the transfer of signal between the cells.

Morphogenetic Receptors

Signals culminating in morphological rearrangements (see Table 1.1) are mediated by specific classes of receptor proteins, introduced in chapter 2. Table 3.1 lists the major categories of morphogenetic receptors and the cytoplasmic protein families which mediate their interactions with cytoskeletal elements. Each class of these integral membrane receptors initiates the assembly of largely unique membrane-cytoskeletal arrays, and as a consequence, specific pathways are activated which play roles in initiating and/or maintaining polarity. These receptors interact with cytoskeletal structures by the assembly of different filament-binding complexes. The membrane skeletal proteins of the complexes (Table 3.2) dictate the subsequent events which proceed temporally as a consequence of receptor engagement. For example, activation of a growth factor receptor by its ligand triggers an organization and a morphology of the cell which is quite different from that which occurs during the remodeling of cellular architecture after stimulation by a differentiation factor.

Growth factors and their *receptors* comprise two groups of morphogenetic molecules (Table 3.1). Growth factors may act by paracrine, autocrine or juxtacrine mechanisms (discussed in chapter 2) and may be present in association with the ECM (9) as a mechanism for maintaining an available reserve. Other cell surface proteins, some of which bind to specific ECM proteins and to proteins on adjacent cells, are also involved in morphological reorganizations. These receptors are responsible for *anchorage to the ECM* and *to adjacent cells*, morphogenetic processes which are *major determinants of polarity and cell function.* Some of these surface proteins stabilize cell-cell interactions and others form junctions which allow the cells in a tissue to communicate by permitting the selective passage of small molecules between cells. Cell adhesion proteins include four broad classes, integrins, cadherins, immunoglobulin-like proteins and selectins. The first two of these form complexes which interact with microfilaments in stable structures which transmit signals from cell interactions with matrix and adjacent like cells, respectively.

Several of the morphogenetic receptors listed in Table 3.1 comprise three families of cell *junctions* (Table 3.2) defined on the basis of their cellular function (10):

1) anchoring or *adherens* junctions, which include junctions which anchor the cell to both extracellular matrix and to adjacent cells;
2) occluding (*occludens* or tight) junctions; and
3) communicating junctions, responsible for the transfer of chemicals between cells which is required for signaling and for tissue function.

Integrins, the best studied of the receptor classes, are ubiquitous and have been implicated in morphogenetic processes in many cells (11). Receptors which mediate interactions of cells with *ECM* and with *other cells* are *crucial to the development and maintenance of polarity in polarized cells* (12, 13). The interactions of

Table 3.1. Classes of receptors responsible for establishing morphogenic signals

Receptor class	Ligands	Membrane skeletal proteins	Cell function
Receptors for cell growth modulators	Growth factors, cytokines; differentiation factors; apoptosis factors	Direct and indirect interactions with microfilaments	Transduction of signal to the cytoskeleton and nucleus to modulate transcription culminating in growth, differentiation, apoptosis
Integrins	Extracellular matrix components	Vinculin, talin, α-actinin, tensin	Anchorage of cells to extracellular matrix proteins
Cadherins	Cadherins on adjacent cells	Catenins: α, β and γ	Initiation of cell-cell interactions in *adherens junctions*; assembly of microfilaments at the sites of contact and initiation of first stages of epithelial cell polarity
		Catenins: β	Assembly of *zonula occludens*; linkage of cadherins to intermediate filaments; stabilization of polarized state
Immunoglobulin-like cell adhesion molecules	Immunoglobulin-like (CAMs)	??	Cell-cell adhesion
Selectins	Oligosaccharides on cell surface glycoproteins	α-actin, vinculin	Transendothelial migration

Table 3.2. Receptor-cytoskeleton cellular interaction sites

Site	Transmembrane component	Cytoskeletal component	Signal transduction proteins
Erythrocyte membrane skeleton	Band 3 (anion channel) Glycophorin	Actin protofilaments (via ankyrin, spectrin, band 4.1, adducin)	None?
Focal adhesion	Integrin	Microfilaments	See Table 4.3
Platelet membrane skeleton	Integrin GPIIb/IIIa, GPIb-IX	Actin protofilaments?	See Table 4.5
Adherens junctions	Cadherin	Microfilaments (via α-,β -catenins)	Src, Yes, Lyn, Others?
Desmosome	Desmoglein, desmocollin (cadherins)	Intermediate filaments	None?
Tight junction (*occludens* junction)	Occludin	Intermediate filaments	None?
Communicating junctions Connexons (gap junctions)	Connexins	None	None?
Neuromuscular junction	Acetylcholinesterase Dystroglycan	Microfilaments	MuSK, PKC, others
Mitogenic signaling complexes	Growth factor/cytokine receptors Transmembrane complex gp's	Microfilaments, ?	Mitogenic pathway proteins Ras to MAPK pathway
Microvillar complex	ErbB2, ErbB3 E-cadherin	Microfilaments, α-actin, ezrin α, β, γ-caterin	Src; Abl,PLCγ, PI-3-kinase

integrins with matrix proteins and their assembly into focal adhesions mediate the *initial phase of morphogenesis,* imposing a horizontal or two-dimensional polarity on the cell. Homophilic interactions between cadherins on adjacent cells initiate the *secondary phase,* leading to three-dimensional or vertical polarity. Chapter 8 discusses the order of establishment of different adhesive complexes and their roles in further ligand-induced perturbations and ultimate cell fate. Integrins are αβ-heterodimeric receptors which recognize specific sequences, such as the RGD tripeptide (chapter 4.2), in extracellular matrix proteins (14, 15). Engaged integrins assemble complexes which form *focal adhesions* comprising microfilaments linked to the receptors by an assembly of cytoplasmic proteins unique to these adhesion sites (16, 17). Focal adhesions anchor cells stably but reversibly to the matrix and are sites of bidirectional signal transduction (18) which regulate a number of aspects of polarized cell function, including gene expression (19). They contain signaling proteins such as the focal adhesion kinase (FAK) (20, 21, 22) and the small G-protein Rho (17) (chapter 2). A discussion of focal adhesions and a model for their assembly is presented in chapter 4. Integrins in platelets associate with microfilaments and a different set of cytoplasmic membrane skeletal proteins to facilitate the gross morphological changes which occur upon platelet aggregation. These shape changes result in increased adhesivity and are critical to the platelet's role in early stages of wound repair. Unlike the reversible morphological changes which occur after a stimulation event in most cells, platelet shape changes leading to aggregation are both global and irreversible. The resultant morphology is not the major functional consequence of platelet activation; the important consequence is the creation of efficient adhesive sites for binding platelets to the extracellular matrix in vessel walls and to each other for creation of a contractile mass in clotting. The mechanisms and consequences of platelet activation are discussed in chapter 4.

The other class of adherens junction is that formed between two like cells and is mediated by a homotypic interaction between Ca^{2+}-sensitive *cadherin molecules.* Unlike the ubiquitous integrin-mediated ECM interactions, cadherin-containing junctions are more restricted in distribution. They will be discussed further in the section on epithelial cells, in which they are a major determinant of polarized morphology and differentiated function. The *communicating junctions* are also more tissue-specific.

An important general paradigm developing in signal transduction, *"crosstalk" between signaling pathways,* attempts to explain the sometimes unpredicted cellular effects seen in response to stimulation by a morphogenetic ligand. For example, stimulation of a growth factor receptor culminates in *pleiotypic responses* in the organization of the cytoskeleton, leading to morphological alterations as well as mitogenic stimulation. As a conseqence of ligand binding, a cascade of signaling events is initiated, leading to the assembly of ligand-specific complexes linked to the cytoskeleton. The composition of the proteins in the complexes determines the cellular pathways involved in crosstalk, or cross-modulation, between pathways. The *points of intersection* of the mitogenic pathway with pathways regulating the assembly and disassembly of cytoskeletal elements are critical to crosstalk between the pathways necessary for full cellular response to the growth factor. Further, *the cellular responses to engagement of a receptor is modulated in complex manners by other membrane receptors and their signaling assemblies* (23, 24). Examples of this "crosstalk" between growth factor receptors and other surface

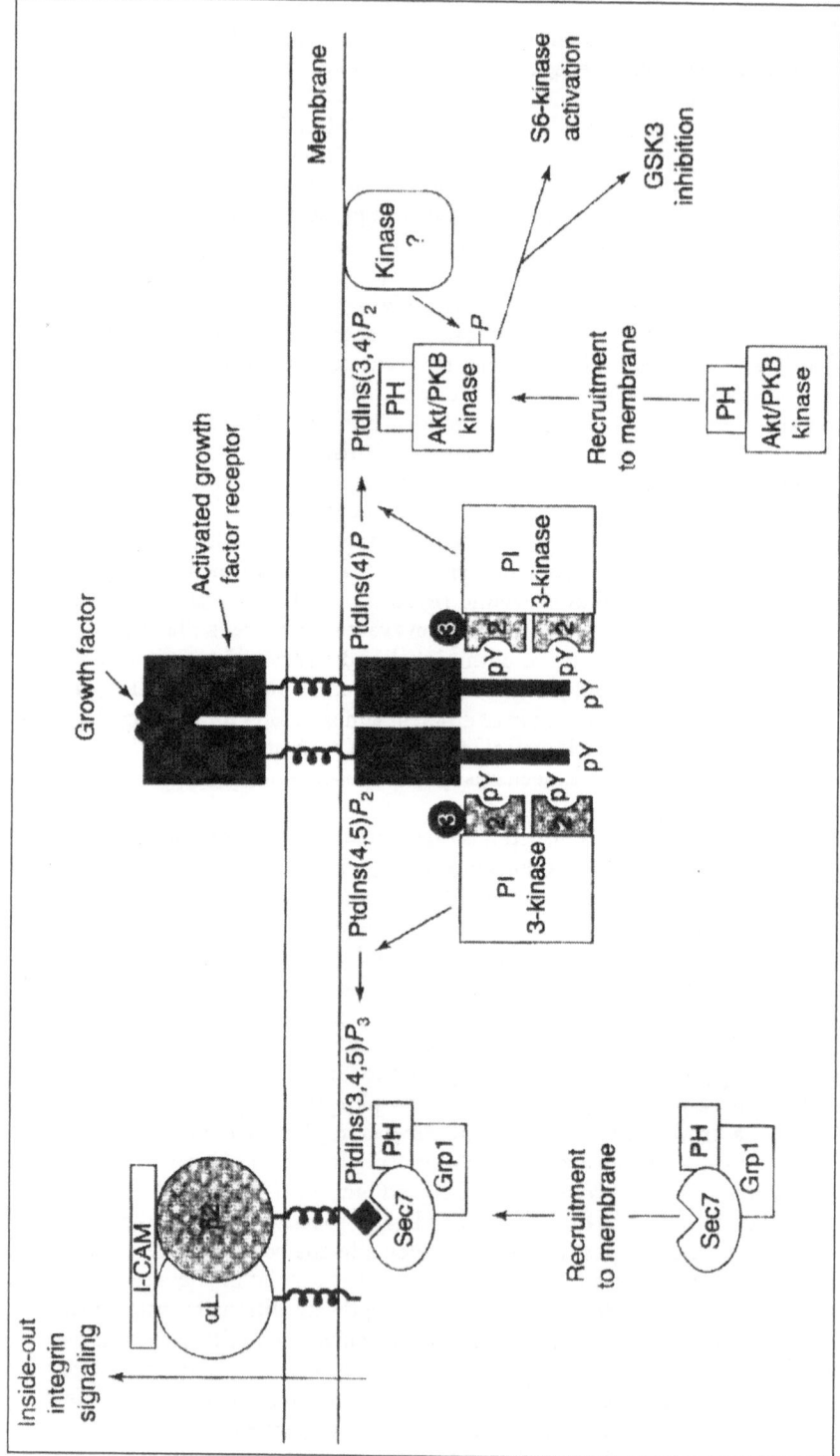

Fig. 3.2. Model showing mechanisms for crosstalk between pathways activated via a growth factor receptor and a cell adhesion receptor. Figure from ref. 25 (Lemmon MA, Falasco, M, Ferguson, KM, Schlessinger J. Trends Cell Biol 1997: 7, 237–242), with kind permission of Elsevier Science - NL, Sara Burgerhartstraat 25, 1055 KV Amsterdam, The Netherlands.

receptors are described in chapter 5, and complexities in crosstalk are explored further in chapter 8. Figure 3.2 depicts a model for the crosstalk in which a ligated growth factor receptor modulates integrin ligand binding by an inside-out signaling mechanism (25). A growing recognition of the importance of crosstalk between pathways is leading to the elucidation of these complex linkages between signaling pathways for a number of morphogenetic receptors. Integration of these pathways with those regulating the assembly of the cytoskeleton are even more complex and will be discussed further in chapter 8. Inside-out signaling is discussed in chapter 4.

Model Systems in the Study of Cell Polarity

In veretbrate organisms, there are six major types of tissue: connective, blood, lymphoid, nerve, muscle and epithelial. Among these tissues there are significant differences in the relative contributions of cell-matrix and cell-cell interactions to polarity and function. Connective tissue, which is largely matrix material containing relatively few cells, relies much more heavily on cell-matrix than on cell-cell interactions, whereas the opposite is true for epithelial cells which interact in sheets. Cell types in use as model systems for studies on molecular aspects of the generation and maintenance of polarity are of two broad general types:

1) cells which interact with extracellular matrix or with other cells in the formation of tissues or specialized structures; and

2) cells which do not require anchorage for the initiation of morphogenesis.

The first category includes cells such as epithelial cells, which not only require interactions with the ECM, but also with adjacent cells for both establishing and maintaining polarity. Neurons require interactions with matrix components and with other cells to serve as a guidance or tracking function for the unidimensionally polarizing axon. Cells in this category require receptors on the cell surface which interact with specific proteins in the matrix and/or on another cell to establish the site for initiation. The second class of cells, typified by yeast, do not require an insoluble ligand to initiate the morphogenetic process, but respond to either external or internal cues or both. The first class is said to be anchorage dependent and the second anchorage independent. Tumor cells lose their anchorage dependence, providing these cells greater freedom for motility and for proliferation as a consequence of release from the constraints of cell-cell and and cell-matrix interactions. This concept is discussed further in chapters 5 and 7.

Early studies on cultured fibroblasts showed that these cells require adhesion to a substratum for cell proliferation (26) and that malignantly transformed cells are less dependent on anchorage for growth (27-29). Binding of integrins to ECM and assembly of membrane-microfilament complexes into *focal adhesions* play a major role in attachment and thus in anchorage dependence (29, 30). Prevention of attachment with RGD peptides or growth in suspension prevents proliferation and induces anoikis, or attachment deprivation-induced apoptosis (30). Apoptosis is the normal mechanism whereby cell homeostasis is maintained in different tissues, and mechanisms for this critical cellular pathway and its aberrant function in neoplasia and other disease states are discussed in chapters 5 and 7. In many respects, growth and differentiation in most cells are opposing effects. As a general rule any event which triggers cell differentiation tends to slow mitogenesis.

Some of the basic concepts involved in the establishment of morphology arose in studies of motility in fibroblasts which will be discussed in more detail in chapter

4. Foremost among these is the attachment of cells to the extracellular matrix via integrin association with ECM components. Engagement of integrins by their ligands initiates the assembly of microfilament-associated focal adhesions (23), culminating in the most basic step in the generation of polarity. In fibroblasts focal adhesions define the sites at which cells attach transiently in their movement across the substratum. The focal adhesion in the fibroblast, described in chapter 4, is the prototype for these adhesive structures; the structures, integrin isoforms and other components, and major functions are somewhat different in highly differentiated cells such as epithelial or neural cells. A cell which has established focal contacts is polarized vertically into a dorsal and a ventral aspect, leading to a relatively "two-dimensional," horizontal polarity. In more polarized cells such as epithelial cells, the establishment of this initial level of polarity must be followed by a second stage of polarization leading to a vertical polarization required for the differentiated function of the cell.

Epithelial Cells

Early work on polarity focused on the epithelial cell (31), which in its differentiated form is relatively rigidly polarized. A simple differentiated epithelial cell is an asymmetric, vertically polarized cell characterized by structurally and functionally specialized domains, an apical and a basal-lateral, or *basolateral*, domain. The *apical* domain faces the lumen of an epithelial tissue and the basolateral domain interacts with extracellular matrix components in the basal lamina or basement membrane. Figure 3.3 depicts a polarized epithelial cell and the components known to associate with each domain (32). Specific cell surface proteins are partitioned into each of the domains, and each domain contains specialized lipids and proteins and assembles specific cytoskeletal structures. The *apical* domain is more specialized, expressing the proteins required for organ-specific function. The apical surface of most cells, but particularly epithelial cells, is covered by numerous microvilli, which greatly increase the surface area and are involved in specific interactions with environment. Additionally, glycolipids as well as glycosylphosphatidylinositol-linked proteins are targeted to the apical domain, although the mechanism for targeting and the functional significance of the presence of these glycoconjugates at the apical surface are poorly understood.

The basolateral surface of the simple epithelial cell expresses proteins responsible for carrying out basic cellular functions, many of which are common to cells other than epithelial cells. These include the Na^+/K^+-ATPase, the anion channel and receptors for growth factors, hormones and neurotransmitters. The Na^+/K^+-ATpase associates with the cytoskeleton by interacting with an analog of membrane skeletal protein spectrin (3). The basal segment which makes direct contact with the ECM contains integrins. Each type of receptor in both the apical and the basolateral domains assembles a large cytoskeletal-associated transduction system comprising a distinct set of membrane skeletal proteins (Table 3.1). These assemblies are an integral part of the cellular machinery for the establishment and maintenance of polarity.

In the fully polarized cell, lateral diffusion of receptors in the plane of the membrane is severely restricted, as described in the following discussion, and maintenance of the domains is essential for the vectorial transport of ions and solutes across the epithelium (33, 34). Polarity and the differentiated state in epithelial cells are controlled in a complex multi-stage manner in which responses to cell interactions with the ECM, other cells and growth factors are coordinately regulated with

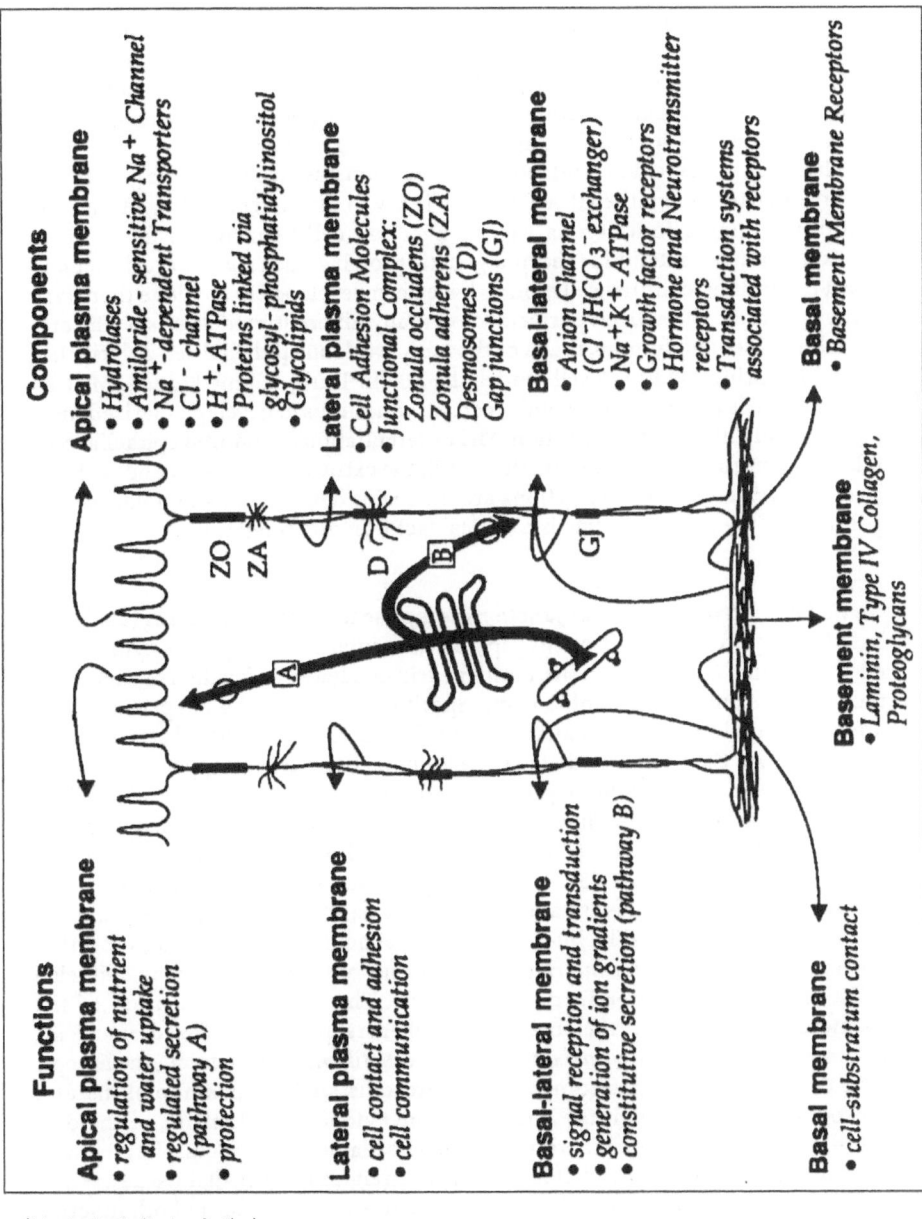

Fig. 3.3. Distribution of components in polarized epithelial cell. Figure from ref. 32 (Rodriguez-Boulan E, Nelson WJ. Science 1989; 245: 718-725), reprinted with permission. Copyright 1989 American Association for the Advancement of Science.

Components

Apical plasma membrane
• Hydrolases
• Amiloride-sensitive Na⁺ Channel
• Na⁺-dependent Transporters
• Cl⁻ channel
• H⁺-ATPase
• Proteins linked via glycosyl-phosphatidylinositol
• Glycolipids

Lateral plasma membrane
• Cell Adhesion Molecules
• Junctional Complex:
  Zonula occludens (ZO)
  Zonula adherens (ZA)
  Desmosomes (D)
  Gap junctions (GJ)

Basal-lateral membrane
• Anion Channel (Cl⁻/HCO₃⁻ exchanger)
• Na⁺-K⁺-ATPase
• Growth factor receptors
• Hormone and Neurotransmitter receptors
• Transduction systems associated with receptors

Basal membrane
• Basement Membrane Receptors

Basement membrane
• Laminin, Type IV Collagen, Proteoglycans

Functions

Apical plasma membrane
• regulation of nutrient and water uptake
• regulated secretion (pathway A)
• protection

Lateral plasma membrane
• cell contact and adhesion
• cell communication

Basal-lateral membrane
• signal reception and transduction
• generation of ion gradients
• constitutive secretion (pathway B)

Basal membrane
• cell-substratum contact

cytoskeletal organization (24). Mutation or inappropriate expression of any one of an array of morphogenetic proteins can result in an alteration in the morphogenetic potential or capability of the cell to maintain a stable polarized state, often producing altered cellular functions. Alterations in morphology and function which occur upon neoplastic transformation are discussed in chapter 7.

The study of the cellular and molecular mechanisms involved in the generation and maintenance of epithelial cell polarity is key to the understanding of the normal differentiated function as well as aberrant cell behavior resulting from misexpression or mutation of morphogenetic proteins. Such studies are difficult in intact tissue; thus, cultured cell models have been developed to facilitate them (31). Although studies have been performed on many different models of specific epithelial cell lines and tumors, much of the basic work on epithelial cell polarity has been done using a cultured cell model, the Madin-Darby canine kidney (MDCK) cell. These cells can be induced under certain conditions to develop polarized morphology and differentiated function. The differentiation of cultured epithelial cells is largely controlled by cell interactions with the extracellular matrix and with adjacent cells. These same interactions are the major determinants of polarity, suggesting that the polarized state is a critical factor in differentiation.

Junctions

Cell-cell interaction sites or *junctions* comprise three general classes (Table 3.2) based on their function. All except gap junctions interact with specific cytoskeletal structures. In *adherens junctions*, both integrin-mediated and cadherin-mediated, receptors are linked to microfilaments. Specific isoforms of the integrin subunits are being identified for the mediation of the interaction of epithelial cells with the matrix (11, 35). Cadherins mediate cell-cell interactions at both types of adhesion sites via Ca^{2+}-dependent, homophilic interactions of the receptors on one cell to those on an adjacent cell. These adhesion receptors are *linked to microfilaments* through a specialized class of membrane skeletal proteins called *catenins* and catenin-related proteins (8). In *occludens junctions both desmosomes* and *hemidesmosomes* are linked to and stabilized by bands of intermediate filaments. *Desmosomes* are button-like intercellular connections which staple cells together and are *anchoring sites for intermediate filaments* of the keratin type in epithelial cells, or the desmin type in cardiac muscle. The adhesion receptors of the desmosomal junctions are members of the cadherin superfamily called desmogleins or desmocollins. They are *linked to intermediate filaments* by lateral attachments to a cytoplasmic plaque composed of proteins such as desmoplakins and plakoglobin (36). Through the binding of the cytoplasmic domains of the desmosomal membrane proteins with intermediate filaments, a structural framework linking cells is generated which lends tensile strength to epithelial tissues. *Hemidesmosomes* are morphologically similar structures which differ in both function and composition. They form stablilizing connections between the cell and basal lamina, a specialized mat of extracellular matrix proteins between the extracellular matrix and connective tissue. Hemidesmosomes also interact with intermediate filaments, but unlike the lateral attachments in desmosomes, the keratin filaments in hemidesmosomes are frequently buried within the plaques. The transmembrane proteins in hemidesmosomes are integrins rather than cadherins. Establishment of these cell-cell interaction sites is a requisite early step in the polarization of epithelial cells to form differentiated epithelial tissue. Both types of interaction sites are clearly important in providing stable mechanical support, yet both are quite dynamic structures easily perturbed by environmental stimuli (10). The third type of junction is the *tight junction* which is composed of the membrane protein occludin linked to cytoplasmic plaque proteins ZO-1, ZO-2, ZO-3 and cingulin (37). The tight junctions are also *anchored to intermediate filaments*.

When epithelial cells in culture approach confluence, they can be induced to undergo morphological and functional differentiation simply by the addition of calcium. The calcium ion is required for the mediation of the *calcium-dependent homotypic interaction of cadherin molecules* on cells which provides a key stabilizing event in the polarization of the cells. This binding event triggers the assembly of microfilamentous structures at the engaged cadherin molecules, leading to alterations in both membrane and cytoskeletal protein arrangements in the cells. Once the cells are associated into these cadherin-based *adherens junctions* which are largely found in epithelial cells, the interactions between the cells are further secured by the assembly of another specialized structure, the *zonula occludens*, or tight junction. The tight junction, unique to epithelial cells, provides the permeability barrier required for the separation of components in the lumen from those accessible to the bloodstream. They are impermeant to water-soluble components including salts. They comprise narrow bands of membrane proteins which completely encircle epithelial cells and make contact with like regions on adjacent cells. In epithelia of the intestine, the tight junctions are located on the apico-lateral face just below the microvillar surface. These junctions prevent the diffusion of proteins in the plane of the cell membrane, allowing the maintenance of the apical and basolateral domains. Interestingly, they also provide a barrier to diffusion of lipids in the exoplasmic, but not the endoplasmic face of the epithelial cell membrane (38). A second mechanism of segregation of membrane components in epithelial cells is by linkage to specific cytoskeletal elements of the particular domain. For example, Na,K-ATPase is restricted to the basolateral domain by its formation of a complex with ankyrin and spectrin linked to the basolateral cytoskeleton (5, 37).

In order to function normally as epithelial tissues, clusters of interacting epithelial cells must be able to "communicate" their intracellular information by limited diffusion of components, primarily ions and small molecules. This transfer is accomplished through *gap junctions* which are formed when specialized regions of the plasma membrane, called *connexons*, on adjacent cells are aligned (39). Connexons are composed of rings of six identical subunits of unique integral membrane proteins called *connexins*, each containing four putative transmembrane-spanning α-helices. The connexon contains a large central aqueous pore formed by an α-helix from each subunit. Connexins are members of a highly related multigene family comprising at least 13 members, from which different connexons can be assembled.

Formation of the cadherin complexes at the adherens junction is one of the key stages in epithelial polarization. The crystal structure of cadherin extracellular domains suggests that they form a linear cell adhesion zipper stabilizing the junctions through cooperative interactions. Cadherins are linked to microfilaments and other cytoplasmic proteins by dynamic associations with the catenins. Pulse-chase studies indicate that the initial event is association of β-catenin with cadherin (8). Alternatively, plakoglobin may associate with cadherin, but not both. The initial event is followed by binding α-catenin to β-catenin/plakoglobin. α-Catenin has homologies to vinculin and is proposed to be the linkage site to microfilaments associated with the junction, though the molecular nature of that association needs to be clarified. A complicating factor in this scheme is the involvement of β-catenin in associations with other cellular components, including transcription factors in the Wnt pathway (40) and the actin-binding protein fascin (41), as described in chapter 7. Thus, β-catenin may have a role as a master regulator coordinating adhesive and gene expression events (40).

Cadherin junctions are dynamic structures which change during processes such as cell migration, mitosis and neoplastic transformation (42, 43). Tyrosine kinases have been implicated in these changes by localization of phosphotyrosine-containing proteins (44), inhibition of tyrosine phosphatases (43, 45) and viral transformation with transforming, but not nontransforming mutants of v-*src* (46, 47). Changes in cell and junction structure were accompanied by increased tyrosine phosphorylation of cell-cell junctions. Treatment of these transformed cells with tyrosine kinase inhibitors resulted in the reformation of junctions (43). Both cadherin and catenins are tyrosine-phosphorylated in the transformed cells. The level of phosphorylation of catenin is much higher than that found for other Src-phosphorylated proteins at membrane-microfilament interfaces, such as vinculin, talin, integrin and calpactin. Finally, three members of the Src family (Yes, Src and Lyn) were found associated with isolated liver cell-cell junctions (45). These combined results suggest that Src family tyrosine kinases located at the intercellular junction plaque play an important role in the dynamics of cell-cell interactions via a phosphorylation-dephosphorylation cycle. Not surprisingly, losses and modifications of cadherins and catenins have been implicated in neoplasia (chapter 7).

Apical Domains

Apical domains are the most specialized feature of epithelial cells and more often contain specifically localized components. Glycolipids are targeted entirely to the apical domain, as are GPI-linked proteins; the significance of the partitioning of the lipids between domains is not well understood. However, roles for glycosphingolipids (GSLs) in adhesion and signal transduction have been proposed (48-50). GSLs are cell type-specific lipids which are altered significantly in development and neoplastic transformation. In multicellular organisms they appear to have roles in cellular interactions by homotypic interactions between cells or matrix or heterotypic interactions with lectins. They are implicated in control of cell proliferation by modulation of transmembrane signaling through integrins and other tyrosine kinase-linked receptors. Modulation of Ca^{2+} signaling, PKC and mitogenic pathway components by GSLs can cause changes in cellular functions such as motility, proliferation, differentiation and apoptosis.

The intestinal brush border is a specialization of the epithelial apical cell surface which exhibits a highly structured elaboration of densely packed microvilli (51). It has been a favorite system for the study of membrane-cytoskeleton interactions because of the ease of isolation and the amenability of the highly structured cytoskeleton to ultrastructural analyses. Together these factors have permitted correlated biochemical and morphological observations. Microvillar microfilament cores are highly organized and oriented with their + ends at the microvillus tip in an ill-defined dense plaque apparently associated with the plasma membrane and their–ends in a perpendicular band of microfilaments called the terminal web (52). The terminal web consists of two regions. The region into which the core is inserted is composed of fine filaments bridging the rootlets of the cores and contains myosin and nonerythroid spectrin (51). The second region contains circumferential bundles of microfilaments of mixed polarity perpendicular to the rootlets

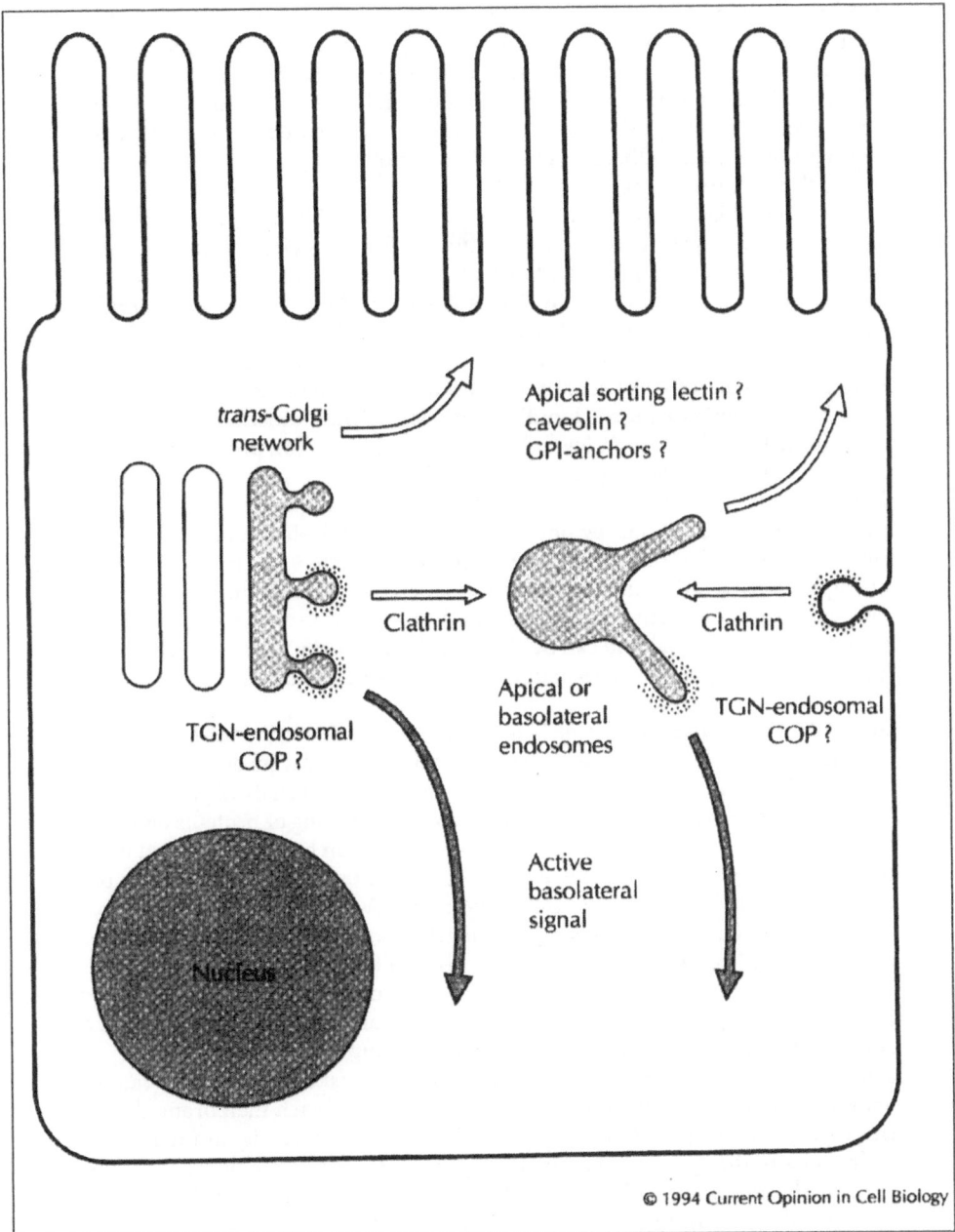

Fig. 3.4. Membrane protein sorting mechanisms in epithelial cells. Figure from ref. 56 (Matter K, Mellman I. Curr Opin Cell Biol 1994; 6: 545-554), courtesy of Current Biology Ltd.

associated with myosin, α-actinin and tropomyosin, all of which are absent from the microfilaments in the tubular regions of the microvilli. The terminal web microfilaments insert into the membrane at adhesion plaques at the level of the zonula adherens which contains α-actinin and vinculin. Below the zonula adherens, a dense meshwork of cytokeratin-containing intermediate filaments is associated with the macula adherens. These filaments often appear to interact with the microvillar core rootlets (53).

The microvillar cores of the brush border have highly organized bundles of microfilaments (53), formed and maintained by specific distributions of cytoskeleton-associated proteins. Microfilament cores contain three major proteins which are not present on the filaments located in the terminal web region: fimbrin (68 kDa), villin (95 kDa) and a myosin I (51). In addition, tropomyosin is present in the rootlet regions of the core which extend into the terminal web. Fimbrin is a ubiquitous protein present in a wide variety of tissues. Purified fimbrin assembles microfilaments into bundles. Villin is a calcium-sensitive protein which has a limited distribution. At low calcium concentrations (0.1 μM), villin bundles microfilaments. At higher calcium concentrations, villin acts as a severing agent, a filament-capping protein and a nucleator of filament formation. Myosin I, which is found in only the tubular regions of the microvilli of the brush border, has been implicated in the lateral association of microfilaments with the membrane as a complex with calmodulin. Its mode of association with the membrane is not well understood. Myosin I molecules in other systems are also associated with cellular membranes (54) and can be shown to bind directly to phospholipids. The role of these phospholipid associations has not been established.

Generation and Maintenance of Polarity: Sorting of Proteins and Lipids

The receptors and their cytoskeleton-associated signal transducing arrays are major mediators of the generation and maintenance of epithelial cell polarity. However, these structures alone are insufficient for the targeting of proteins and lipids to the appropriate domains. In addition to the diffusion barrier presented by the tight junctions, specific mechanisms are required for the polarized delivery of proteins (55) and lipids (38) to the cell surface (37). Two general mechanisms have been described for sorting proteins in epithelial cells, a direct and an indirect mechanism (Fig. 3.4) (56). In MDCK cells, the proteins of both domains are synthesized in the ER and routed together through a common Golgi compartment. Sorting to domain-specific vesicles occurs in the trans-Golgi network (TGN). In some MDCK strains, basolateral domain targeting of the Na^+/K^+ATPase was achieved by selective stabilization, possibly by the cytoskeleton (57) after random delivery to both domains (58). Further, the proteins at each membrane domain can be continuously exchanged *(transcytosis)* by endocytosis and delivery in endosomes to the opposite domain. In hepatocytes apical proteins are first delivered to the basolateral domain and then targeted to the apical domain by transcytosis (59, 60). In Caco-2 and primary intestinal cells, apical proteins utilized both the direct pathway from the TGN as well as the indirect pathway through the basolateral domain. Thus, epithelial cells use a cell type- and protein-dependent selection of two different pathways for sorting of newly synthesized apical proteins.

Cells are constantly endocytosing material from both apical and basolateral domains, requiring mechanisms for re-establishing polarized expression. Since

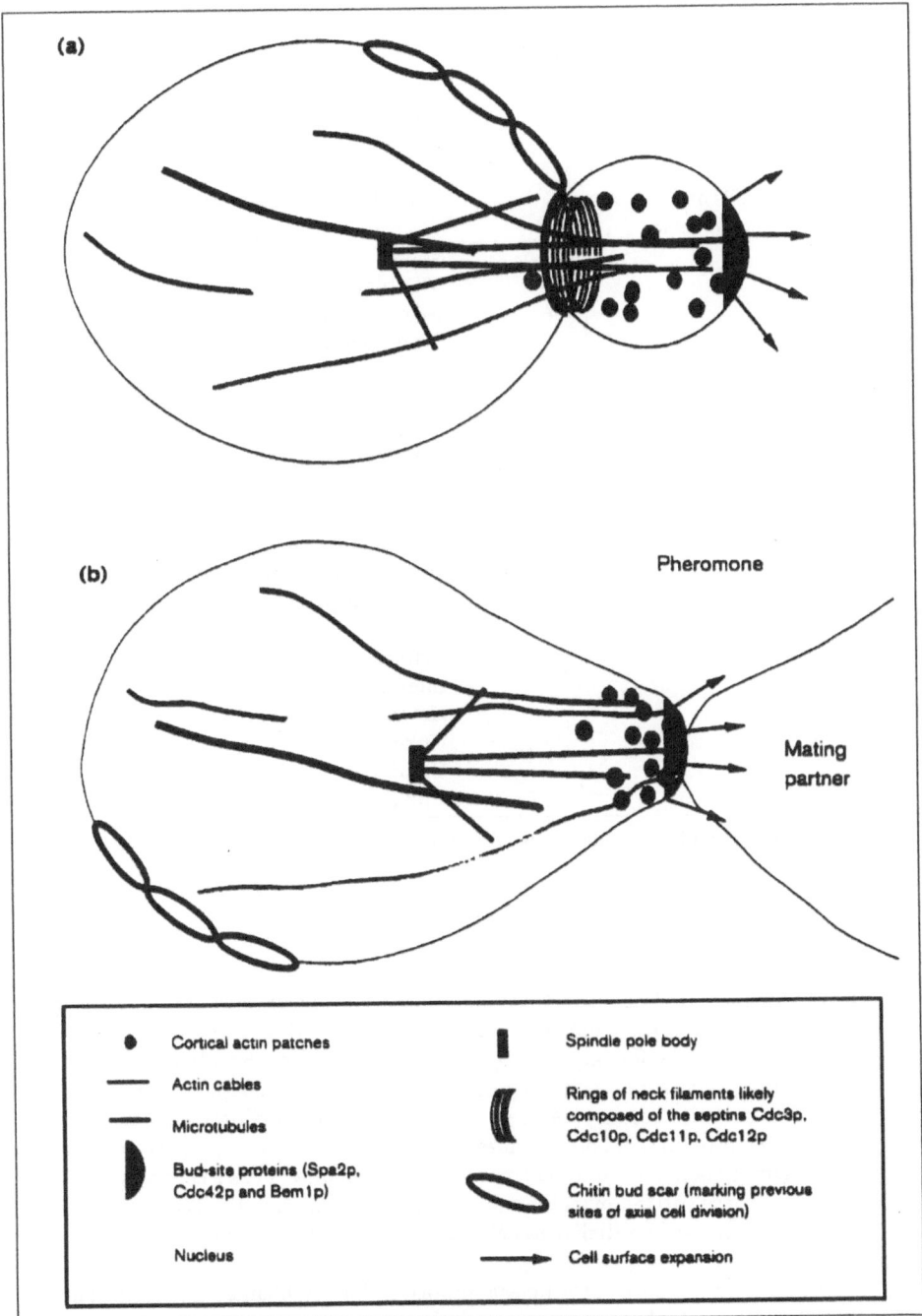

Fig. 3.5. Polarized yeast cell. Figure from ref. 62 (Chant J. Curr Opin Cell Biol 1996; 8: 557-565), courtesy of Current Biology Ltd.

Table 3.3. Examples of components important for bud-site selection and budding

Genes	Gene product
Septin-ring genes	
CDC3, CDC10, CDC11, CDC12	10-nm filament ring components
Axial bud-site selection genes	
BUD3	Novel
BUD4	GTP-binding domain
AKL1	a-factor protease
AXL2 (BUD10)	Type-1 plasma membrane glycoprotein
General bud-site selection genes	
RSR1 (BUD1)	Ras-related protein
BUD2	GTPase-activating protein (GAP)
BUD5	GDP-GTP exchange factor (GEF)
Polarity-establishment genes	
CDC24	GEF for Cdc42p
CDC42	Rho/Rac GTPase
BEM1	Src homology 3 (SH3) domains containing protein
Diploid bud-site selection genes	
ACT1	Actin
SPA2	Internal repeats, coiled-coil domain
RVS161, RVS167	Homologues, SH3 domain in RVS161
BNI1, BUD6, BUD7	Unknown
BUD8	Unknown
BUD9	Unknown

Modified from ref. 65 (Roemer T, Vallier LG, Snyder M. Trends Cell Biol 1996; 6: 434-441), courtesy of courtesy of Current Biology Ltd.

both transcytosed and recycling receptors can be found in a common pool of endosomal vesicles, a mechanism is necessary for targeting. The GPI-anchor is proposed to be a signal for targeting to the apical membrane. Basolateral domain proteins contain sequences in their cytoplasmic domains which form a β-turn targeted to that domain. Transcytosis requires an additional sorting mechanism for returning each protein to its proper domain. These proteins accumulate in a tubular compartment beneath the apical membrane which may act as a central sorting station (60). These sorting principles are applicable to neuronal cells as well, with exogenously expressed epithelial apical proteins transported to the axonal plasma membrane (55).

Two additional facets of sorting are important for establishing and maintaining the polarized epithelial structure: regulation of the sorting pathways and movement of membrane components within the cell. As expected, the former is a function of the cytoskeleton and its associated motor proteins. In the epithelial cell, microtubules form parallel tracks extending from the basal (microtubule + end) to the the apical side (microtubule − end) of the cell (5). These permit trafficking of intracellular vesicles in either direction, depending on the motor used, as described in chapter 6. The cytoplasmic surfaces of the plasma membrane domains are associated with cortical actin filaments which appear to be involved in vesicle movements near the membrane, probably using myosin I as a motor protein (61). Much less is known about the factors controlling the pathways for sorting membrane molecules in epithelial cells (60). Heterotrimeric G-proteins were implicated along with cyclic AMP and PKA by inhibitor experiments. Similarly, phorbol ester effects suggested a role for PKC, which is probably activated normally through stimulation of a PLC as described in chapter 2.

Yeast Cells

Because of their amenability to genetic analyses, yeast provides an important model for the study of cell polarization (4). For example, budding yeast become polarized as part of their normal cell division cycle (budding) and as a response to mating pheromone (projection or schmoo formation) as shown in Figure 3.5 (62). Four primary morphogenetic events are required during the cell division cycle in budding yeast: bud site selection, bud emergence, bud growth and cytokinesis (63). Each of these is coupled to particular phases of the cell cycle and triggered by the activation of the cyclin/cdk1 complexes which are described in chapter 5. As the yeast cell initiates the cell cycle or responds to mating pheromone, its internal structures, including the secretory apparatus and cytoskeleton, assume an anisotropic distribution (Fig. 3.5) (64). Although many aspects of cell polarization are similar in yeast and animal cells, the fact that yeast cells have cell walls creates some additional complexity. To grow and divide, yeast cells must add additional cell wall. Cell wall assembly must be coupled to the processes which regulate morphogenesis during cell division and to the cell cycle (63). Thus, at least four different types of components participate in the budding process: membrane/structural, cytoskeletal, signaling and biosynthetic. A compilation of examples of components is listed in Table 3.3.

The site of bud formation is established from spatial cues remaining from the previous bud site and may occur adjacent to the previous bud site (axial budding) or at the opposite end of the cell (bipolar budding), depending on the genotype and nutrition conditions (4). Mutation studies indicate the involvement of a complex of filamentous proteins called septins in the localization process in axial budding; actin is more important in bipolar budding events (Table 3.3) (4). Axial bud site selection also depends on Bud3, Bud4, Axl1 and Axl2 (Bud10) (65), all of which are associated with the septin ring filament complex during the establishment and extension of the bud. Bud6, Bud7, Bud8, Bud9 and actin are essential for bipolar budding. Bud1, Bud2 and Bud5 are required for both. Loss of function mutations in Bud3, Bud4 and Axl1 lead to bipolar budding, indicating that site selection is an hierarchical process. Axl1 serves as a switch between the axial and bipolar mechanisms. Information from the spatial cues for both axial and bipolar budding is transmitted to the cytoskeleton through a small G-protein cascade containing Bud1 and cdc42 (62). These two G-proteins are linked by cdc24, which serves as a target of Bud1 GTPase and an exchange factor (GEF) for cdc42. How the GTPase cascade is assembled and organized and how cdc42 is coupled to the cytoskeleton in the budding process are unclear.

Complex formation and linkage to the cytoskeleton have been best characterized for the mating response in which a defined extracellular signal triggers the extension of the process along a chemoattractant (pheromone) gradient, a

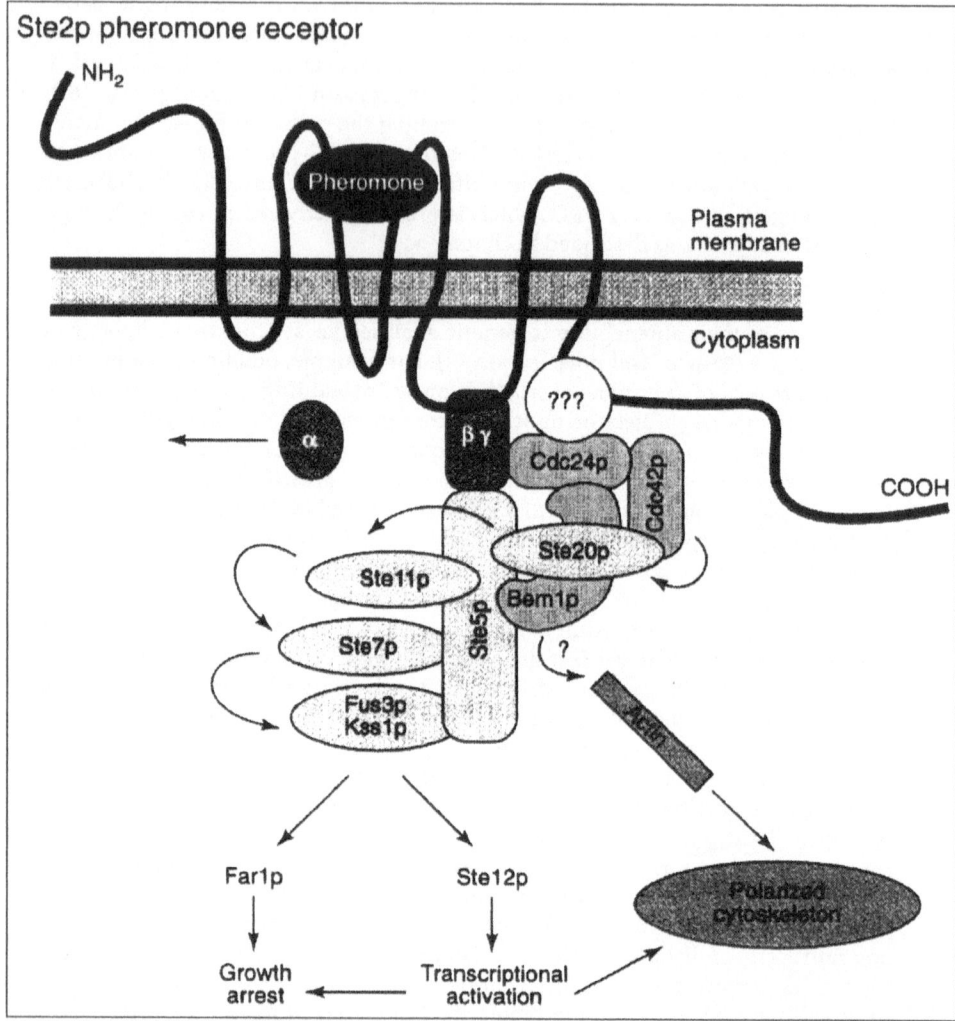

Fig. 3.6. Complex of components regulating polarization and cell cycle arrest. Figure from ref. 65, reprinted from Roemer T, Vallier LG, Snyder M, Selection of polarized growth sites in yeast, Trends Cell Biol 1996; 6: 434-441, with kind permission of Elsevier Science - NL, Sara Burgerhartstraat 25, 1055 KV Amsterdam, The Netherlands.

phenomenon similar to that described in chapter 4 for the first stage of chemotaxis (65). Pheromone treatments deplete cells of Axl2 and Bud4 which should reduce axial budding. In yeast cultures, cells of one mating type produce small peptide pheromones which bind and conformationally activate G-protein-linked (serpentine) receptors (Ste2/3) (66). Coupling of the receptor to G-protein results in its dissociation into α and βγ subunits. G-protein dissociation is regulated by two types of GAP proteins which promote GTP hydrolysis and subunit reassociation

Table 3.4. Genes related to the actin cytoskeleton

Gene (synonym)	Encoded protein	Function
ACT1	Actin	Structural; essential for polarized secretion and endocytosis
ACT2 (ARP2)	Actin-like protein	Cytokinesis?
ACT3 (ARP3)	Actin-like protein	?
SAC6	Fimbrin	Actin filament bundling; essential for polarized secretion and endocytosis
SAC7	?	Essential for cytoskeleton assembly
MYO1	Conventional myosin heavy chain	Cytokinesis?
MYO2	Type V unconventional myosin	Essential for polarized secretion
MYO4	Type V unconventional mysoin	?, nonessential
ANC1	Related to human ENL and AF-9	Transcriptional regulation?
ABP1	SH3 and cofilin-like domains	?, nonessential
SI	SH3 domains	Important for actin assembly?
SI	Talin-like	Important for actin nucleation?
PPY1	Profilin	Actin monomer sequestering; actin assembly regulation
COF1	Cofilin	Actin severing; essential
CAP1	Actin-capping protein α-subunit	Actin cytoskeleton assembly
CAP2	Actin-capping protein β-subunit	Actin cytoskeleton assembly
TPM1	Tropomyosin	Actin filament bundling; essential for polarized secretion
TPM2	Tropomyosin	?
RAH3	?	Regulation of actin assembly under osmotic stress?
RVS167	SH3 domains	Actin cytoskeleton assembly?
END3	Putative Ca^{2+}- and PIP_2-binding protein	Important for actin assembly and endocytosis?
VRP1	Verprolin	Actin polarization?
SMK	Kinesin-like protein	Polarized secretion?
SP	?	Phosphatidylinsitol metabolism?

Modified from ref. 63 (Cid VJ, Duran A, del Rey F, Snyder MP, Nombela C, Sanchez M. Microbiol Rev 1995; 59: 345-386), courtesy of American Society for Microbiology.

as described in chapter 2. Mutations which dissociate the G-protein heterotrimer can constitutively activate the pheromone response pathway. Conformationally activated receptors can be down-regulated by phosphorylation and ubiquination of its C-terminal cytoplasmic domain, the latter required for proper endocytosis and degradation.

Pheromone receptor activation is coupled via the G-protein $\beta\gamma$ complex (β/γ = Ste4/18) to a specific MAPK cascade (Ste20, Ste11, Ste7, Fus3/Kss1) (see Fig.

2.7). The MAPK signaling pathway is physically organized as a multimeric complex on the scaffolding protein Ste5, which is associated with all four components of the MAPK cascade as well as the βγ complex from the G-protein, as illustrated in Figure 3.6 (65). The key component for biochemically coupling the G-protein to the MAPK cascade is Ste20, the founding member of the PAK family of Ser/Thr kinases (67). These kinases share not only homologous kinase domains, but also binding sites for the small G-proteins Rac and cdc42 which have been implicated in cytoskeletal rearrangements for the formation of cellular protrusions (chapters 1 and 4). Thus, they provide a potential link between pheromone receptor activation and the cytoskeleton. In support of that idea, cdc42 has been shown to be present at regions of polarized growth and at the tips of mating protrusions. Cdc42 also binds directly to Ste20 (68). Although cdc42 and Ste20 associate, cdc42 does not appear to be required for Ste20 activation, the mechanism of which is unclear.

For extension of protrusions the pheromone response pathway must be coupled to the cytoskeleton. Three different filamentous elements are implicated in budding: septin filaments, actin and microtubules. Septins were first associated with cytokinesis because of their distribution during the cleavage process (63). However, recent studies suggest that they also play a role in bud site selection during budding, coupling of cell cycle signals to polarization, localized synthesis of chitin for cell walls and morphogenesis during the pheromone response. Budding yeast is a good organism for studying actin because only one actin gene is present (69). Actin in yeast is observed in two forms, cortical patches and microfilament cables, as well as at the bud site. The key component in the organization of actin at the bud site appears to be Bem1, an SH3 domain-containing protein which binds Ste5, Ste20 and actin (Fig. 3.6) (66). A second actin-binding- protein implicated in the complex at the bud site is Bni1 which binds cdc42, Bud6 and profilin. Thus, the structure of the bud site complex(es) may rival that of adhesion complex(es). Unraveling the molecular nature of this complex will be a challenge, but also an opportunity for understanding an important aspect of cytoskeletal control and the linkage between signaling and cytoskeletally regulated cellular processes.

The possibility of profilin as a bud site complex component is a reminder of the likely importance of actin polymerization in protrusion formation as described in chapter 4. The key components for actin rearrangements are available in these yeast as shown in Table 3.4. Of particular importance in forming and stabilizing protrusions based on studies in other systems are the capping and severing proteins, profilin, myosins, tropomyosin and crosslinking proteins. How these proteins are organized and regulated for the participation of microfilaments in the polarization process remains unclear. Yeast actin cables are similar to stress fibers whose assembly is regulated by Rho. Four different Rho analogs have been implicated in budding in addition to cdc42. One or more of these Rho proteins may be involved in cable formation as part of the process of stabilizing the bud. Interestingly, analogs of Ste20 kinase have been shown to phosphorylate and activate myosin I which has been implicated in formation of protrusions in other cell types (chapter 4). However, the role of myosin I in these cases is unclear. It may either participate as a contractile element directly in the protrusion process, or it may be involved in the vesicular traffic necessary for bringing membrane to the site of the protrusion.

As mentioned previously, yeast must not only provide new membrane for the protrusion, but also provide the materials for cell wall synthesis at the appropriate site. Cell wall growth is linked to cortical actin (63). Electron microscopy indicates that actin patches are associated with sites of membrane invaginations which are attached to microfilament cables. Furthermore, actin cables appear to be directly involved in the polarization of the secretory pathway for movement of materials in vesicles to the plasma membrane for cell wall synthesis (4). Vesicular traffic also requires polarized microtubules whose formation is dependent on actin function, suggesting an interaction between the microtubules, actin cytoskeleton and secretory pathway. How these components are organized at the molecular level in yeast is unclear. As shown in later sections of this chapter and in chapter 4, there are common features between yeast polarization, neuronal axon formation and cell motility, but each of them has its own special traits. Undoubtedly, the power of yeast genetics, the availability of the sequence of the complete yeast genome and the application of biochemical and cell biological approaches should yield interesting contributions from this organism to the understanding of the general question of the relationships between the structure and functions of dynamic cell projections.

Sensory Receptor and Mechanosensitive Cells

Sensory Receptor Cells

One of the most important functions of multicellular organisms is the ability to sense and respond to the environment. During evolution, animals have developed increasingly complex mechanisms and structures for assessing visual, tactile, auditory, olfactory and other environmental information (70). A common feature of these sensory cells is the use of gated ion channels for signal transduction (71). However, because of the extreme variability of the input signals, the cell structures involved in signal reception are quite varied and include such elaborations as the rod and cone cells of the eye and stereocilia of the receptor cells in the ear. Since cell polarization and the formation of specialized cytoskeleton-based cell surface features play an important role in sensory reception in the auditory and olfactory responses, we will consider those in more detail. The sensor cells for these responses are superficially similar, but differ substantially in the organization of the sensing structures and in the signaling mechanisms coupled to those structures. Both cell types are polarized and have a ciliated surface, though the nature of the cilia differ in both structure and cytoskeletal components. The sensor cell of the olfactory epithelium is a neuron which extends an apical dendrite into the mucus bathing the epithelial surface (72). This dendrite has 10 or more immotile microtubule-based cilia. Its axon extends from the basal surface of the epithelium and conveys electrical signals to the brain. The auditory sensor cell is the hair cell, so called because of the microfilament-based stereocilia at the apical surface (73). The basal surface of the hair cell conveys signals via a synaptic connection to neurons from the brain.

The key to understanding these sensory mechanisms is the role of the organization of the sensing structures in transducing these extracellular signals. Olfaction is a "classical" signal transduction mechanism in which the signals are small molecules which bind to specific G-protein-linked receptors localized to the cilia of the sensor cells (71). The receptors which provide the specificity for odorant recognition (74) are coupled to a G-protein transduction system to activate adenylate cyclase and produce cAMP. The cAMP acts directly on gated cation channels to produce a current carried by Na$^+$ and Ca^{2+}. How the signaling system is organized and localized to the cilia is unclear, though some G-proteins have been shown to associate with microtubules (75). Interestingly, the visual system also uses a G-protein signaling system, but the cellular structures used for collecting the signal and the second messenger used are completely different.

The critical feature of the auditory sensor cells is the apical bundle of sensory hairs composed of a microtubule-based kinocilium and an array of hexagonally-packed stereocilia (tens to hundreds) in rows of varied heights (73). The stereocilia contain a core of highly crosslinked microfilaments, providing structural rigidity, which is important for their signaling mechanism. The stereocilia extend into the fluid of the cochlea. Pressure waves in this fluid generated by sound impinging on the ear cause a deflection of the stereocilia bundles at their bases (76). These mechanical deformations activate stretch-sensitive cation channels (77) in the stereocilium membrane to generate the electrical signal which is transmitted to the brain. A critical question is how the mechanical deformation is transmitted to the channel. An interesting aspect of hair cell structure is that the tips of the shorter stereocilia are linked by cell surface fibers to the sides of adjacent taller stereocilia (73). These "tip link" structures are proposed to act as a gating spring to open nearby channels (76). This model and localization data suggest that the channels are located near the tips of the stereocilia (76), though this interpretion has been questioned (73). Obviously, an identification and characterization of the specific channel molecules involved in the signaling would help to resolve the mechanism.

Mechanosensitive Cells

In addition to sensory mechanisms, cells must often respond to direct mechanical stimuli such as gravity, touch, shear forces and osmotic forces. The most similar to the sensory mechanisms is the touch sensation. Investigations of touch at the molecular level have been facilitated by studying the genetics of touch insensitive mutants of the nematode *Caenorhabditis elegans* (78). In this organism the touch sensation is mediated by six neurons. Isolation of mechanosensory deficient (Mec) mutants has led to the identification of 12 genes, two of which encode proteins similar to amiloride-sensitive sodium channels and are presumed to be mechanosensitive ion channels similar to those of the tip-link model for hair cells (76). Other members of this superfamily of cation transporters are involved in ion transport in epithelia and are known to be composed of three subunits (79). Two of the *mec* genes encode unique α and β tubulins which form large diameter microtubules found only in neuronal processes of the touch cells. In contrast to the conventional microtubules of the neurons, these specialized microtubules are oriented parallel to the length of the process and may be part of the sensory apparatus (78). How the stimulus gets transferred to the channels is yet unclear. However, the recent observation that a putative subunit of amiloride-sensitive channel interacts with a collagen IVa analog provides one possible link from cell surface structures to the channel (80).

Volume regulation is a common cell function, particularly in polarized, secretory cells. A number of cellular signaling components have been implicated in these responses, including calcium/calmodulin, eicosanoids, phosphoinositides and stretch-activated channels. Two types of sensors have been suggested, the cytoskeleton and intracellular solutes (81). One simple mechanism suggests that osmotically-induced swelling causes a cytoskeletal reorganization which activates stretch-activated channels and calcium influx. An alternative mechanism proposes that changes in the concentration of intracellular solutes activate G-protein-linked phospholipase A2 to release arachidonic acid which is then converted to eicosanoids. These lipids are released from the cell and stimulate receptor-linked PLC to hydrolyze phosphoinositides, potentially altering microfilament organization, and produce DAG and IP3 which can stimulate PKC. The diversity of responses in different cells suggests that multiple mechanisms may be involved.

Endothelial cells are located between the flowing blood and vascular wall and are thus subjected to much greater fluid forces than most tissue cells (82). Fluid shear stress modifies the shape of endothelial cells from polygonal to ellipsoidal (83). This shape change is accompanied by the redistribution of microfilaments into stress fibers (84) and remodeling of focal adhesions (82) with a concomitant shift of vinculin into focal adhesions (85). Although organizational changes of microtubules and intermediate filaments are also observed, microfilaments appear to be the primary agent for transmission of shear stress effects, such as changes in gene expression, because the effects can be blocked by drugs which interfere with microfilament turnover (86). Numerous signaling elements have been implicated in the changes which are induced by shear stress, including mechanosensitive ion channels, G-proteins, tyrosine kinases and Ser/Thr kinases (82). Temporal aspects of the effects suggest that the earliest changes occur in mechanosensitive ion channels which can regulate calcium and pH changes in cells. In addition to effects induced via gated cation channels, endothelial cells may also be responding to perturbations through cell surface receptors such as integrins (87). Integrin clustering can induce changes in inositide synthesis, which, as described in chapter 1, can lead to changes in the cytoskeleton. Furthermore, integrins may act directly as mechanoreceptors to transduce signals to the cytoskeleton (88) which may be transmitted via microfilaments and intermediate filaments to the nucleus (89). Whether such effects are important to transcriptional changes activated by shear stress is unknown. The identification of a shear stress responsive element on endothelial cell DNA and a shear stress-related transcriptional factor (90) suggests involvement of a signaling pathway as part of the response. One possibility would be the MAPK pathway.

Neurons

The ultimate horizontally polarized cell is the neuron which may attain lengths in meters in some species. This unusual shape allows it to transmit information via electrical and chemical signals between cells through cellular extensions (*neurites*) and cell-cell connections (*synapses*). The specificity of the transmission is determined by the localization of the neuron in the organism, the type of extension and the nature of the synapse. The complexity of the information transfer can be illustrated by the fact that a large neuron may connect to as many as 10^5 other cells. Neurons are nondividing cells which migrate to their final position in the organism after generation from precursor cells (91). Differentiation into a functioning neuron occurs with the growth of an *axon* and multiple *dendrites* from the cell body (92). Both types of extensions form via the activity of a motile cell surface projection called the *growth cone*. Axons and dendrites differ substantially in structure and function since the axon must carry signals over long distances, while the dendrites receive and process these signals (93).

Axons are long thin processes which extend from the neuron body and form branches to make multiple connections with other neurons (via the dendrites) or other cell types, such as muscle cells (via neuromuscular junctions). Freeze-etch electron microscopy has shown a highly detailed structure of axons (94). A dense network of filaments, presumably microfilaments, is associated with the axonal membrane and linked to two domains containing longitudinal filaments more centrally located in the axon. One of the longitudinal elements consists of crosslinked neurofilament bundles; the other domain contains microtubule bundles suspended in a loose granular matrix. The microtubule domains are proposed to provide channels within which organelles are transported. Three types of membranous structures are observed within axoplasm, axoplasmic reticulum, mitochondria and vesicles (94), the latter presumably including materials which are undergoing microtubule-dependent transport down the axon (95). Dendrites are more diverse in structure than axons, exhibiting branched configurations and special branchlets known as *dendritic spines.* Although microtubules in axons are of uniform polarity, those in dendrites are mixed (93). Axons and dendrites may also differ in caliber, myelinated axons being the largest (96). In the large axons, caliber is related to neurofilament content; in smaller axons and dendrites microtubules appear to determine the caliber. Differences in size and structure may relate to MAPs associated with the two types of these neurites, MAP-2A and MAP-2B in axons, tau in dendrites and MAP-1A and MAP-1B in both (97), as indicated in chapter 1. Neurite extension is initiated by microfilament-driven motility of the growth cone as described in chapter 4. However, neurite elongation is primarily dependent on microtubules, while guidance depends on microfilament dynamics in the growth cone (98). At the distal ends of neurites individual microtubules of the bundles splay into the growth cone (99). The neurite extension is stabilized by microtubule assembly in the growth cone and by microtubule bundling with different MAPs serving as stabilizing agents for the complex neurite cytoskeleton formed during the development of the two types of processes (96).

The critical aspect of neuron function is the formation of the correct neuronal connections, often requiring long axonal traverses to the target cell (100). This axonal specificity develops in three stages: pathway selection, which decides the route along which the axon travels; target selection, which involves recognition of the appropriate cell set for the formation of connections; and address selection, which determines the particular cell(s) to which the connections are made. Pathway selection occurs before the onset of neuronal electrical activity via three types of guidance mechanisms, *differential adhesion, chemotropism* and *repulsion,* which are also fundamental to target selection. Address selection results from the formation of specialized junctions, called *synapses,* and requires electrical activity. A more detailed description of the formation of one type of synapse, the neuromuscular junction, is provided in the following section. The formation of a network via these three stages is only the first phase of neuronal development and is followed by a refinement phase in which synapses are eliminated and neurites retract (101). This network maturation is dependent on the type of neuron, the type of neurite and environmental factors and may be triggered by signaling through calcium channels, leading to an increase in intracellular calcium and disruption of the axonal cytoskeleton.

Differential adhesion to extracellular matrix molecules has long been recognized as an important contributor to cell sorting into tissues and to directed cell migration. This mechanism for neurons requires that specific adhesion receptors are expressed on the surface of the growth cones which recognize a pattern of extracellular matrix or cellular ligands along the intended pathway (100). The ligands include collagens, laminins and fibronectin, which can all bind specific integrin receptors and specific cell-cell adhesion molecules, such as cadherins and immunoglobulin (Ig)-related cell adhesion molecules (CAMs). These provide positive guidance over or through adjacent cells. The ligand-receptor interactions not only define steps in a path, but can also activate processes in the signaling mechanisms for *growth cone motility* which is described further in the next chapter. Interestingly, growth cones distinguish between fibronectin and laminin by forming different types of adhesions (102). Integrin-based adhesions resemble focal contacts and contain paxillin and phosphotyrosine. These results imply that the different adhesion mechanisms can also induce different signaling pathways for influencing growth cone guidance.

In addition to this positive mechanism for guidance, growth cone advances may be inhibited by secreted molecules of nearby cells which cause *growth cone collapse*. Two families of these growth cone repellants have been described, *semaphorins* and *netrins* (103). As the growth cone travels toward and approaches its target region, a second type of repulsion mechanism is observed (104). Neuronal growth cones carry members of the Eph family of tyrosine kinase receptors (see Fig. 2.2), while cells away from the target area or nontarget cells in the target region may express cell surface ligands for these receptors. The ligand-receptor interaction signals a change in the cytoskeleton of the growth cone leading to its withdrawal. The large number of Eph family receptors may provide the diversity of signals necessary for specific recognition in the nervous system, while the cell surface location and spatial distribution of the ligands presents the necessary gradients for guidance of growth cones. Finally, the ECM molecule *tenascin-C* may be able to provide both positive and negative guidance cues since different recombinant domains of the proteins can confer either adhesion or anti-adhesion changes to growth cones and cell bodies (105), implying the presence of different cell surface receptors and the potential for different signaling mechanisms for controlling the cytoskeleton and guidance.

As mentioned above, the final stage of development is the death of a substantial fraction of the neurons, often more than half, depending on the species. This overproduction and *apoptosis* of excess neurons is regulated by polypeptide factors (*neurotrophic factors*) secreted by the target tissues, the best known of which is nerve growth factor. Other factors are similar in structure to NGF, but are specific to different groups of neurons and thus stabilize different types of neuronal connections.

The Neuromuscular Junction

Much of the intercellular communication in developing and adult nervous systems occurs through chemical synapses. Synapses are highly specialized structures that mediate the transfer of information from a neuron to its target cell. In the central nervous system, synapses are responsible for conveying information from one neuron to another, while in the peripheral nervous system they mediate communication between motor neurons and muscle cells. Because these structures are responsible for the transmission and storage of information, the biochemical and structural changes that accompany synaptic signaling likely underlie the biological processes of behavior, learning and memory. Rapid and efficient transmission at the chemical synapse requires that signaling components be concentrated within these structures. Hence, the presynaptic neuron must deliver its chemical signal to the precise site of contact with its target cell. Likewise, signal transmission is most efficient when receptors in the postsynaptic cell are localized to the point of contact with the neuron. Synaptic transmission generally results in a change in the electrical activity of the postsynaptic cell by stimulating ion channel activity. Since local membrane polarization/depolarization is responsible for the propagation of the electrical signal by the action potential mechanism, the clustering of channels at specific sites in the postsynaptic membrane is essential for efficient signal transmission. For these reasons numerous proteins and mechanisms have evolved to specifically localize channels and other signaling molecules at synapses.

The neuromuscular junction (NMJ), the site at which a nerve cell communicates with a muscle fiber, is the most well-studied example of chemical synapses (106). Historically, the accessibility and relative simplicity of this synapse has facilitated its detailed structural and biochemical analysis. In addition, muscle fibers are reinnervated following nerve damage, allowing the process of synaptogenesis to be studied in the adult animal. Numerous synapse-specific proteins have been identified and the roles of these components in the structure and maintenance of the NMJ characterized. More recently, genetic studies have shed light on the mechanisms by which the NMJ develops, and the roles of specific proteins in mediating intercellular communication between the motor neuron and the muscle fiber. From these studies is a picture emerging of a highly specialized structure that mediates rapid and focal intercellular communication, highlighting the critical importance of the localization of signaling components in signal transduction. Moreover, these studies underscore the importance of cooperative and reciprocal signaling between different cell types during tissue development.

Structure and Function of the Neuromuscular Synapse

The NMJ is specifically adapted to propagate an electrical signal from the motor neuron to the muscle fiber through the neurotransmitter acetylcholine (ACh). There are several steps involved in synaptic transmission at the NMJ. An action potential propagated along the axon of a motor neuron to its nerve terminal stimulates the opening of voltage sensitive calcium channels in the presynaptic membrane. The influx of calcium ions into the nerve terminal triggers the fusion of intracellular neurotransmitter-containing vesicles with the plasma membrane (see chapter 6), resulting in the secretion of acetylcholine into the synaptic cleft. ACh diffuses across the cleft to the postsynaptic muscle fiber, and binds to receptors in the plasma membrane. Neurotransmitter binding then opens the sodium ion channel associated with the receptor, and the local depolarization of the membrane results in an action potential within the muscle fiber. ACh releases from the receptor and is cleared by uptake or acetylcholinesterase-catalyzed hydrolysis.

At the microscopic level, a number of specific structures at both the nerve terminal and the muscle fiber innervation site contribute to the concentration of the components involved in these synaptic transmission events. Nerve terminals are rich in synaptic vesicles that carry ACh. These vesicles are particularly concentrated near regions of the presynaptic membrane called active zones, where the vesicles fuse with the plasma membrane and release their contents. As expected, active zones contain calcium channels and docking proteins involved in membrane-vesicle fusion. Nerve terminals are also rich in mitochondria which supply the energy required for ACh synthesis and secretion, and microfilaments which can constrain vesicles as part of the regulated secretion pathway (see chapter 6). In the region of contact with the nerve terminal, the postsynaptic membrane of the muscle fiber is thickened relative to the surrounding membrane and rich in acetylcholine receptors (AChRs). Junctional folds in the membrane of the fiber precisely align with the active zones of the nerve terminal. Also, the nuclei immediately adjacent to the site of innervation, or the synaptic nuclei, may be considered as part of the NMJ. These nuclei are morphologically distinguishable from their extrasynaptic counterparts, and as described below, are involved in the expression of synapse-specific genes. Between the pre- and postsynaptic membranes is a basal lamina which is involved in cementing the nerve terminal to the muscle fiber. The basal lamina also acts as a repository for acetylcholinesterase, an enzyme that inactivates ACh, as well as for growth factors and other molecules involved in the development of the NMJ.

The key factor in the fidelity of synaptic transmission at the NMJ is the enrichment of AChRs in the postsynaptic membrane. The AChR consists of five homologous polypeptide subunits that form a sodium ion channel which opens upon binding of neurotransmitter. The high concentration of AChRs in the postsynaptic membrane ensures a focal depolarization of the plasma membrane upon stimulation which is ultimately responsible for the muscle response. AChRs are concentrated at the crests of the junctional folds of the postsynaptic membrane, but their levels fall off dramatically within the junctional folds and in the extrasynaptic membrane. This distribution puts the receptors immediately adjacent to the active zones in the presynaptic membrane where ACh is released. Hence, the cytoskeletal elements responsible for the clustering of AChRs in the postsynaptic membrane, the factors that contribute to vesicle fusion in the active zones, as well as the elements that are involved in the alignment of the presynaptic active zones with the postsynaptic junctional folds, are as much a part of the signaling mechanism as are the neurotransmitter ligand and its ion channel receptor.

Development of the Neuromuscular Synapse

Since the concentration of AChRs in the postsynaptic membrane is triggered by the innervating motor neuron during the process of synapse development, an examination of the steps involved in NMJ formation leads to insight into the molecular mechanisms underlying the specific localization of signaling components. The formation and establishment of the NMJ involves a series of events mediated by reciprocal intercellular signaling between the motor neuron and the target muscle fiber. During development, motor neurons of the spinal cord extend axons that branch to innervate multiple fibers in a single muscle. As the axon makes contact with its target muscle fiber, it loses its myelin sheath, settles into shallow gutters on the fiber surface, and becomes capped by a Schwann cell. Upon arrival the neuron sends a signal to the muscle fiber indicating that this region will develop into a synapse and that postsynaptic signaling components should be concentrated in the immediate area. Likewise, the fiber sends a retrograde signal to the neuron indicating that it is undergoing innervation at this site and that this branch of the axon needs to extend no further. The result of these reciprocal signaling events is that each fiber is innervated via a single synaptic site, so that each fiber receives a single action potential upon stimulation.

The innervating axon stimulates the concentration of AChRs at the NMJ by three distinct mechanisms (107). The first two involve the transcriptional regulation of AChR subunit genes expressed by specific nuclei. Upon innervation, nuclei immediately adjacent to the NMJ transcribe AChR genes at a higher rate than prior to innervation. In addition, the transcription of AChR genes by extrasynaptic nuclei appears to be suppressed after innervation. The result is an accumulation of

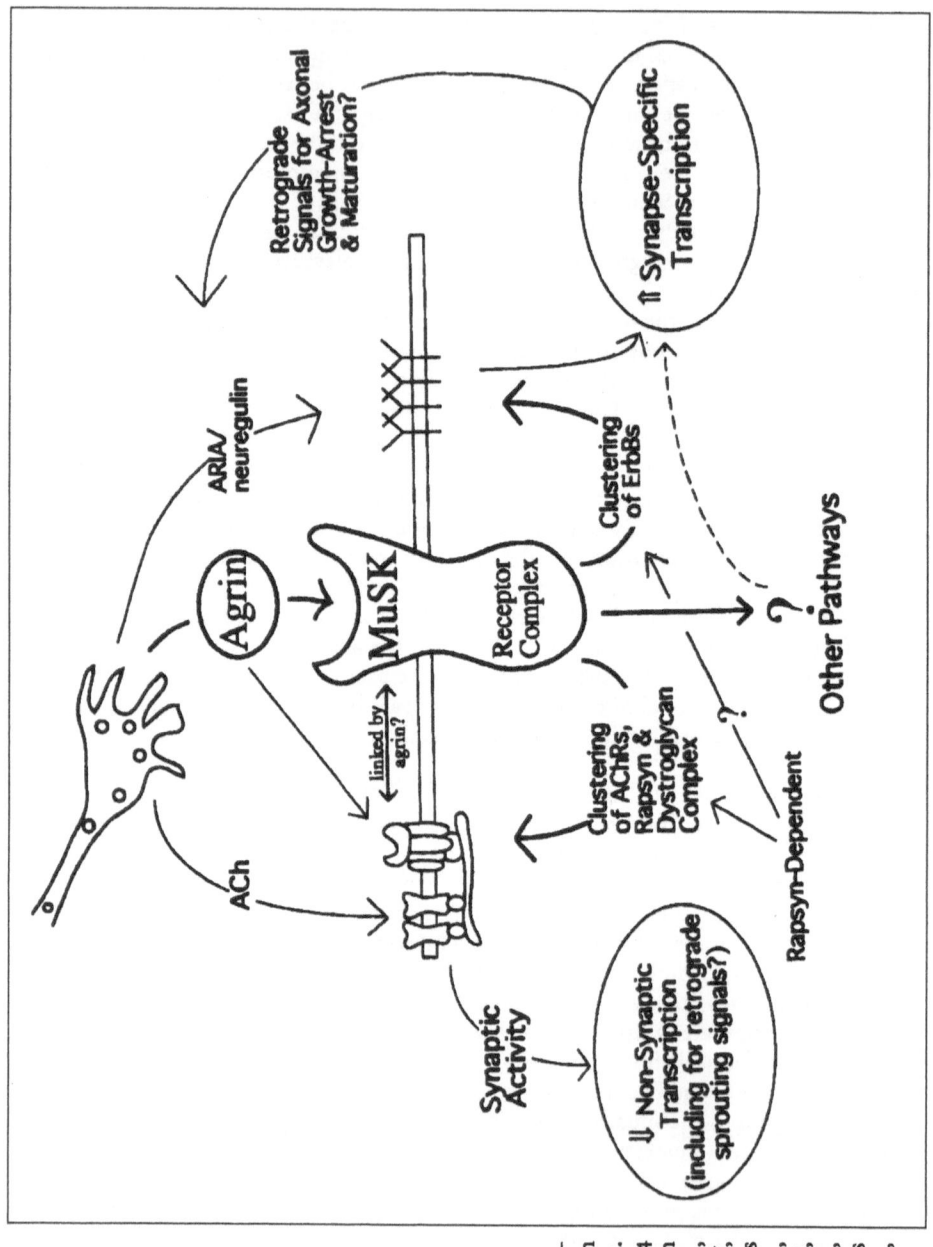

Fig. 3.7. Agrin/MuSK signaling system in formation of neuromuscular junction. Figure from ref. 114 (DeChiara TM, Bowen DC, Valenzuela DM, Simmons MV, Poueymirou WT, Thomas S, Kinetz E, Compton DL, Rojas E, Park JS, Smith C, DiStefano PS, Glass DJ, Burden SJ, Yancopolous GD. 1996. Cell 85, 501-512), courtesy of Cell Press.

AChR subunit message at the NMJ, and local protein synthesis from this message pool contributes to the concentration of receptors at the synapse. However, the more prevalent mechanism for concentrating receptors at the synaptic site involves the neuron-stimulated recruitment of diffusely distributed AChRs from the extrasynaptic membrane upon innervation, and the clustering of newly synthesized receptors at the synapse.

The molecular events underlying the three neuron-regulated AChR concentration mechanisms are still in the process of being elucidated, but great strides have been made in the past few years. The dissection of these pathways has been facilitated by an accumulation of knowledge concerning the identity of proteins present at the NMJ, and more recently by genetic studies in which the genes encoding some of these components have been deleted in mice and the effects on NMJ development and structure determined. The results of these studies reveal some familiar themes, as well as some unexpected and novel signaling mechanisms.

The two mechanisms for AChR transcriptional regulation at extrasynaptic and synaptic nuclei are quite disparate. The suppression of AChR expression by extrasynaptic nuclei is a secondary event that follows innervation, and appears to be mediated by cholinergic transmission. An axonal action potential results in the release of ACh at the NMJ, eliciting an action potential in the muscle fiber. This action potential is then propagated along the muscle fiber and delivered to voltage-sensitive calcium channels in the extrasynaptic membrane. The influx of calcium that accompanies stimulation is thought to be a major contributor to the regulation of genes expressed at extrasynaptic nuclei. The neuron-mediated increase in AChR subunit expression at synaptic nuclei also appears to be a secondary event following muscle fiber innervation, but in this case the signal goes through a growth factor receptor/tyrosine kinase pathway at the NMJ. In recent years it has become clear that one of the major paracrine factors expressed by neurons to influence the growth and behavior of surrounding glial cells is a series of growth factors called the neuregulins. Neuregulins expressed in the axons of neurons in the peripheral nervous system influence the proliferation and differentiation of Schwann cell precursors during development, and the survival of Schwann cells in the adult (108). In the central nervous system, the neuregulins influence axonal guidance and innervation of the developing hindbrain (109). At the NMJ the neuregulins are secreted at the nerve terminal and are deposited on the basal lamina. The neuregulins bind to and activate members of the ErbB family of receptor tyrosine kinases, the prototype of which is the epidermal growth factor (EGF) receptor. These receptors undergo an extensive series of heterodimerization events in response to ligand binding, and it appears that two different members of the receptor family are required for normal signal transduction in many cell types (110). Like AChR, ErbB receptors are also clustered in the post-synaptic membrane following muscle fiber innervation. Hence, neuron-derived neuregulin associated with the basal lamina binds to and activates ErbB receptors in the post-synaptic membrane. The signal is propagated to synaptic nuclei by a mechanism that includes activation of the MAPK and PI3K pathways, resulting in the stimulation of AChR subunit genes.

Early work suggested that a neuronally-derived factor present in the basal lamina is responsible for the clustering of pre-existing AChRs in the post-synaptic membrane. More recent studies suggest that this factor, called agrin, is probably the initial signal delivered from the incoming axon to the muscle fiber that sets in motion the entire process of NMJ development (see Fig. 3.7). Agrin is synthesized by motor neurons and transported down axons to be secreted at the nerve terminal and deposited on the basal lamina. Purified agrin is capable of inducing the clustering of AChRs in cultured myotubes, and neutralizing antibodies to agrin inhibit neuron-induced AChR concentration in co-cultures. Such studies laid the groundwork for the agrin hypothesis which states that agrin released at motor axon terminals interacts with receptors on the surface of myofibers to initiate postsynaptic differentiation (111).

Recent studies with mice deleted in agrin expression strongly support the agrin hypothesis (112). Homozygous mutants developed normally, but were stillborn. No gross abnormalities were observed in any tissues of these animals, including skeletal muscle, but the embryos were incapable of movement, suggesting a defect in signaling at the NMJ. Closer examination revealed serious defects in NMJ formation. Most axon terminals displayed no clustering of AChRs, and all other components of the postsynaptic apparatus, including basal laminal, membrane and cytoskeletal proteins were disrupted. In addition to the observed defects in postsynaptic differentiation, defects in the properties of motor neurons were also observed in homozygous agrin knockouts. Axons appeared to overrun their target fibers ending with no apparent specializations and presynaptic vesicles and active zones were rarely observed. Taken together, the observations made with knockout mutants indicate that agrin is a very potent initiator of postsynaptic differentiation, and that an agrin-stimulated retrograde signal from the muscle fiber to the axon is required for presynaptic differentiation. Hence, the dissection of NMJ development reveals that reciprocal communication between cell types is critical for the proper tissue development.

Although the effects of agrin on postsynaptic differentiation have been known for several years, the molecular target for agrin at the surface of muscle fibers remains to be clearly defined. A number of candidate receptors for agrin have been proposed over the years, including integrins, neural cell adhesion molecule and *dystroglycan*. Of these, dystroglycan has received a good deal of attention because it was demonstrated to interact directly with agrin. Moreover, dystroglycan acts as a postsynaptic transmembrane linker between the basal lamina, by interacting with laminins, and the muscle cytoskeleton, by interacting with the spectrin-like dystrophin homolog *utrophin*. This dystroglycan complex interacts with AChRs, and is thought to contribute to their clustering in the postsynaptic membrane. However, some splice variants of agrin that possess no AChR clustering activity bind avidly to dystroglycan; likewise some fragments of agrin active in receptor clustering activity do not bind dystroglycan.

One important clue pointing to the identity of the agrin receptor was the observation that agrin stimulates the tyrosine phosphorylation of specific proteins in muscle cells including subunits of AChR. About the same time, a receptor tyrosine kinase that showed muscle specific expression was cloned and characterized (113). This protein, called MuSK for muscle-specific kinase, is expressed very early in the muscle lineage and becomes highly concentrated at NMJs as the muscle matures. Moreover, mice homozygous for the deletion of MuSK display a very similar phenotype as those deleted in agrin (114). These animals are incapable of movement and die at birth from an inability to breathe. Like agrin mutants, no gross abnormalities in skeletal muscle are observed, but no AChR clustering is found at synapses and axons overrun their targets. Together with data indicating that agrin stimulates MuSK tyrosine phosphorylation in cultured myotubes and that radiolabeled agrin may be covalently crosslinked to MuSK in these cells, these observations strongly implicate the MuSK receptor tyrosine kinase as a binding receptor for agrin. However, agrin is incapable of interacting with purified MuSK extracellular domain and when MuSK is ectopically expressed in fibroblasts, it is incapable of interacting with or becoming stimulated by agrin (115). The requirement that MuSK be expressed in myotubes strongly suggests that another component present in these cells is required for agrin stimulation. The molecular identity of this component, referred to as MASC for myotube-associated specificity component, remains to be determined.

While most of the signaling events following MuSK activation remain to be elucidated, the ultimate effector for agrin in the postsynaptic membrane may be a protein called 43K or *rapsyn*. Rapsyn is a myristoylated zinc finger-containing protein present at the synapse in a roughly one-to-one ratio with AChR. Its localization in the postsynaptic membrane precisely matches that of AChR clusters in developing embryos and in adult animals. When co-expressed with AChR subunits in non-muscle cells, rapsyn induces the clustering of otherwise diffuse receptors into small aggregates, strongly suggesting that rapsyn plays a major role in AChR clustering in the postsynaptic membrane. This hypothesis is confirmed by the phenotype of mice in which rapsyn expression has been deleted (116). Rapsyn knockout animals are capable of some movement and even shallow breathing, but still die within a few hours of birth. As in agrin and MuSK knockouts, the overall phenotype of skeletal muscle appears normal in rapsyn-deleted animals. In contrast, much of the NMJ develops normally. Synaptic nuclei preferentially express AChRs and receptors accumulate in normal quantities, but they do not cluster in the post-synaptic membrane. Other proteins of the postsynaptic membrane and cytoskeleton, including utrophin, dystroglycan and the ErbB receptors are also diffusely distributed in rapsyn-deficient animals, suggesting a prominent role for this protein in specifically localizing signaling components at the NMJ. However, some components of the muscle fiber remain clustered, such as laminins and acetylcholinesterase. Hence, rapsyn may link a subset of components of the membrane to postsynaptic scaffolding elements, while other components are linked directly or through unidentified adaptor proteins. Although microfilaments have been identified as the major cytoskeletal element associated with the TMJ, the molecular mechanism of linking this complex junction to the cytoskeleton is still unclear (117). A novel isoform of spectrin (β-*spectrin*) has been implicated from its localization and concentration; however, a number of other membrane skeletal and focal adhesion complex components, including vinculin, talin, paxillin, filamin, α-actinin, tropomyosin 2, dystrophin and ankyrin have been localized to the postsynaptic NMJ and probably contribute to the organization of the membrane complex and its linkage to microfilaments.

Summary

Development of polarity is a key feature of many cell functions. Polarization may be local or global, transient or permanent. The initiating signal and the site of origin may be imposed either externally or internally. A critical factor is the activation of morphogenetic receptors to establish membrane-cytoskeletal signaling complexes whose formation leads to subsequent cellular reorganizations. Investigations of polarization mechanisms have focused primarily on epithelial, yeast and neural cells. Epithelial cells provide an impermeant barrier by cell-cell associations at tight junctions which separate their more specialized apical membranes from their basolateral surfaces. Formation of the epithelial cell layer involves integrin-mediated adhesion to basal lamina and the establishment of cell-cell, cadherin-containing adherens junctions. Cadherin is linked to microfilaments via associations with catenins. Adherens junctions are dynamic structures; the association of this complex with Src family kinases provides a regulatory mechanism. The compositions of the apical and basolateral domains are maintained by intracellular sorting of components through vesicle sorting pathways. Genetic analysis has made substantial contributions toward characterizing mechanisms for polarization in budding yeast. A heterotrimeric G-protein-linked pathway couples mating pheromone to gene expression via a MAPK cascade. The mating response is integrated with the cytoskeleton and cell division through a multimeric complex containing the G-protein βγ subunits, MAPK cascade components and an actin-binding protein, all associated with a scaffolding protein, providing an excellent example of how multiple functions in a cell can be coordinated.

Environmental information is gathered in higher animals by a variety of sensory cells with different mechanisms for collecting inputs. A common mechanism for transducing that information into signal is via gated channels, activated by chemical or mechanical signals. Such signals are transmitted by neurons, the ultimate polarized cells which have specialized processes for making electrical connections, axons and dendrites. Establishment of proper connections for these processes is critical to proper neural function and relies on three types of guidance mechanisms: differential adhesion, chemotropism and repulsion. The mechanisms depend on coupling of specific types of cell surface receptors to the cytoskeleton in the growing tip (growth cone) of the neurite. A particularly interesting case is the Eph receptor whose contact with cellular ligands results in withdrawal of the growth cone. Chemical synapses provide the structure for some types of neural communication. Neuromuscular junctions are a well-studied example in which the acetylcholine receptor is the ion channel transducing the signal. Development of the junction requires the clustering of receptors on the muscle cell induced by an extracellular matrix component agrin secreted by the neuron. Agrin appears to act through a specific tyrosine kinase receptor MuSK. An association with actin filaments involving the myristoylated, zinc finger protein rapsyn has been implicated, but the coupling mechanism is unknown.

References

1. Glotzer M, Hyman AA. Cell polarity. The importance of being polar. Curr Biol 1995; 5:1102-1105.
2. Drenckhahn D, Jons T, Kollert-Jons A, Koob R, Kraemer D, Wagner S. Cytoskeleton and epithelial polarity. Ren Physiol Biochem 1993; 16:6-14.
3. Mays RW, Nelson WJ, Marrs JA. Generation of epithelial cell polarity: roles for protein trafficking, membrane-cytoskeleton, and E-cadherin-mediated cell adhesion. Cold Spring Harb Symp Quant Biol 1995; 60:763-773.
4. Drubin DG, Nelson WJ. Origins of cell polarity. Cell 1996; 84:335-344.
5. Mays RW, Beck KA, Nelson WJ. Organization and function of the cytoskeleton in polarized epithelial cells: a component of the protein sorting machinery. Curr Opin Cell Biol 1994; 6:16-24.
6. Nicolson GL. Transmembrane control of the receptors on normal and tumor cells. II. Surface changes associated with transformation and malignancy. Biochim Biophys Acta 1976; 457:57-108.

7. Bennett V, Gilligan DM. The spectrin-based membrane skeleton and micron-scale organization of the plasma membrane. Annu Rev Cell Biol 1993; 9:27-66.
8. Aberle H, Schwartz H, Kemler R. Cadherin-catenin complex: protein interactions and their implications for cadherin function. J Cell Biochem 1996; 61:514-523.
9. Taipale J, Keski-Oja J. Growth factors in the extracellular matrix. FASEB J 1997; 11:51-59.
10. Gumbiner BM. Cell adhesion: the molecular basis of tissue architecture and morphogenesis. Cell 1996; 84:345-357.
11. Sheppard D. Epithelial integrins. Bioessays 1996; 18:655-660.
12. Schmidt JW, Piepenhagen PA, Nelson WJ. Modulation of epithelial morphogenesis and cell fate by cell-to-cell signals and regulated cell adhesion. Semin Cell Biol 1993; 4:161-173.
13. Rosales C, Juliano RL. Signal transduction by cell adhesion receptors in leukocytes. J Leukocyte Biol 1995; 57:189-198.
14. Hynes RO. Integrins: versatility, modulation, and signaling in cell adhesion. Cell 1992; 69:11-25.
15. Ruoslahti E, Pierschbacher MD. New perspectives in cell adhesion: RGD and integrins. Science 1987; 238:491-497.
16. Turner CE, Burridge K. Transmembrane molecular assemblies in cell-extracellular matrix interactions. Curr Opin Cell Biol 1991; 3:849-53.
17. Burridge K, Chrzanowska-Wodnicka M. Focal adhesions, contractility, and signaling. Annu Rev Cell Dev Biol 1996; 12: 463-518.
18. Dedhar S, Hannigan GE. Integrin cytoplasmic interactions and bidirectional transmembrane signalling. Curr Opin Cell Biol 1996; 8:657-669.
19. Ben-Ze'ev A. Animal cell shape changes and gene expression. BioEssays 1991; 13: 207-212.
20. Otey CA. pp125FAK in the focal adhesion. Int Rev Cytol 1996; 167:161-183.
21. Schlaepfer DD, Hunter T. Signal transduction from the extracellular matrix—a role for the focal adhesion protein-tyrosine kinase FAK. Cell Struct Funct 1996; 21:445-450.
22. Hanks SK, Polte TR. Signaling through focal adhesion kinase. Bioessays 1997; 19: 137-145.
23. Yamada K, Miyamoto S. Integrin transmembrane signaling and cytoskeletal control. Curr Opin Cell Biol 1995; 7: 681-689.
24. Juliano R. Cooperation between soluble factors and integrin-mediated cell anchorage in the control of cell growth and differentiation. BioEssays 1996; 18: 911-917.
25. Lemmon MA, Falasca M, Ferguson KM, Schlessinger J. Regulatory recruitment of signalling molecules to the cell membrane by pleckstrin-homology domains. Trends Cell Biol 1997; 7:237-242.
26. Folkman J, Moscona A. Role of cell shape in growth control. Nature 1978; 273:345-349.
27. Boone CW, Takeichi N, Paranjpe M, Gilden R. Vasoformative sarcomas arising from BALB/3T3 cells attached to solid substrates. Cancer Res 1976; 36: 1626-1633.
28. Juliano R. Signal transduction by integrins and its role in the regulation of tumor growth. Cancer Metastasis Rev 1994; 13:25-30.
29. Ruoslahti E, Reed JC. Anchorage dependence, integrins, and apoptosis. Cell 1994; 77:477-478.
30. Frisch SM, Vuori K, Ruoslahti E, Chan-Hui P-Y. Control of adhesion-dependent cell survival by focal adhesion kinase. J Cell Biol 1996; 134:793-799.
31. Simons K, Fuller SD. Cell surface polarity in epithelia. Annu Rev Cell Biol 1985; 1:243-288.

32. Rodriguez-Boulan E, Nelson WJ. Morphogenesis of the polarized epithelial cell phenotype. Science 1989; 245:718-725.

33. Berridge MJ, Oshman JL. Transporting Epithelia. New York: Academic Press, 1972:91-108.

34. Almers W, and Stirling C. Distribution of transport proteins over animal cell membranes. J Membr Biol 1984; 77: 169-186.

35. Quaranta V. Epithelial integrins. Cell Differ Dev 1990; 32:361-365.

36. Garrod DR. Desmosomes and hemidesmosomes. Curr Opin Cell Biol 1993; 5:30-40.

37. Le Gall AH, Yeaman C, Muesch A, Rodriguez-Boulan E. Epithelial cell polarity: new perspectives. Semin Nephrol 1995; 15:272-284.

38. van Meer G, van't Hof W, van Genderen I. Tight junctions and the polarity of lipids. in Cereijido M, ed. Tight Junctions. Boca Raton, FL: CRC Press Inc., 1992: 187-201.

39. Goodenough DA, Goliger JA, Paul DL. Connexins, connexons, and intercellular communication. Annu Rev Biochem 1996; 65:475-502.

40. Hunter T. Oncoprotein networks. Cell 1997; 88:333-346.

41. Tao YS, Edwards RA, Tubb B, Wang S, Bryan J, McCrea PD. β-catenin associates with the actin-bundling protein fascin in a noncadherin complex. J Cell Biol 1996; 134:1271-1281.

42. Takeichi M. Cadherin cell adhesion receptors as a morphogenetic regulator. Science 1991; 251:1451-1455.

43. Volberg T, Zick Y, Dror R, Sabanay I, Gilon C, Levitzki A, Geiger, B. The effect of tyrosine-specific protein phosphorylation on the assembly of adherens-type junctions. EMBO J 1992; 11: 1733-1742.

44. Maher PA, Pasquale EB, Wang JYJ, Singer SJ. Phosphotyrosine-containing proteins are concentrated in focal adhesions and intercellular junctions in normal cells. Proc Natl Acad Sci USA 1985; 82:6576-6580.

45. Tsukita S, Oishi K, Akiyama T, Yamanashi Y, Yamamoto T, Tsukita S. Specific proto-oncogenic tyrosine kinases of src family are enriched in cell-to-cell adherens junctions where the level of tyrosine phosphorylation is elevated. J Cell Biol 1991; 113:867-879.

46. Hamaguchi M, Matsuyoshi N, Ohnishi Y, Gotoh B, Takeichi M, Nagai Y. p60[v-src] causes tyrosine phosphorylation and inactivation of the N-cadherin-catenin cell adhesion system. EMBO J 1993; 12:307-314.

47. Volberg T, Geiger B, Dror R, Zick Y. Modulation of intercellular adherens-type junctions and tyrosine phosphorylation of their components in RSV-transformed cultured chick lens cells. Cell Regulat 1991; 2:105-120.

48. Hakomori S. Bifunctional role of glycosphingolipids. Modulators for transmembrane signaling and mediators for cellular interactions. J Biol Chem 1990; 265: 18713-18716.

49. Hakomori S. Functional role of glycosphingolipids in cell recognition and signaling. J Biochem 1995; 118:1091-1103.

50. Zeller CB, Marchase RB. Gangliosides as modulators of cell function. Am J Physiol 1992; 262:C1341-1355.

51. Bretscher A. Microfilament structure and function in the cortical cytoskeleton. Annu Rev Cell Biol 1991; 7:337-374.

52. Mooseker MS, Tilney LG. Organization of an actin filament-membrane complex. Filament polarity and membrane attachment in the microvilli of intestinal epithelial cells. J Cell Biol 1975; 67:725-743.

53. Mooseker MS. Organization, chemistry, and assembly of the cytoskeletal apparatus of the intestinal brush border. Annu Rev Cell Biol 1985; 1:209-241.

54. Mooseker MS, Cheney RE. Unconventional myosins. Annu Rev Cell Dev Biol 1995; 11:633-675.

55. Rodriguez-Boulan E, Powell SK. Polarity of epithelial and neuronal cells. Annu Rev Cell Biol 1992; 8:395-427.

56. Matter K, Mellman I. Mechanisms of cell polarity: sorting and transport in epithelial cells. Curr Opin Cell Biol 1994; 6: 545-554.

57. Nelson WJ. Renal epithelial cell polarity. Curr Opin Nephrol Hypertens 1992; 1:59-67.

58. Hammerton RW, Krzeminski KA, Mays RW, Ryan TA, Wollner DA, Nelson WJ. Mechanism for regulating cell surface distribution of Na+,K+-ATPase in polarized epithelial cells. Science 1991; 254:847-850.

59. Hubbard AL. Targeting of membrane and secretory proteins to the apical domain in epithelial cells. Semin Cell Biol 1991; 2:365-374.

60. Mostov KE, Cardone MH. Regulation of protein traffic in polarized epithelial cells. BioEssays 1995; 17:129-138.

61. Fath KR, Mamajiwalla SN, Burgess DR. The cytoskeleton in development of epithelial cell polarity. J Cell Sci Suppl 1993; 17:65-73.

62. Chant J. Generation of cell polarity in yeast. Curr Opin Cell Biol 1996; 8: 557-565.

63. Cid VJ, Duran A, del Rey F, Snyder MP, Nombela C, Sanchez M. Molecular basis of cell integrity and morphogenesis in *Saccharomyces cerevisiae*. Microbiol Rev 1995; 59:345-386.

64. Chant J. Cell polarity in yeast. Trends Genet. 1994; 10: 328-333.

65. Roemer T, Vallier LG, Snyder M. Selection of polarized growth sites in yeast. Trends Cell Biol 1996; 6:434-441.

66. Leberer E, Thomas DY, Whiteway M. Pheromone signalling and polarized morphogenesis in yeast. Curr Opin Genet Devel 1997; 7:59-66.

67. Sells MA, Chernoff J. Emerging from the Pak: the p21-activated protein kinase family. Trends Cell Biol 1997; 7:162-167.

68. Peter M, Neiman A, Park H-O, Van Lohuizen M, Herskowitz I. Functional analysis of the interaction between the small GTP-binding protein cdc42 and the Ste20 protein kinase in yeast. EMBO J 1996; 15:7046-7059.

69. Ayscough KR, Drubin DG. Actin: general principles from studies in yeast. Annu Rev Cell Dev Biol 1996; 12:129-160.

70. Shepherd GM. Sensory transduction: entering the mainstream of membrane signaling. Cell 1991; 67:845-851.

71. Torre V, Ashmore JF, Lamb TD, Menini A. Transduction and adaptation in sensory receptor cells. J Neurosci 1995; 15: 7757-7768.

72. Reed RR. How does the nose know? Cell 1990; 60:1-2.

73. Hackney CM, Furness DN. Mechanotransduction in vertebrate hair cells: structure and function of the stereociliary bundle. Am J Physiol 1995; 268:C1-C13.

74. Buck LB. The olfactory multigene family. Curr Opin Neurobiol 1992; 2:282-288.

75. Yan K, Greene E, Belga F, Rasenick MM. Synaptic membrane G-proteins are complexed with tubulin in situ. J Neurochem 1996; 66:1489-1495.

76. Pickles JO, Corey DP. Mechanoelectrical transduction by hair cells. Trends Neurosci 1992; 15:254-259.

77. Watson PA. Function follows form: generation of intracellular signals by cell deformation. FASEB J 1991; 5:2013-2019.

78. Garcia-Anoveros J, Corey DP. Mechanosensation: touch at the molecular level. Curr Biol 196; 6:541-543.

79. Corey DP, Garcia-Anoveros J. Mechanosensation and DEG/ENaC ion channels. Science 1996; 273:323-324

80. Liu J, Schrank B, Waterston RH. Interaction between a putative mechanosensory membrane channel and a collagen. Science 1996; 273:361-364.

81. Hoffmann EK, Dunham PB. Membrane mechanisms and intracellular signalling in cell volume regulation. Int Rev Cytol 1995; 161:173-262.

82. Davies PF. Flow-mediated endothelial mechanotransduction. Physiol Rev 1995; 75:519-560.

83. Dewey CF Jr, Bussolari SR, Gimbrone MA Jr, Davies PF. The dynamic response of vascular endothelial cells to fluid shear stress. J Biomech Eng 1981; 103:177-184.

84. Franke RP, Grafe M, Schnittler H, Seiffge D, Mittermayer C, Drenckhahn D. Induction of human vascular endothelial stress fibers by fluid shear stress. Nature 1984; 307:648-650.

85. Girard PR, Nerem RM. Shear stress modulates endothelial cell morphology and F-actin organization through the regulation of focal adhesion-associated proteins. J Cell Physiol 1995; 163:179-193.

86. Morita T, Kurihara H, Maemura K, Yoshizumi M, Yazaki Y. Disruption of cytoskeletal structures mediates shear stress-induced endothelin-1 gene expression in cultured porcine aortic endothelial cells. J Clin Invest 1993; 92:1706-1712.

87. McNamee HP, Liley HG, Ingber DE. Integrin-dependent control of inositol lipid synthesis in vascular endothelial cells and smooth muscle cells. Exp Cell Res 1996; 224:116-122.

88. Wang N, Butler JP, Ingber DE. Mechanotransduction across the cell surface and through the cytoskeleton. Science 1993; 260:1124-1127.

89. Maniotis AJ, Chen CS, Ingber DE. Demonstration of mechanical connections between integrins, cytoskeletal filaments, and nucleoplasm that stabilize nuclear structure. Proc Natl Acad Sci USA 1997; 94:849-854.

90. Ando J, Kamiya A. Flow-dependent regulation of gene expression in vascular endothelial cells. Jpn Heart J 1996; 37:19-32.

91. Rakic P. Principles of neuronal cell migration. Experentia 1990; 46:882-891.

92. Diaz-Nido J, Ulloa L, Sanchez C, Avila J. The role of the cytoskeleton in the morphological changes occurring during neuronal differentiation. Sem Cell Devel Biol 1996; 7:733-739.

93. Black MM, Baas PW. The basis of polarity in neurons. Trends Neurosci 1989; 6:211-214.

94. Schnapp BJ, Reese TS. Cytoplasmic structure in rapid-frozen axons. J Cell Biol 1982; 94:667-679.

95. Vale RD, Schnapp BJ, Reese TS, Sheetz MP. Different axoplasmic proteins generate movement in opposite directions along microtubules in vitro. Cell 1985; 40:559-569.

96. Schoenfeld TA, Obar RA. Diverse distribution and function of fibrous microtubule-associated proteins in the nervous system. Int Rev Cytol 1994; 151:67-137.

97. Hirokawa N. Microtubule organization and dynamics dependent on microtubule-associated proteins. Curr Opin Cell Biol 1994; 6:74-81.

98. Sobue K. Actin-based cytoskeleton in growth cone activity. Neurosci Res 1993; 18:91-102.

99. Gordon-Weeks PR. Growth cones: the mechanism of neurite advance. BioEssays 1991; 13:235-239.

100. Goodman CS, Shatz CJ. Developmental mechanisms that generate precise patterns of neuronal connectivity. Neuron 1993; 10 (Suppl.):77-98.

101. Neely MD, Nicholls JG. Electrical activity, growth cone motility and the cytoskeleton. J Exper Biol 1995; 198:1995.

102. Gomez TM, Roche FK, Letourneau PC. Chick sensory neuronal growth cones distinguish fibronectin from laminin by making substratum contacts that resemble focal contacts. J Neurobiol 1996; 29:18-34.

103. Dodd J, Schuchardt A. Axon guidance: a compelling case for repelling growth cones. Cell 1996; 81:471-474.
104. Tessier-Lavigne M. Eph receptor tyrosine kinases, axon repulsion, and the development of topographic maps. Cell 1995; 82:345-348.
105. Dorries U, Taylor J, Xiao Z, Lochter A, Montag D, Schachner M. Distinct effects of recombinant tenascin-C domains on neuronal cell adhesion, growth cone guidance, and neuronal polarity. J Neurosci Res 1996; 43:420-438.
106. Hall ZW, Sanes JR. Synaptic structure and development: the neuromuscular junction. Cell 1993; 72:99-121.
107. Sanes JR. Genetic analysis of postsynaptic differentiation at the vertebrate neuromuscular junction. Curr Opin Neurobiol 1997; 7:93-100.
108. Carraway KL III, Burden, SJ. Neuregulins and their receptors. Curr Opin Neurobiol 1995; 5:1-7.
109. Gassmann M, Lemke G. Neuregulins and neuregulin receptors in neural development. Curr Opin Neurobiol 1997; 7:87-92.
110. Carraway KL III, and Cantley LC. A neu acquaintance for ErbB3 and ErbB4: a role for receptor heterodimerization in growth signaling. Cell 1994; 78:5-8.
111. Bowe MA, Fallon JR. The role of agrin in synapse formation. Annu. Rev. Neurosci. 1995; 18:443-462.
112. Gautam M, Noakes, PG, Moscoso L, Rupp F, Scheller RH, Merlie JP, Sanes JR. Defective neuromuscular synaptogenesis in agrin-deficient mutant mice. Cell 1996; 85:525-535.
113. Valenzuela DM, Stitt TN, DiStefano PS, Rojas E, Mattsson K, Compton DL, Nunez L, Park JS, Stark JL, Gies DR, Thomas S, Le Beau MM, Fernald AA, Copeland NG, Jenkins NA, Burden SJ, Glass DJ, Yancopolous GD. Receptor tyrosine kinase specific for the skeletal muscle lineage: expression in embryonic muscle, at the neuromuscular junction, and after injury. Neuron 1995; 15: 573-584.
114. DeChiara TM, Bowen DC, Valenzuela DM, Simmons MV, Poueymirou WT, Thomas S, Kinetz E, Compton DL, Rojas E, Park JS, Smith C, DiStefano PS, Glass DJ, Burden SJ, Yancopolous GD. The receptor tyrosine kinase MuSK is required for neuromuscular junction formation in vivo. Cell 1996; 85:501-512.
115. Glass DJ, Bowen DC, Stitt TN, Radziejewski C, Bruno J, Ryan TE, Gies DR, Shah S, Mattsson K, Burden SJ, DiStefano PS, Valenzuela DM, DeChiara TM, Yancopoulos GD. Agrin acts via a MuSK receptor complex. Cell 1996; 85: 513-523.
116. Gautam M, Noakes PG, Mudd J, Nichol M, Chu GC, Sanes JR, Merlie JP. Failure of postsynaptic specialization to develop at neuromuscular junctions of rapsyn-deficient mice. Nature 1995; 377:232-236.
117. Carraway CAC, Carraway, KL. In: Hesketh HE, Pryme IF, eds. Treatise on the Cytoskeleton, Greenwich, CT: JAI Press, 1996:207-238.

CHAPTER 4

Cell Adhesion and Motility

Introduction

One of the defining characteristics of living organisms and cells is movement. At the cellu-lar level motility can be defined as directed shape changes. As described in chapter 1, cell shape depends on two factors, adhesion and the cytoskeleton. Thus, cells move by a cycle which is initiated with a cytoskeleton-dependent protrusive activity (1). This protrusion defines the direction of movement, sometimes in response to some extracellular force such as an attractant. Directionality is set through formation of adhesions by the protrusive element (Fig. 4.1) (2). Translocation of the cell then occurs through contractile and detachment phases, and the cycle is set to repeat. One of the important lessons in studying cell motility, illustrated in Figure 4.2, is that there are wide variations in motility among different types of cells. However, the fundamental molecular mechanisms appear similar in most cases, so the differences may reside in the organization and control of those mechanisms.

Cell Motility

Cell motility is accomplished by a "crawling" mechanism initiated in response to cell surface stimuli and involving a cycle of cellular shape changes (1, 3). Although different mechanisms have been proposed for various aspects of the cycle, it can be readily described by four phases: *protrusion, adhesion, traction/contractility and detachment* (2, 4). Two different types of cell motility have been described: amoeboid and fibroblast-like. These differ in the rates of movement of the cells and the details of the individual steps. However, they may simply represent extremes of a motility continuum with variations in the mechanisms of the phases and the balance of forces acting on the moving cell (see Fig. 4.3). For example, amoeboid movement is the more rapid and exhibits different types of protrusive and contractile processes (Fig. 4.2). Similarly, although motility requires adhesion, motility of fibroblasts is inversely related to adhesivity and stress fiber formation (5). It is the integration of these molecular processes which ultimately determines the motile behavior (4). Finally, motility is necessarily a directed process, involving polarization of the cell, discussed in chapter 3. How that polarization is achieved is an essential aspect of motility, particularly for cells which move in response to external signals, as in chemotaxis, described later in this chapter.

Protrusion

Cell movement is absolutely dependent on protrusive activity, but the mechanism(s) for development of the protrusions is still unclear (2). Different cells

Signaling and the Cytoskeleton, by Kermit L. Carraway, Coralie A. Carothers Carraway and Kermit L. Carraway III. © 1998 Springer-Verlag and R.G. Landes Company.

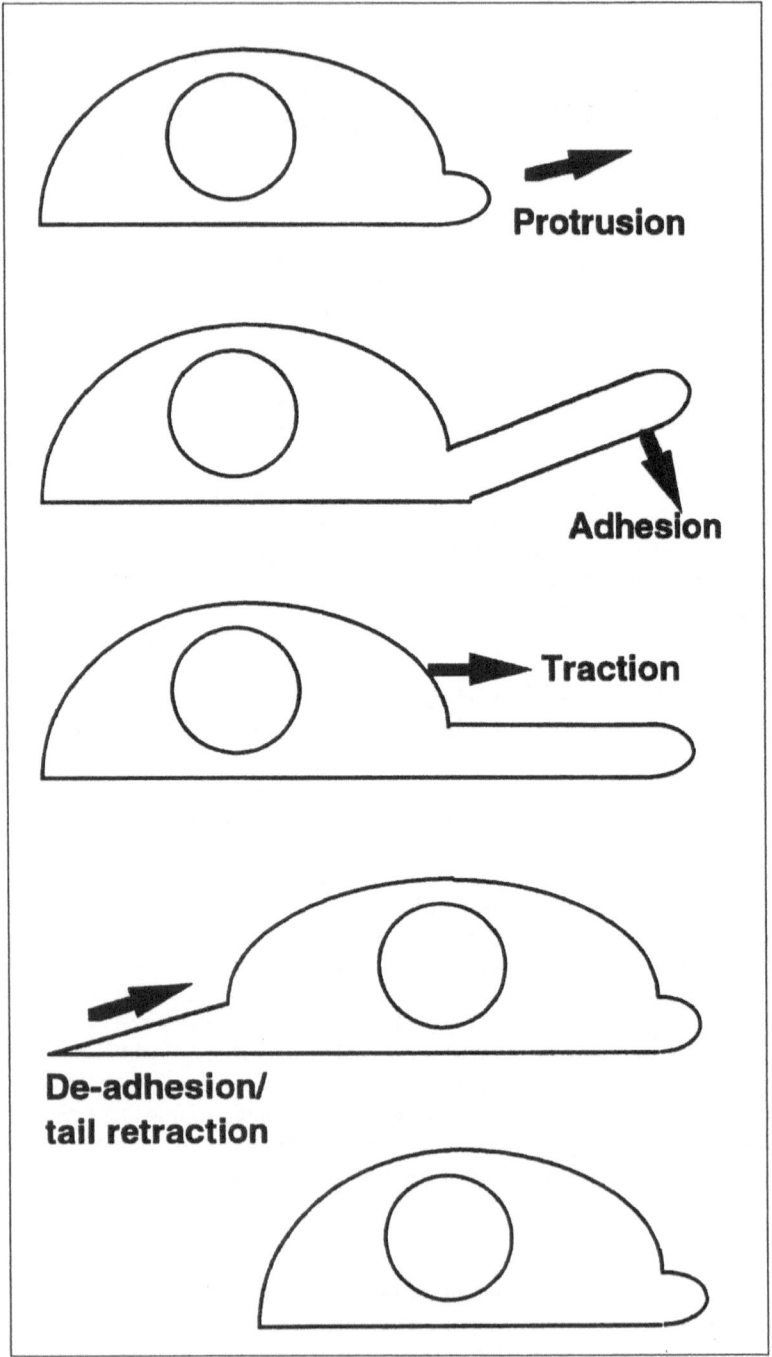

Fig. 4.1. Stages in migration of a motile cell. Figure from ref. 2 (Mitchison TJ, Cramer LP. Cell 1996; 84: 371-379), courtesy of Cell Press.

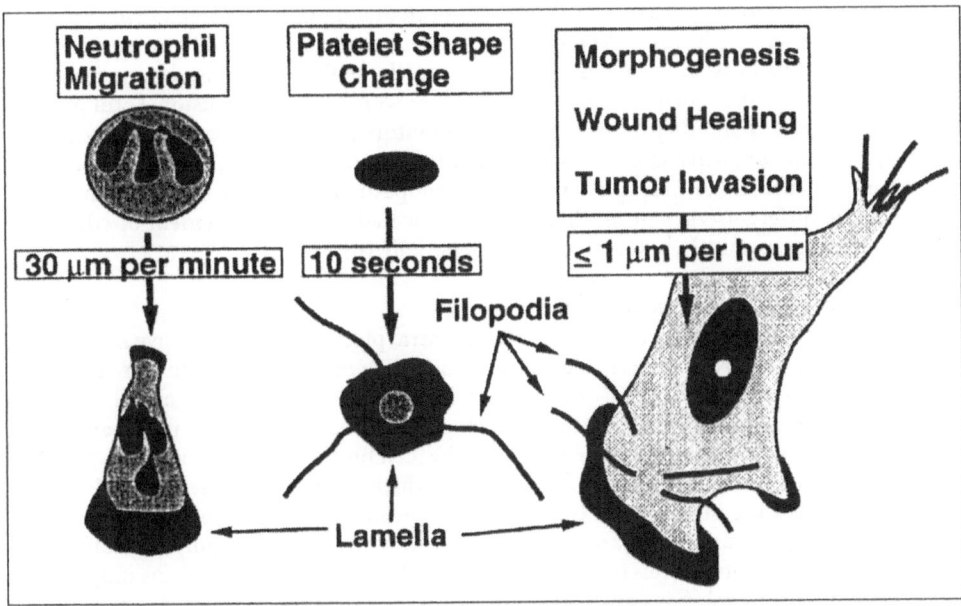

Fig. 4.2. Types and speeds of cell crawling movements. Figure from ref. 1 (Stossel TP. Blood 1994; 84: 367-379), courtesy of W.B. Saunders Co.

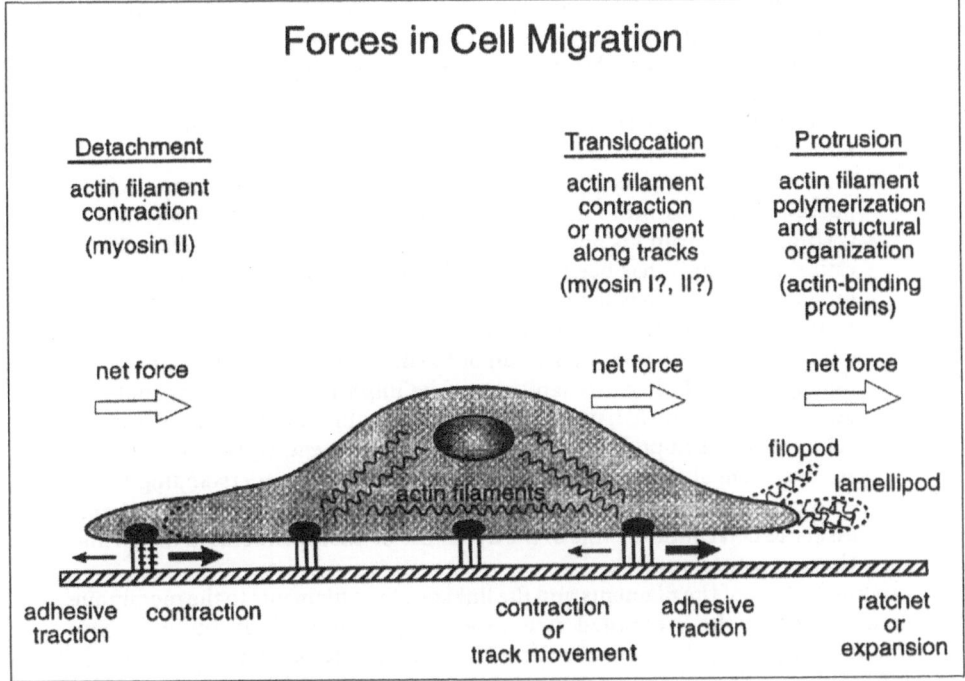

Fig. 4.3. Model for forces involved in cell migration. Figure from ref. 4 (Lauffenburger DA, Horwitz AF. Cell 1996; 84: 359-369), courtesy of Cell Press.

may use different types of processes and mechanisms, but the mechanisms may have overlapping features. Amoeboid cells advance by pseudopodal protrusions, whereas fibroblast-like cells use filopodia and lamellipodia (2). Several general mechanisms have been proposed for the forces which cause protrusion: actin polymerization, osmotic pressure and myosin motors (6, 7). These are not necessarily mutually exclusive and may operate at different times and/or places in the protrusive process. For example, extension of filopodia and lamellipodia appear to require different types of mechanisms as indicated by the involvement of cdc42 and Rac, respectively, and may occur sequentially (6).

Microfilaments and Membranes

Since actin polymerization occupies a central place in most protrusive mechanisms (8), it has received much of the attention. Extension of protrusions is correlated temporally and spatially with actin polymerization in a number of cell types (6). The orientation of filaments in many protrusions requires that polymerization occur at the "membrane" rather than the "cell body" end of the filament. This requirement raises two questions, neither of which has been completely resolved. What drives the membrane expansion which permits the extension of filaments near the membrane? What are the nucleation sites involved in the polymerization, existing filaments or sites of severing activity? Two types of mechanisms may account for the membrane expansion, motor-driven and polymerization-driven (Fig. 4.4):

1) In the *first* mechanism, a membrane-attached myosin moves toward the barbed ends of filaments (9). Myosin I, the first membrane myosin described, is an obvious candidate and is appropriately localized (10). Moreover, genetic ablation of myosin I isoforms in *Dictyostelium* results in reduced cell motility (11). However, this observation does not necessarily imply a direct role of myosin I in membrane movement since it might instead be required for vesicular transport to move materials to the site of the protrusion (12) rather than a direct role in the protrusive force. Ablation of myosin I in double mutants of *Dictyostelium* (see chapter 6) causes conditional defects in fluid phase pinocytosis without loss of the ability to extend processes (13), suggesting that myosin I is not required for protrusion in this organism. Interestingly, myosin V has been implicated in filopodial extension in neuronal growth cones as described below (14).

2) In the *second* mechanism, the protrusion force results from local osmotic effects due to actin polymerization or to simple thermal fluctuations of the membrane (2). Either mechanism may be important in different contexts. Two different models have been suggested for the rapid turnover of actin subunits in lamellipodia: treadmilling of long filaments (15) and nucleation/release of small filaments (16). In particular, a better understanding of the sites of polymerization and the factors controlling filament turnover is necessary to determine the role of the dynamics of microfilaments in the formation of protrusions (17).

Crosslinking of the filaments and the linkage of the filaments to the membrane should stabilize the protruded structure (4, 6). A number of microfilament crosslinking proteins have been localized in protrusive structures (18). Obviously, different types of crosslinking are required for filopodia, which have tighly bundled microfilaments, and lamellipodia, which have networks. Several types of experi-

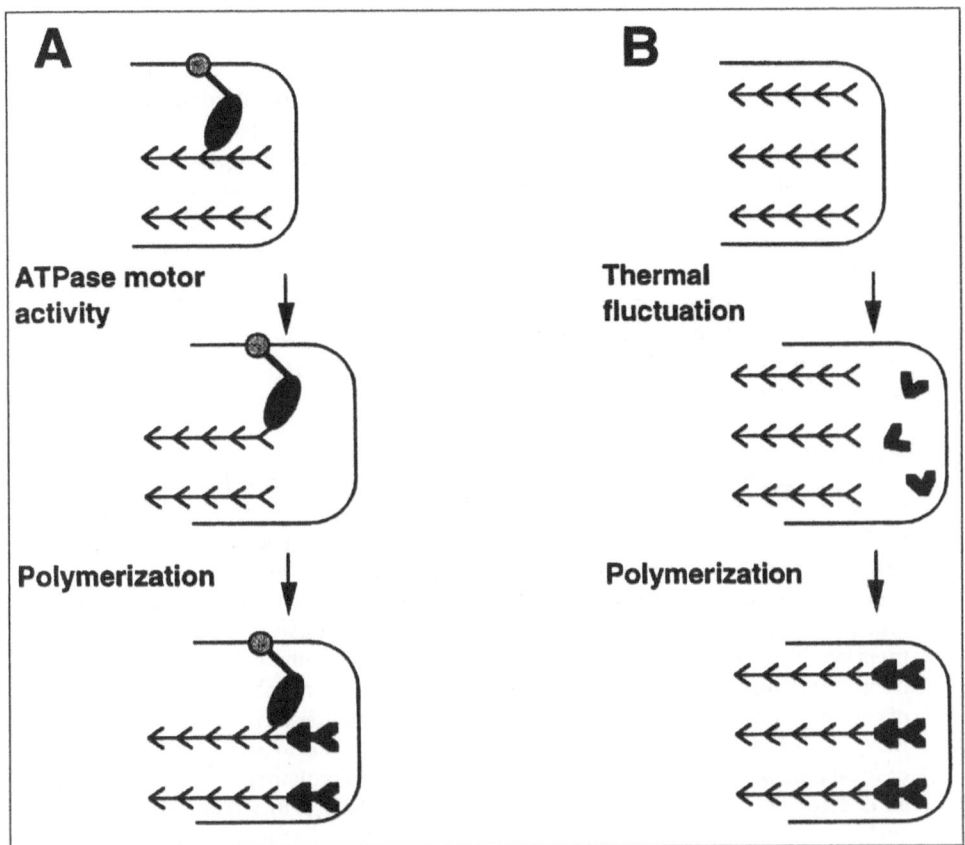

Fig. 4.4. Models for generation of protrusive force. Figure from ref. 2 (Mitchison TJ, Cramer LP. Cell 1996; 84: 371-379), courtesy of Cell Press.

ments indicate that crosslinking is important for motility. Melanoma cells lacking filamin (ABP-280) do not extend normal pseudopods or migrate normally, an effect that is reversed by expression of the protein in the cells by transfection (19). Likewise, disruption of the expression of ABP-120, a similar crosslinking protein in *Dictyostelium*, results in an inability to extend pseudopods and to locomote normallly (20). Ezrin, which has been implicated in crosslinking microfilaments to membranes (21), appears to be involved in formation of cell protrusions (22). These combined results suggest that stabilization of membrane-microfilament interactions and microfilament networks is an important aspect of the formation of lamellipodia and pseudopodia.

An important question concerning the development of cell protrusions is the source of the actin and membrane for the protrusive structure. Actin synthesis occurs at the leading edge of migrating fibroblasts due in part to the isoform-specific localization of β-actin mRNA (23). One hypothesis concerning the source of membrane is that it is provided by an endocytotic recycling pathway which inserts membrane at the leading edge of the protrusion (24). For example, newly synthesized viral glycoproteins are preferentially inserted at the leading edge (25). However, physical measurements of membrane lipid and protein diffusion indicate that these are moving forward rather than backward as required by this model (8). Moreover, the site of bulk endocytotic membrane fusion appears to be nearer the interface of the cell body and protrusion than the leading edge of the protrusion. In contrast, particles attached to dorsal cell surfaces and membrane receptors of protrusions definitely exhibit a backward movement. One explanation for the latter is that these components are attached to submembrane microfilaments which are also undergoing a rearward movement (2, 10, 26). However, other models for explaining these movements cannot be excluded. The mechanism for the movements of the microfilaments is also uncertain (2, 15), complicated by the complexities of the actin polymerization/depolymerization cycle and microfilament crosslinking in the cell (27).

Regulation

Any analysis of the role of protrusions in motility must consider how formation of those protrusions is regulated. Motility of cells occurs in response to cell surface stimuli. Receptors for these stimuli include the serpentine type coupled to heterotrimeric G-proteins, tyrosine kinase receptors and other receptor types (3). A critical question is how these cell surface receptors are coupled to cytoskeletal reorganizations such as those which occur with the extension of protrusions. As mentioned in chapter 1, small G-proteins of the Rho family have been shown by microinjection studies in some cell types to play important roles in the formation of specific types of protrusions (see Fig. 1.3). Cdc42 promotes formation of filopodia, Rac promotes formation of lamellipodia and ruffles and Rho promotes formation of focal adhesions and stress fibers (28). These three regulatory elements provide an ordered sequence of events related to cell motility (17, 28). An external signal, such as bradykinin, can activate cdc42, which can induce the formation of filopodia as exploratory processes at the cell migratory front. Cdc42 activation or external signals such as growth factors can activate Rac to produce lamellipodia. Activation of Rho by Rac or by extracellular factors, such as LPA or bombesin, which act through trimeric G-protein-linked receptors, triggers the formation of adhesions and stress fibers as the next stage of the migration cycle. However, cdc42 may also antagonize Rho-mediated effects on the cytoskeleton (29). The molecular mechanisms for the roles of these small G-proteins in protrusion formation remains uncertain, though PI3K has been implicated in Ras-stimulated membrane ruffling (30), presumably as an intermediate between Ras and Rac. Interestingly, cdc42 in cultured fibroblasts appears to be localized primarily in the Golgi (31). Its sensitivity to the Golgi-disrupting drug brefeldin A and to the ADP-ribosylation factor (ARF) implicated in intracellular membrane trafficking led to the suggestion that the role of cdc42 in filopodia formation is in regulating vesicle delivery to the plasma membrane.

As described for Ras in chapter 2, three types of proteins have been characterized which can regulate Rho family members as described in chapter 2: GEFs, GAPs and guanine nucleotide dissociation inhibitors (GDIs) (32). RhoGDI can dissociate Rho family members from membranes which should prevent their involvement in membrane-associated morphological changes. GAP proteins which interact with specific Rho family members have been identified and may contribute to their specific effects on cell behavior. The GAP Bcr has specificity for Rac and also contains a domain with homology to the exchange factor Dbl (33), which has been identified as an oncogene. *A number of proteins implicated in small G-protein regulation have multiple domains with different potential specificities, suggesting that they could be involved in multimeric protein complexes whose activities are regulated by their associations as well as by other mechanisms.* A number of downstream signaling kinases have been described for these small G-proteins, but none of these has been demonstrated to provide the linkage to protrusive activity via actin reorganization necessary for explaining effects on this phase of cell motility.

As mentioned previously, one of the key processes in protrusive activity is actin polymerization, probably regulated by the availability of nucleation sites (6). A likely participant in actin polymerization in protrusions is PIP$_2$ which can regulate actin binding to both gelsolin and profilin. Gelsolin has been implicated in cell motility by transfection (19), antisense (34) and knockout (35) experiments. Synthesis of PIP$_2$ is stimulated in permeabilized platelets by Rac which also increases the availability of free barbed ends, thus providing sites for actin polymerization (36). Such observations may permit a direct linkage of cell surface activation through the small G-proteins to microfilament regulation in protrusions. A second possible link is through the interaction of cdc42 with the Wiscott-Aldrich Syndrome protein (WASP) (37). Expression of this protein in cells yields clustered structures containing both the protein and F-actin. Moreover, WASP contains proline-rich domains similar to those in vasodilator-stimulated phosphoprotein (VASP) which has been implicated in actin polymerization. For example, VASP appears to play a role in the filament formation and motility of the bacterium *L. monocytogenes* in infected eukaryotic cells (27) (see chapter 1). A neural analog of WASP (N-WASP) has been identified as a Grb2 binding protein. Like WASP, it has a pleckstrin homology domain and cofilin-related domain (38). It can associate with actin, calmodulin, PIP$_2$ and Grb2 and depolymerize microfilaments. N-WASP-expressing cells form microspikes when treated with EGF, consistent with its association with cdc42 and an ability to link to EGF receptors via Grb2. Thus, these observations provide evidence for a direct pathway for linking cortical actin reorganization to receptor tyrosine kinase activation through a Rho family member.

Focal Adhesions

Advancement of the cell by a crawling mechanism requires that the protrusion be attached by an adhesion to a substratum. The formation of focal adhesions provides both stability and directionality to the cell motility process. These close associations with the substratum in cultured cells were first identified by electron microscopy (39), but are most readily observed by interference reflection microscopy (40). Microscopic studies have also identified these sites as the termini of stress fibers (41). Thus, they constitute the site of the dynamic membrane-microfilament interactions involved in regulating cell shape and cell motility (18). Understanding these phenomena requires a knowledge of the components, organization and dynamics of the focal adhesions. Most of the research in this area has been devoted to identifying the components and their interactions, though some progress has been made in recent years in delineating factors which contribute to regulating their dynamics.

Components and Structure

Focal adhesion components can be divided into three classes: transmembrane, extracellular and cytoplasmic (Table 4.1) (42). The major family of proteins which mediates formation of focal adhesions and other associations with the extracellular

Table 4.1. Summary of focal contact components

Extracellular side	Transmembrane	Cytoplasmic side		
		Structural	Enzymatic	Other
vitronectin	integrin	α-actinin	v-Src	paxillin
fibronectin		actin	PKC	zyxin
heparan sulfate		fimbrin	FAK	foculin
		filamin	calpain II	cCRP
proteoglycan		VASP	FRNK	
		MARCKS	tenuin	
		tensin		
		dystrophin		
		vinculin		
		talin		
		radixin		
		ezrin		
		moesin		

From ref. 42 (Lo SH, Chen LB. Cancer Metas Rev 1994; 13:9-24), with kind permission of Kluwer Academic Publishers.

matrix is the integrins. Integrins are heterodimeric transmembrane glycoprotein complexes composed of α and β subunits (Table 4.2) (43). Integrins generally contain a large extracellular domain for both α and β subunits, a transmembrane domain for each subunit and a small cytoplasmic domain (except for β4) (44). Since there are 8 β and 14 α subunits, a large repertoire of heterodimers is possible, but the actual number is much more restricted. Some α subunits can associate with only one β; others are more promiscuous. Moreover, some heterodimers have more than one specificity and overlapping specificities. Molecular analyses indicate that the β subunit cytoplasmic domain is responsible for integrin localization to focal adhesions. Both β1 and β3 heterodimers have been so located, and the localization depends on the ligation of the integrin. For example, α5β1, but not αvβ3, localizes to focal adhesions on fibronectin substrates, while αvβ3, but not α5β1, localizes to focal adhesions on vitronectin substrates (45). Since localization appears to result from association of the β cytoplasmic domain with a cytoskeletal complex, the implication is that the α subunit can modulate that association.

Substantial progress has been made in characterizing the interactions of integrins at the cell surface, some of which are shown in Table 4.2 (43). The most common feature of the major ligands is short peptide sequences, exemplified by the RGD sequence in matrix proteins such as collagens, laminin, fibronectin and vitronectin. Both subunits and divalent cations are essential for integrin associations with its ligands (44). Moreover, the affinity of integrins for extracellular ligands depends on additional factors and may be regulated by *inside-out signaling* (44, 45), as described for IIbIIIa in platelets in a later section. This inside-out signaling can be triggered by activation of platelet G-proteins and involves changes in

Table 4.2. The integrin receptor family

Subunits		Ligands and Counterreceptors
β_1	α_1	Collagens, laminin
	α_2	Collagens, laminin
	α_3	Fibronectin, laminin, collagens
	α_4	Fibronectin (V25), VCAM-1
	α_5	Fibronectin (RGD)
	α_6	Laminin
	α_7	Laminin
	α_8	?
	α_V	Vitronectin, fibronectin (?)
β_2	α_L	ICAM-1, ICAM-2
	α_M	C3b component of complement (inactivated), fibrinogen, factor X, ICAM-1
	α_X	Fibrinogen, C3b component of complement (inactivated)?
β_3	α_{IIb}	Fibrinogen, fibronectin, von Willebrand factor, vitronectin, thrombospondin
	α_V	Vitronectin, fibrinogen, von Willebrand factor, thrombospondin, fibronectin, osteopontin, collagen
β_4	α_6	Laminin ??
β_5	α_V	Vitronectin
β_6	α_V	Fibronectin
$\beta_7(=\beta_P?)$	α_4	Fibronectin (V25), VCAM-1
	α_{IEL}	?
β_8	α_V	?

Subunits and interactions of major vertebrate integrins are listed. Several subfamilies exist, each with 2-9 α subunits and a common shared β subunit (β_1, β_2, β_3, or β_7). In addition, several of the α subunits can interact with other β subunits (β_4-β_8). Each αβ receptor recognizes one or more extracellular ligands or counterreceptors on other cells. It should be noted that the ligand specificity of a given receptor can be markedly affected by its environment or state of activation. Modified from ref. 43 (Hynes RO. Cell 1992; 69: 11-25), courtesy of Cell Press.

intracellular pH, Ca^{2+}, phosphatidylinositides and protein (including integrin) phosphorylation (46). A role for inside-out signaling in extracellular fibronectin fibrillogenesis has been demonstrated using CHO cells expressing recombinant IIbIIIa (47). Significantly, an intact actin cytoskeleton was required in addition to integrin activation for the fibronectin matrix assembly.

Despite numerous investigations, the organization of proteins at the cytoplasmic surface of the focal adhesion is still poorly understood. Since this structure links microfilaments to the extracellular matrix through integrins, four questions have been paramount in this area of research. What components are associated with the cytoplasmic domain of the integrins? What components are associated with microfilaments? What components provide the linkage between those elements? What mechanisms and molecules modulate the dynamics of those linkages?

A major problem throughout much of this research has been the difficulty in isolating sufficient amounts of focal adhesions for biochemical analyses (41, 48). Thus, it has not been possible to enumerate the components or describe their stoichiometries. The use of magnetic beads containing an integrin ligand to induce focal adhesion complexes for isolation (49) provides an approach which should solve at least part of that problem and should also provide valuable information about the dynamics of adhesion formation (50). Interestingly, the complexes isolated contain an array of signaling molecules, including tyrosine kinases, enzymes of the inositide lipid pathways and ion channels (50). However, to date much of the information about the structural components of the focal adhesion has been obtained by immunolocalization studies rather than direct biochemical analyses, using antibodies against proteins isolated from other sources, such as smooth muscle (41). These components have then been isolated or prepared as recombinant proteins for biochemical studies of their interactions. Thus, a catalog of focal adhesion proteins and their potential associations in the adhesion complex has been developed. The caveat to this approach is that the *interactions in the cell, with its multiplicity of interacting components, may not always be identical to those found by* in vitro *studies.*

Two prominent components of focal adhesions have been shown to bind to the cytoplasmic domain of the integrin subunit: α-actinin and talin. Since α-actinin also binds microfilaments, it can provide a direct link between the cytoskeleton and the membrane. The link to talin is less direct. Talin binds vinculin, and vinculin is reported to bind microfilaments via a site which can be masked by intramolecular interactions (51). This intramolecular blocking mechanism, similar to that found in some protein kinases (52), provides a possible means of regulating formation of the membrane-microfilament interaction during formation of the adhesion. Interestingly, phosphoinositides have been found to dissociate the intramolecular linkage in vinculin, providing another mechanism by which these phospholipids may regulate cytoskeletal organization (53, 54). Other proteins may be involved in the basic organization of the membrane-microfilament linkage. Of particular interest is tensin, a 215-235 kDa protein with multiple binding domains. Tensin has three actin binding domains, one of which is a weak barbed-end-capping site (42). This domain might allow regulated barbed end actin polymerization at the membrane, which is necessary for addition of G-actin to the filament near the membrane, as described for protrusion formation. Tensin also has an SH2 domain, by which it could link microfilaments to tyrosine phosphorylated proteins in the focal adhesion complex, providing a mechanism for regulating the structure of the complex. Although a large number of highly specific models of the focal adhesion complex has been presented, one of which is shown in Figure 4.5, the details of the structure are still uncertain. The general picture is of a complex organization containing numerous interacting proteins with substantial redundancies in the linkages. It should not be surprising that there are redundant linkage mechanisms in the focal adhesions since other similar membrane complexes linked to the cytoskeleton have been found to exhibit multiple association mechanisms, as exemplified by the erythrocyte membrane skeleton described in chapter 1.

Formation and Regulation

A critical feature in the regulation of cell motility is the assembly and stabilization of focal adhesions and stress fibers. Two recent types of studies have been particularly informative in addressing focal adhesion formation (55): integrin-mediated recruitment of focal adhesion components (56) and growth factor and

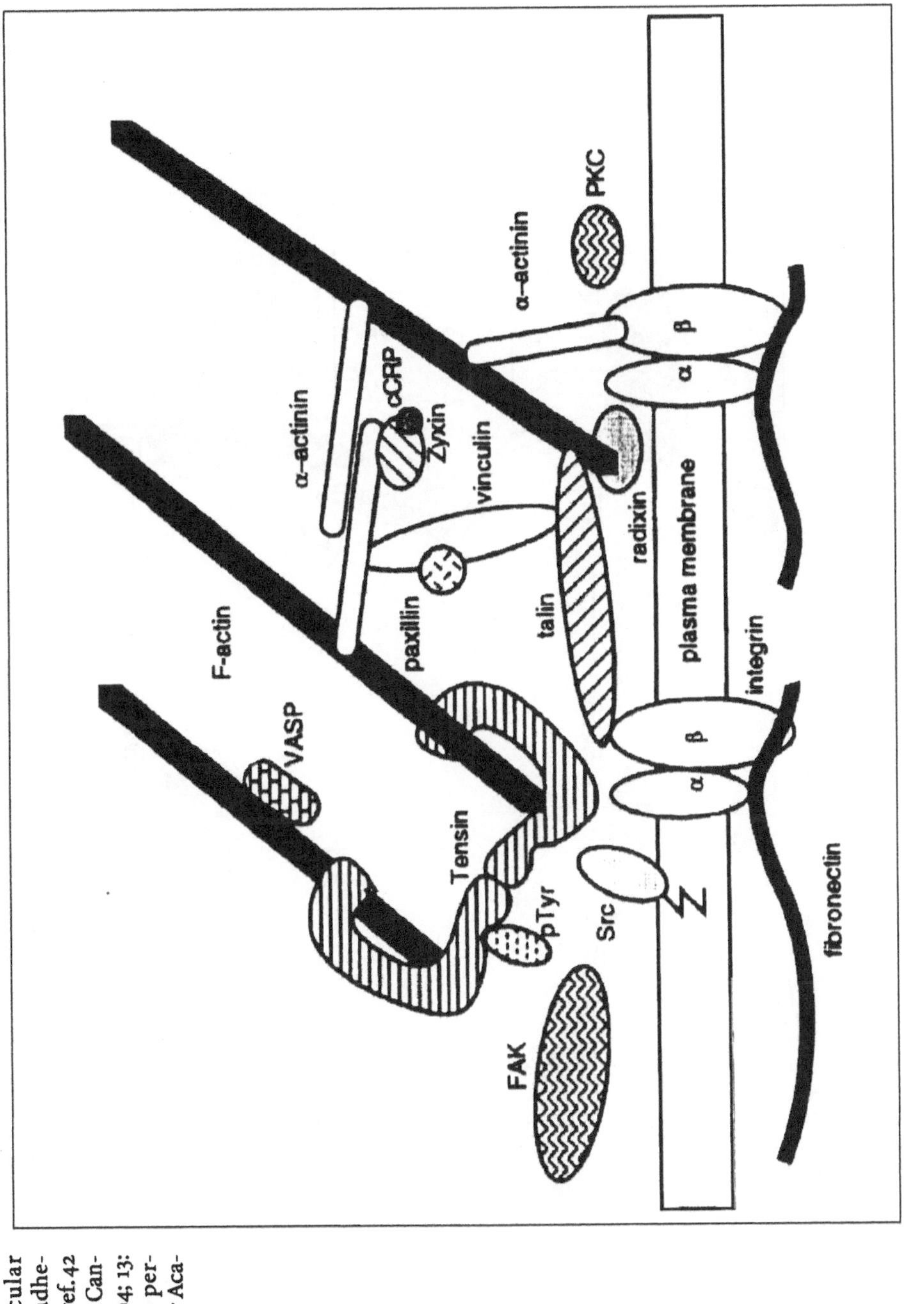

Fig. 4.5. Molecular model for focal adhesion. Figure from ref. 42 (Lo SH, Chen LB. Cancer Metas Rev 1994; 13: 9-24), with kind permission of Kluwer Academic Publishers.

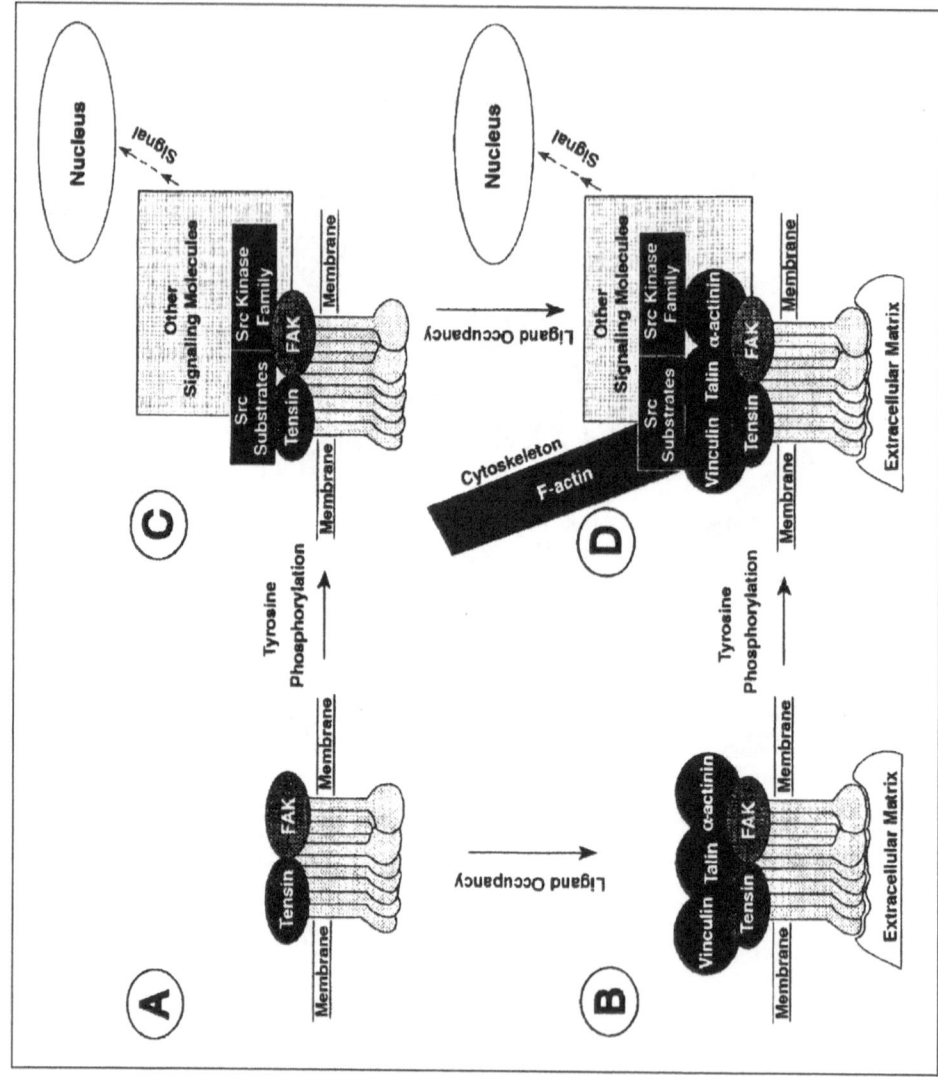

Fig. 4.6. Models for integrin-mediated recruitment and organization of signaling and cytoskeletal components. Figure from ref. 57 (Miyamoto S, Teramoto H, Coso OA, Gutkind JS, Burbelo PD, Akiyama SK, Yamada KM. J Cell Biol 1995; 131: 791-805), courtesy of Dr. Ken Yamada.

Table 4.3. Molecular hierarchies of transmembrane redistribution of cytoplasmic proteins by integrins

	A	B	C	D
				F-actin
				Paxillin
Cytoskeletal			Filamin	
molecules		Talin		Talin
		α-actinin		α-actinin
		Vinculin		Vinculin
Src	Tensin	Tensin	Tensin	Tensin
substrates	FAK	FAK	FAK	FAK
			Cortactin	Cortactin
			pp120	pp120
			c-Src	c-Src
Src Kinase			c-Yes	c-Yes
family			c-Fyn	c-Fyn
			c-Csk	c-Csk
Signaling			Gap	Gap
molecules			PLC-γ	PLC-γ
			PI3-kinase	PI3-kinase
			PTP-1D	PTP-1D
			RhoA	RhoA
			Rac1	Rac1
			Grb2	Grb2
			Raf1	Raf1
			MEKK	MEKK
			MEK1	MEK1
			ERK1	ERK1
			ERK2	ERK2
			JNK1	JNK1

A, Integrin aggregation; B, Integrin aggregation plus integrin occupancy; C, Integrin aggregation plus tyrosine phosphorylation; D, Integrin aggregation plus integrin occupancy, tyrosine phosphorylation, and actin cytoskeletal integrity. From ref. 57 (Miyamoto S, Teramoto H, Coso OA, Gutkind JS, Burbelo PD, Akiyama SK, Yamada KM. J Cell Biol 1995; 131: 791-805), courtesy of Dr. Ken Yamada.

lysophosphatidic acid induced assembly of focal adhesions and stress fibers (57). Such studies emphasize the congruity and complementarity of certain aspects of integrin- and growth factor-induced signaling responses and demonstrate the complexity of the pathways involved.

Beads coated with fibronectin or adhesive peptides can induce rapid integrin aggregation and assembly of focal adhesion cytoplasmic proteins in a hierarchical process (Fig. 4.6) (56, 57). Use of a nonaggregating ligand, such as an RGD peptide, results in movement of the integrin into focal adhesions. In contrast, integrin clustering in the absence of ligand results in recruitment of the specific

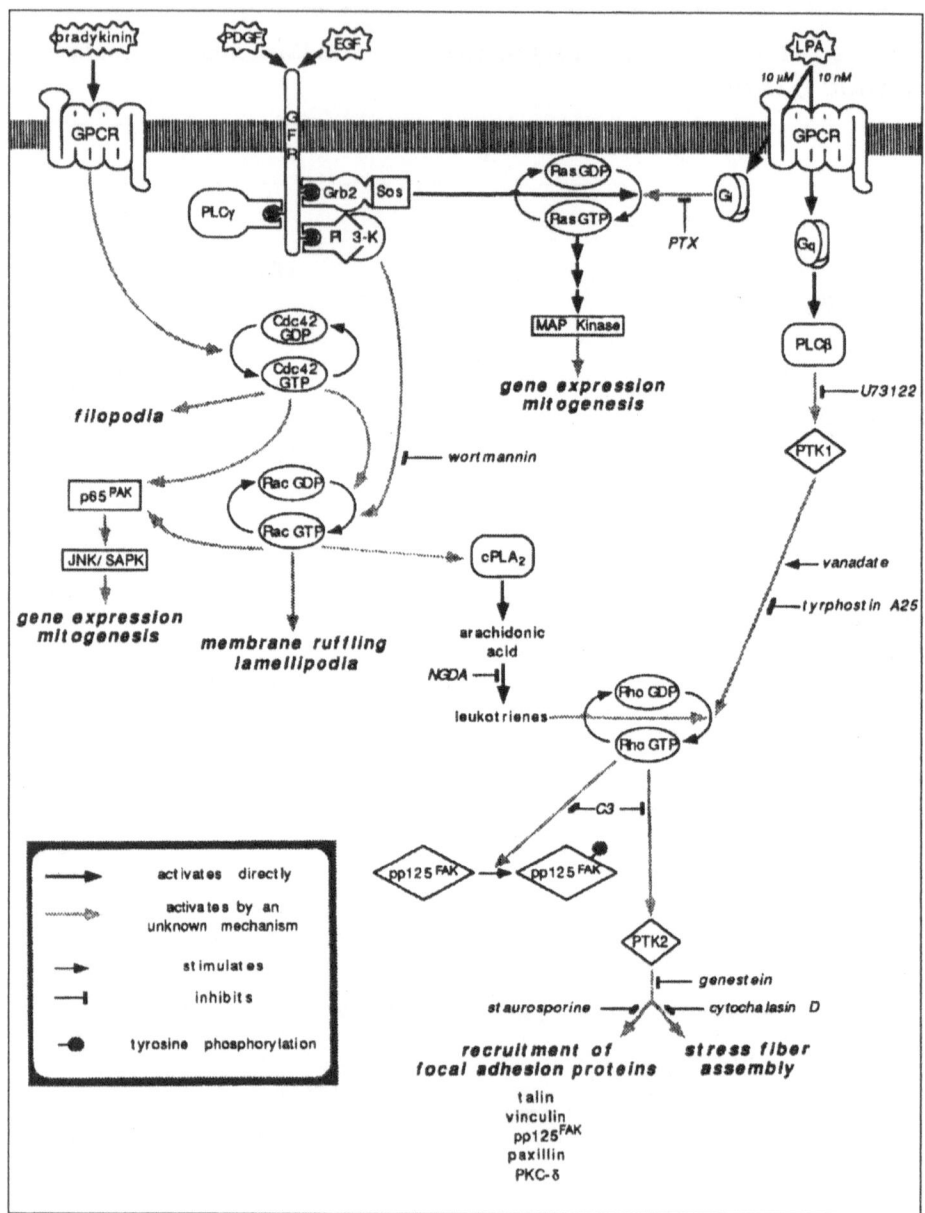

Fig. 4.7. Pathways involved in growth factor- and LPA-stimulated assembly of focal adhesions and stress fibers. Figure from ref. 55 (Craig SW, Johnson RP. Opin Cell Biol 1996; 8: 74-85), courtesy of Current Biology Ltd.

cytoplasmic components FAK and tensin. Tyrosine phosphorylation induced by the integrin clustering is necessary for recruitment of additional components, including Src family kinases and substrates and members of other signaling pathways. Clustering in the presence of ligand recruits α-actinin, talin and vinculin, major structural components of the focal adhesion. Tyrosine phosphorylation is required for assembly of microfilaments on these complexes. Based on these results, a minimum of six different transmembrane responses to integrin interactions with fibroblast cell surfaces can be demonstrated which are described in part in Figure 4.6 (57). These involve recruitment to focal adhesions of over 30 different signaling and cytoskeletal molecules (Table 4.3), culminating in the stabilization of the adhesions by F-actin. One curious aspect of focal adhesion formation is the inhibition of the association of "early" molecules, such as FAK and tensin, by cytochalasin. Since F-actin does not form at this stage, other cellular events dependent on cellular actin polymerization may be required for the initiation of the formation of adhesions.

A critical issue in understanding the assembly of the focal adhesion is how the individual processes are regulated. As with the structural organization of focal adhesions, many of the regulatory components have been recognized, but their temporal and spatial contributions have not yet been delineated. Cell attachment to extracellular matrix can activate at least five classes of signaling pathways: tyrosine kinases, serine/threonine kinases, inositide turnover, ion transporters and G-proteins (50). Extracellular agents, including growth factors, neuropeptides and lysophosphatidic acid, can promote adhesion and stress fiber formation (58). These membrane signals stimulate actin reorganizations via small G-proteins (Fig. 4.7) (55). As described above, activation of the small G-protein Rho via trimeric G-protein-linked receptors can stimulate formation of focal adhesions and stress fibers (59). Both matrix attachment and Rho activation are necessary (60). Two questions arise concerning the role of Rho. What is required for activation of Rho? What are the downstream events from Rho that lead to focal adhesion formation? The interdependence of many signaling elements makes answering these questions very complicated. Rho activation can be achieved by two different pathways, Rac-dependent and Rac independent, shown in schematic form in Figure 4.7 (55). The Rac- dependent pathway is involved in the induction of focal adhesions by growth factors, but not LPA, and is mediated by PI3K. Membrane localization of active PI3K by anchoring it to a transmembrane domain will induce filopodia formation, ruffling, focal adhesions and stress fibers in Swiss 3T3 cells by Rac- and Rho-mediated processes. These results indicate the role of PI3K in cytoskeletal changes and suggest that it functions at the membrane (61). Recent studies showing a Gβγ-dependent activation of MAPK through PI3K suggest a mechanism for coupling G-protein activation in this process (62). EGF-stimulated formation of stress fibers is mediated by Rac via arachidonic acid and leukotrienes which can activate Rho (63). The probable target for Rac is phospholipase A2 which hydrolyzes phospholipids to liberate the arachidonic acid. In contrast, activation of Rho via the LPA pathway proceeds through heterotrimeric G-protein activation of phospholipase Cβ and a tyrosine kinase. The details of this pathway remain to be defined.

The downstream elements in the Rho pathway to focal adhesion assembly are also not well defined, but inhibitor studies with genistein suggests an involvement of tyrosine kinases (33). Furthermore, anti-phosphotyrosine prominently stains focal adhesions (64), inhibition of tyrosine phosphatases can stimulate adhesion and stress fiber formation (65), and disruption of adhesion results in reduced phosphotyrosine (66). A major tyrosine kinase of focal adhesions is FAK, whose recruitment by integrin clustering leads to its autophosphorylation on tyrosine and enhanced activity (67, 68). FAK is targeted to focal adhesions by a C-terminal domain which is separate from its integrin binding domain (69). However, microinjection studies of FAK peptides which block FAK incorporation into focal adhesions suggest that it is required for cell migration, but not for focal adhesion formation (70). Similarly, endodermal cells from FAK knockout mice show impaired cell motility but enhanced focal adhesion formation (71). These results suggest a role in microfilament assembly or organization. FAK is noteworthy for its multiple binding domains, including those for integrins and talin which may allow it to play a structural as well as a regulatory role. A natural FAK analog containing the C-terminal, but missing the kinase domain may participate in regulating FAK functions (72). FAK can associate via its phosphorylated tyrosine with Src and other proteins with SH2 domains. Src may then be linked to focal adhesion proteins such as paxillin via its SH3 domains. Alternatively, paxillin can bind directly to FAK (67). As in the case of tensin, paxillin, which associates with the focal adhesion via a LIM domain (73), is thought to play an important role in the organization of the focal adhesion because of its multiple binding domains and reversible phosphorylation (74). An indication of this role and the regulation of focal adhesions is demonstrated by the morphological effects of cAMP on Y1 adrenal cells (75). The rapid rounding of the cells is accompanied by tyrosine dephosphorylation of paxillin, suggesting that cAMP can negatively regulate adhesion stability through effects on tyrosine phosphatases of the adhesion complex. Although many components of the focal adhesions are tyrosine phosphorylated, the molecular role of these phosphorylations in focal adhesion assembly is unknown. Likewise, the role of FAK in the focal adhesion remains uncertain.

A second integrin-binding kinase (integrin-linked kinase, ILK) has been identified by two-hybrid analyses (46). In contrast to FAK, it is a Ser/Thr kinase and can be co-immunoprecipitated with β integrins, indicating an in vivo association. Moreover, overexpression of ILK in epithelial cells induces anchorage-independent growth (76). A second Ser/Thr kinase ROK has been implicated in focal adhesion and stress fiber assembly (77). ROK also contains a pleckstrin homology domain and is translocated from the cytoplasm to the plasma membrane in response to Rho activation. Microinjection of active ROK into HeLa cells promoted formation of stress fibers and focal adhesion complexes, but kinase negative ROK did not. N-terminal truncations induced disassembly of stress fibers and focal adhesion complexes. Thus, ROK may be a factor in a pathway for cell adhesions.

Lipid second messengers may also play important roles in focal adhesion assembly since they are known to be products of enzymes associated with isolated focal adhesion complexes (50), downstream products of Rho activation (78) and important in the dynamics of actin organization (see chapter 1). PIP 5-kinase is activated via Rho (79) and involved in the production of PIP_2, a substrate for PLCγ. One model for the role of this effect in adhesion formation is that Rho recruits PIP 5-kinase to a complex at the plasma membrane for localized production of PIP_2 (80). As described in chapter 2 (Fig. 2.11), hydrolysis of PIP_2 yields diacylglycerol and IP_3 and leads to the activation of PKC. PIP_2 itself may regulate actin polymerization via profilin and gelsolin, F-actin binding by α-actinin and vinculin binding to talin and F-actin (54). All of these factors may contribute to the production and stabilization of stress fibers associated with the focal adhesion. Clustering of integrins by anti-β1 antibodies results in activation of PLA2, production of arachidonic acid and activation of PKC (81). PKC has been localized to focal adhesions and implicated in cell adhesion and FAK activation (68, 82).

PKC appears to play a role in organizing the membrane focal adhesion complex. For example, integrin αvβ5 requires activation of PKC to induce localization of α-actinin, tensin, vinculin and actin to focal adhesions (83). Integrin αvβ3 does not require PKC activation. Phosphorylation, but not localization, of FAK also requires PKC activation for αvβ5, but not αvβ3. Thus, the specificity of integrins for the formation of focal adhesions appears to be determined by other regulatory elements, such as PKC. Arachidonic acid released by PLA2 may be involved in actin polymerization and cell spreading, since inhibition of integrin-mediated leukotriene production inhibits polymerization (63). These combined observations indicate that multiple lipid metabolic pathways are present in focal adhesions and suggest that lipid second messenger contributions should be important to their dynamics. How these effects contribute spatially and temporally to the formation and/or stabilization of focal adhesions remains one of the important questions in the investigations of the relationships between motility and signaling.

The driving force for cytoskeleton assembly at the focal adhesion remains unknown, though recent studies have implicated contractile proteins (48, 84). Inhibition of contractility blocks Rho induction of stress fibers, focal adhesions and tyrosine phosphorylation and leads to integrin dispersal. Stimulation of contractility promotes integrin aggregation into focal adhesions. The contractile protein has not yet been identified, but one candidate may be the novel myosin Myr 5, which is not only a target for Rho, but also a RhoGAP (85, 86). Alternatively, Rho may be acting by regulating myosin II activity through myosin light chain kinase (87). Rho kinase, one of the targets for Rho (88), has been shown to inhibit myosin phosphatase to increase myosin light chain phosphorylation. Microinjection of the active catalytic domain of Rho kinase can induce serum-starved Swiss 3T3 cells to form stress fibers and focal adhesions, while microinjection of an inactive catalytic domain inhibits formation of these structures induced by LPA (89). Interesingly, microtubule depolymerization has also been shown to stimulate contractility and formation of stress fibers (48), apparently by stimulation of myosin light chain kinase phosphorylation (90).

One of the most intriguing aspects of focal adhesions is how their formation is coordinated with the formation of stress fibers. Experiments with cytochalasin have suggested that actin polymerization is required for signaling processes through integrins (66). Microinjection of cells with Rho in the presence of cytochalasin results in the formation of focal adhesions without stress fibers (28). Conversely, staurosporine, a relatively nonspecific tyrosine kinase inhibitor, blocks focal adhesion formation without blocking filament formation, though the filaments are not organized into stress fibers. Rac and cdc42 have also been shown to stimulate the formation of focal adhesions associated with lamellipodia and filopodia, which contain many of the components of the Rho-induced complexes (28). Their formation is independent of Rho. Thus, formation of focal adhesions can occur independently of the organization of actin, but the organized actin filaments are required for the next stage of the motility process and probably for other processes which involve the focal adhesions.

Traction/Contractility

Protrusion and adhesion will advance a cell only if there is a mechanism for moving the cell body and releasing the adhesions once the cell body has advanced over them. Three different types of primary models have been invoked to explain movement of the cell body: surface traction/contractility, cytoplasmic contractility and rolling. As mentioned previously, motility is usually considered in two categories, fibroblast-like and amoeboid, and the first two of these models were developed to describe these types of cells, respectively.

Fibroblast motility is slow and discontinuous. Cell migration speed tends to correlate inversely with adhesion, which must be balanced by the contractile force involved in moving the cell forward (7). The key to understanding this mechanism is the contractile element involved and the spatial and temporal control of its activity. The most likely contractile element is myosin II associated with stress fibers. Stress fibers are directly attached to adhesion sites and oriented with the direction of movement of the cell, as required for traction/contractility. Microfilaments within the stress fibers are oriented in both parallel and antiparallel directions. Thus, contraction by a two-headed myosin can operate as in a muscle fiber (see chapter 1). Contractility of stress fibers can be demonstrated in permeabilized cell models (91). The prime candidate for the motor is myosin II, the conventional muscle myosin. Myosin II is appropriately associated with stress fibers and absent from the protrusive structures in migrating cells (92). How this contractile activity is regulated is unclear. Actomyosin contractility is generally regulated by calcium in muscle, and a calcium gradient from front to rear has been demonstrated for migrating eosinophils (93). One interesting possibility for regulating the calcium influx under these circumstances is stretch-activated calcium channels (94) associated with adhesion sites.

Amoeboid motility is most commonly represented by a cytoplasmic sol to gel transition followed by contraction of the gel (1). The gel contraction pushes water from the cell body against the cell membrane. This mechanism provides both a protrusive activity and, when coupled with weak adhesions, movement of the cell body. Chemotactic activity sometimes governs the directionality of this movement, possibly by signaling a weakening of the cortical cytoskeleton (see later section). The sol to gel transition occurs as a result of crosslinking of cytoplasmic microfilaments by ABP. As noted previously, the formation of orthogonal crosslinks by ABP (95, 96) makes it uniquely suited for gel formation at low crosslinker concentration. Contraction is catalyzed by myosin associated with the filament networks. Completion of this motility cycle requires dissolution of the contracted gel, probably by filament severing, followed by repolymerization of the filaments.

The third mechanism for movement was discovered in studies of fish scale keratocytes (97). These cells move rapidly and smoothly through the advancement of a broad lamella. Thus, they lack both the strong adhesions of the fibroblasts and the pseudopodal protrusions of the amoeba. The question of how they pull their bodies along was answered by a combination of microscopy and microneedle manipulations. Surprisingly, the cell body rolls along behind the advancing front. It will be interesting to see how the organization of the cytoskeletal components at the interfaces and in the cell body are regulated.

Detachment

The discontinuous movement of fibroblasts results from the release of the tail from the substratum and obviously plays an important role in the rate of motility. Two mechanisms have been implicated in the release: contractility and phosphorylation. The contractility provides a mechanical force which can sever the extended tail or rip adhesion sites loose from the substratum. In either case, a fraction of the adhesion sites remain associated with the substratum as the cell moves away. As in the case of movement of the cell body, the contractility appears to be driven by myosin II (98). A role for tyrosine phosphorylation has been invoked from studies of ATP promotion of adhesion release in permeabilized cells (91). Tyrosine phosphorylation correlates with the release of adhesions, and both can be inhibited by tyrosine phosphatases. None of the major phosphorylated proteins correspond to proteins phosphorylated during adhesion assembly. What role these phosphorylations play remains to be determined. Finally, adhesion sites remaining with the cells are internalized at the trailing end. The integrins are cycled back to the front of the cell by a calcium-dependent process involving the tyrosine phosphatase calcineurin (99).

Growth Cone Motility

Motility of the growth cones of growing axons and dendrites, described in chapter 3, can be considered as a special case of cell motility in which two important parameters are changed. *First*, only a part of the cell is moving. Thus, there is no requirement for translation of the cell body. *Second*, the processes are actually growing behind the advancing growth cone. Thus, new membrane and cytoskeletal elements must be added and organized. As noted in chapter 3, a necessary part of neurite extension is the polymerization of microtubules, followed by their stabilization and bundling. Thus, the paradigm for growth cone motility changes from protrusion-adhesion-contraction-retraction to protrusion-adhesion-growth-stabilization. Another important difference is the collapse of growth cones and retraction of neurites in response to interaction of growth cone cell surface receptors with either soluble or cell surface ligands (100). Growth cones act as sensors for directional cues from the matrix and surrounding cells. This guidance function appears to be primarily due to the regulation of actin in the growth cone (101).

As in the case of the fibroblast advancing front, the shape and direction of travel of the growth cone are determined by the microfilament-dependent protrusive activity at the membrane to produce filopodia and lamellae, whose formation is regulated by small G-proteins (101). This regulation appears to operate in very specific ways. In *Drosophila*, Rac has been implicated in axonal growth, whereas cdc42 may be important for both axon and dendrite formation (100). However, Rac also has effects on dendritic growth of Purkinje neurons in transgenic mice. In contrast, Rho appears to participate in growth cone retraction. Since Rho is involved in adhesion and stress fiber formation in other cells, a different role for Rho may reflect formation of different types of adhesions in the advancing growth cone. The opposite nature of the responses through Rac or cdc42 and Rho also suggests that they are coupled to different types of cell surface receptors on the growth cone. For example, Rac or cdc42 could be coupled through adhesion receptors such as integrins, while Rho is coupled to repulsion receptors, such as those of the Eph family tyrosine kinases described in chapters 2 and 3.

The pathways through which these growth cone small G-proteins operate (both upstream and downstream) are largely unknown. A likely candidate as an upstream effector is the FGF receptor whose activation is linked to neurite outgrowth stimulated by cell adhesion molecules L1, N-CAM and N-cadherin (102, 103). Receptor tyrosine kinase signaling often proceeds through recruitment of SH2-containing components. A novel *Drosophila* SH2/SH3 adaptor protein Dock (104) is present in growth cones and implicated in axon targeting, since mutation in the gene *dreadlocks* encoding Dock disrupts photoreceptor cell axon guidance. Members of the Src family of cytoplasmic tyrosine kinases are also present in growth cones (105, 106) as is Abl, whose mutations in *Drosophila* suggest a role in growth cone guidance (107). Not surprisingly, membrane tyrosine phosphatases have been implicated in axon guidance in *Drosophila* (108, 109), though whether they act as phosphatases or adhesion molecules or both (see chapter 2) is unclear. Neurite retraction appears to involve a different pathway. LPA treatment of differentiated PC12 cells causes growth cone collapse and neurite retraction which can be blocked by inhibitors of phospholipase C_β and protein kinase C (110). The results implicate a G-protein-linked mechanism coupled to Rho, similar to that observed for stress fiber assembly in fibroblasts (Fig. 4.7).

Guidance decisions in growth cones require the cooperation of microfilaments and microtubules (101). The presentation of a cue to change direction, such as a positive cell-cell contact, causes a reorganization of the actin in the growth cone, including an increase of the actin in the region of the contact and a decrease in the retrograde actin flow (111) which occurs in the growth cone as it does in the advancing edge of the fibroblast. One interpretation is that growth cone filopodia and lamellae become reoriented through signals from contact receptors which trigger sequential or concomitant filament breakdown and actin polymerization. In contrast, signals leading to growth cone collapse cause a loss of microfilaments at the leading edge of the growth cone (112). The molecular aspects of growth cone turning must be similar to the microfilament changes described for the migrating fibroblast, including a role for myosin V in growth cone filopodial extension (14). Talin and vinculin have been implicated in growth cone filopodial extension (113). Talin appears to be directly involved in extension and retraction, while vinculin plays a role in the structural integrity of the filopodia.

These actin changes are quickly followed by a reorientation of the growth cone microtubules, which exhibit dynamic instability (101). The microtubule changes are dependent on the prior microfilament reorganization (114, 115). How these two cytoskeletal elements are coupled is unclear, though proteins such as ezrin, a microfilament- and microtubule-binding protein whose distribution in the growth cone requires microtubules (116) have been implicated. The final stage in growth cone advance involves bundling of the microtubules and collapse of the membrane of the growth cone to generate the neurite tube (101). MAP-1B has been implicated in bundling in axon extension, but it may be replaced by MAP-1A with maturation (100). The effects of MAP-1B appear to be regulated by phosphorylation at two different sites. Site 1 phosphorylation is implicated in neurite outgrowth and probably catalyzed by cdk5 or MAPK, while site 2 phosphorylation is more abundant in axon shafts, dendrites and cell bodies and is catalyzed by casein kinase 2. Much less is known about dendritogenesis, although MAP-2 is involved (101). MAP-2 phosphorylation has been demonstrated during neuronal development, but little is known about how it is regulated by extracellular signals.

Chemotaxis

Dictyostelium

One of the most compelling questions in cell motility research is how a cell determines its direction of movement, particularly in the case of directed cell movements, such as chemotaxis. Chemotaxis involves orienting the cell motility along a chemoattractant gradient. Cells exhibiting amoeboid type movement develop transient polarity along the gradient, extending pseudopods in the direction of highest concentration (117). The highest sensitivity of the orientation occurs at concentrations near the K_d of the chemoattractant for its cellular receptor, suggesting that the effect correlates with the difference in the number of occupied receptors across the cell (118). The chemotactic response to a directed concentration change can be illustrated for *Dictyostelium*, a well-studied chemotaxing cell system (117). Within a few seconds of exposure to a gradient of chemoattractant, cells undergo rounding, followed by the extension of lamellae, which then become localized into pseudopodia as the cells develop polarity and rapid directed movement (chemokinesis).

The first event in chemotaxis is association of the chemoattractant with its cell surface receptor. cAMP is a commonly studied attractant for *Dictyostelium*, and cAMP receptors have been identified which transmit their signals via trimeric G-proteins and whose affinities are regulated by guanine nucleotides (119). The signal from the G-protein must then be transmitted to the cytoskeleton. Several second messenger pathways are activated in *Dictyostelium* via G-proteins, including adenylyl cyclase, guanylyl cyclase and phospholipase C (120). However, neither adenylyl cyclase nor phospholipase C appears to be involved in chemotaxis. Both guanylyl cyclase and calcium have been implicated in chemotaxis, but the pathways involved are still uncertain (121). Cellular cGMP is regulated not only by guanylyl cyclase activity, but also by calcium and cGMP phosphodiesterase. Interestingly, tyrosine phosphorylation appears not to be involved in cAMP-induced chemotaxis, but does play a role in folate-induced chemotaxis (122). One of the tyrosine-phosphorylated proteins is actin, but this phosphorylation has not been shown to regulate chemotaxis (123).

Chemoattractant treatments lead to rapid changes in the *Dictyostelium* cytoskeleton, including rapid actin polymerization, migration of cytoskeletal proteins such as talin into pseudopodia (123) and the cytoskeletal association, then dissociation of myosin II (124). Myosin II appears not to be involved in cell movement, but has been implicated in polarization via a pathway involving cGMP (121). Phosphorylation of the myosin II heavy chain by a specific protein kinase C reduces its cytoskeletal association (125). A model for the involvement of this enzyme suggests that it is localized to sites of stimulation, where it dissociates myosin II from the cortical cytoskeleton, thus weakening the cortex and permitting pseudopod protrusion. The MHC-PKC is regulated by cGMP (126), but how it is localized remains unknown, though calcium has been implicated in the cell polarization.

Leukocytes

Phagocytic leukocytes, such as neutrophils, play critical roles in eliminating invading organisms and inflammatory debris from animals (127, 128). These cells are attracted to infection and inflammatory sites by chemoattractants produced at the sites. The chemoattactants act as ligands which bind to G-protein-linked

Table 4.4. Selectin family of adhesion molecules

Selectin	Expressed by	Target cell
L-selectin	PMNs, monocytes, lymphocyte subsets	Activated ECs, EVs of PLN and ML
E-selectin	Cytokine-activated ECs	PMNs, monocytes, eosinophils Lymphocyte subsets, some tumor cells
P-selectin	Thrombin-activated platelets and ECs, cytokine-activated ECs	PMNs, monocytes, eosinophils, lymphocyte subsets, some tumor cells

EC, endothelial cell; GMP, granule membrane protein; HEV, high endothelial venule; ML, mesenteric lymphoid tissue; PLN, peripheral lymph node; PMN, polymorphonuclear cell or neutrophil. From ref. 136 (McEver RP. Curr. Opin. Immunol. 1994; 6: 75-84), courtesy of Current Biology Ltd.

(serpentine) receptors on the leukocyte cell surfaces, and stimulate cytoskeletal rearrangements, changes in adhesion and motility. These responses are largely blocked by pertussis toxin, indicating an involvement of the Gi class of heterotrimeric G-proteins. Dissociation of the G-protein complex yields the α and βγ subunits which can then promote activation of a number of signaling events and pathways, including stimulation of Src kinases, activation of the Ras/MAPK cascade, PI3K activation, release of intracellular calcium and activation of phospholipases PLCβ, PLD and PLA2 (128, 129). To prevent damage to the host tissues (130), the multiple pathways must be coordinated to regulate the three major functions of these leukocytes: chemotactic response (directed motility), phagocytosis (see chapter 6) and cell killing via granule secretion and generation of toxic oxidants. Much of this coordination occurs via the small G-proteins of the Ras family: Rab for secretion and phagocytosis, ARF for vesicular trafficking and oxidant production and the Rho family for motility and oxidant production.

A key component in this regulation is Rac. As described previously in this chapter, Rac is involved in the regulation of actin polymerization in membrane protrusions such as ruffles. Signficantly, Rac is also required for the activation of NADPH oxidase for production of toxic oxidants (131). This function appears to be promoted by Rac release from a GDI and association in a complex with other components of the oxidase at the plasma membrane, probably at microfilament attachment sites. How Rac might integrate oxidase and cytoskeleton functions is yet unclear, but formation of a complex of components from both pathways, reminiscent of the budding yeast complex which integrates mating and budding is an intriguing possibility. Consistent with this possibility, receptors for N-formyl-methionyl chemoattractant peptides are proposed to link to a membrane skeleton in stimulated neutrophils (132). In desensitized cells, these receptors are restricted to domains depleted of G-proteins, suggesting that the submembrane cytoskeleton regulates receptor-G-protein interactions.

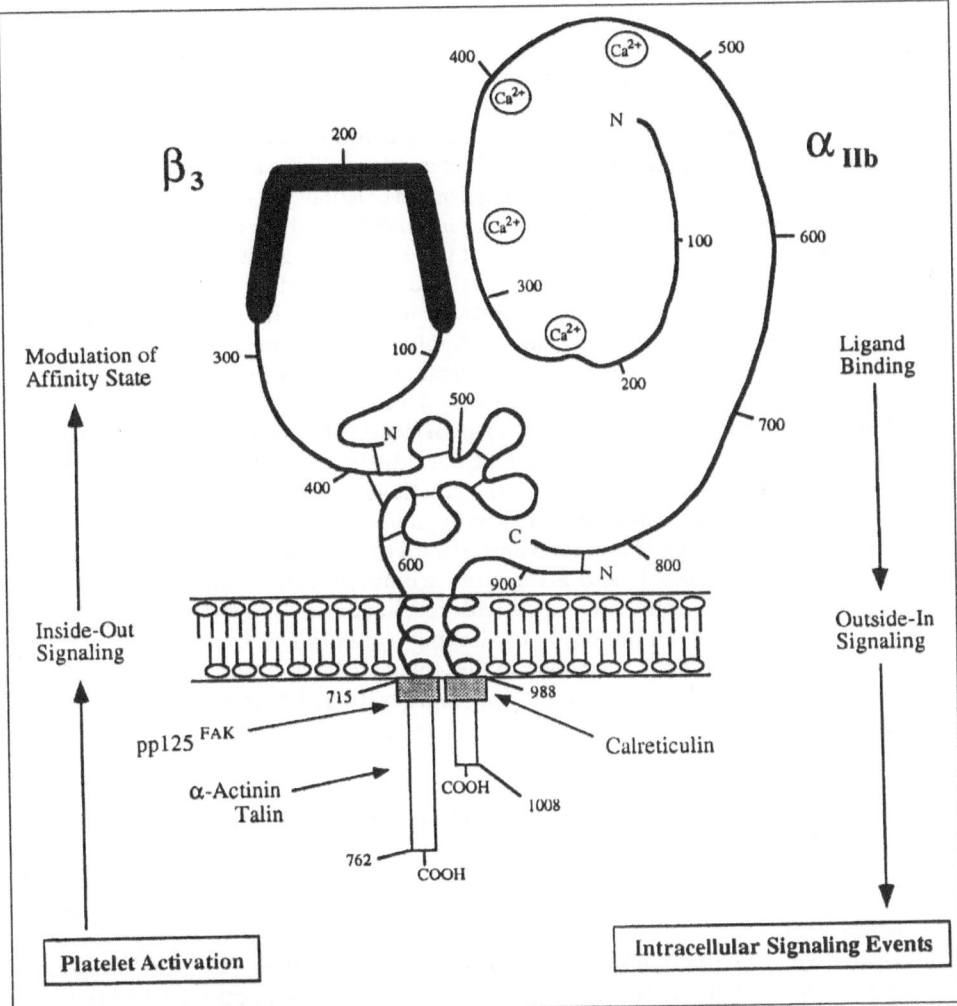

Fig. 4.8. Structure of integrin IIb-IIIa (αIIbβ3), which participates in both inside-out and outside-in signaling. Figure from ref. 143 (Williams MJ, Du X, Loftus JC, Ginsberg MH. Sem Cell Biol 1995; 6: 305-314), courtesy of Academic Press.

    To get to sites of inflammation, leukocytes must migrate through the vascular endothelium (133), a process similar to lymphocyte recirculation and homing (134). The migration involves three discrete processes (135):
1) transient interactions of cell surface selectins to induce rolling of the lymphocytes;
2) integrin-mediated adhesion of chemoattractant-activated leukocytes to endothelial cells; and
3) transendothelial movement of the leukocytes, also thought to be mediated by integrins.

The selectin-mediated processes are the best understood and provide some interesting examples of cooperation between the cell surface, cytoskeleton and signaling pathways. Selectins are transmembrane glycoproteins containing a C-type lectin domain which binds specific carbohydrates (136). Three different classes of selectins (L-selectin, E-selectin, P-selectin) have been described which participate in leukocyte-endothelial cell interactions (Table 4.4) (136). An important aspect of selectin behavior is the localization of leukocyte selectin to microvilli which presumably facilitates the transient interaction of these cells with endothelial cells (137). Moreover, the transience of this selectin interaction is assured by a proteolytic mechanism for its removal from leukocyte cell surfaces upon crosslinking (134). L-selectin is constitutively associated with microfilaments via α-actinin and vinculin. However, although microfilaments are a major structural element of microvilli, the association of selectin with microvilli appears to be independent of linkage to the microfilaments (138). Crosslinking of L-selectin with antibodies results in activation of Lck, phosphorylation of the selectin, association of Grb2/Sos with the selectin, activation of Ras, MAPK and Rac and transient production of toxic oxidants (139). One of the functions of selectin-mediated signaling is likely inside-out activation of integrins for subsequent adhesion events. The relationship between signal initiation and selectin association with the cytoskeleton needs to be clarified. E- and P-selectins are not constitutively linked to the cytoskeleton, but become associated with focal adhesion-like cytoskeletal complexes upon adhesion or antibody crosslinking (137, 140). This anchoring provides both additional mechanical stability to the leukocyte binding and a connection to endothelial signaling pathways preparatory to subsequent steps in the transendothelial movement (140).

Platelet Adhesion

Although strictly speaking, the platelet is not a motile cell, it is among the most thoroughly studied and best understood examples of a cell which undergoes the adhesion, spreading and contractile functions which provide the mechanisms for motility. Platelets play a critical role in hemostasis and thrombosis by their involvement in the maintenance of blood vessel integrity (141). Tissue injury releases agents which activate platelets to undergo shape changes. The disc-shaped unactivated platelet is first converted to a sphere which then puts forth two types of protrusions, flat lamellae and filopodia. Lamellae adhere to extracellular matrix components of injury sites of the endothelial lining. Filopodia bind fibrin strands and other platelets to form a clot. Adhesion initiates a complex set of phenomena, including platelet activation, spreading, aggregation and eventual clot retraction to seal the site of the injury. A number of platelet adhesive receptors have been implicated in these events (142, 143): several integrins, particularly GPIIb-IIIa (also known as αIIbβ3) (Fig. 4.8); GP Ib-IX complex; P- selectin; immunoglobulin-like PECAM-1; CD9; and CD36. The importance of the GPIb-IX and GPIIb-IIIa complexes is indicated by their implication in the inherited clotting disorders Bernard-Soulier Syndrome and Glanzmann's Thrombasthenia, respectively (143). Although association with endothelial components is a critical feature of the platelet response, much of the work on platelet signaling has been done in vitro because of the difficulty of working in animals.

Outside In and Inside Out Signaling

Two different types of platelet signaling have been described: *outside-in* and *inside-out* (Fig. 4.8) (46,143). *Outside-in* signaling is the more traditional type and can occur by two different mechanisms:

1) Platelet agonists, such as thrombin, can bind to receptors and trigger a cascade of cellular changes, including shape change, extension of filopodia, secretion of granular contents, aggregation and contraction, all of which involve the platelet cytoskeleton.

2) Specific adhesion of platelet receptors to insoluble or soluble ligands can trigger specific responses. For example, glycoprotein complex GPIb-IX is particularly important under high shear conditions which occur in arteries, arterioles and capillaries where its interaction with von Willebrand Factor (vWF) exposed in the vessel wall binds platelets to the injury site and triggers platelet activation (143). This signaling results in increased phospholipid metabolism, PKC activation, Ca^{2+} influx and cytoplasmic reorganization, possibly modulated via the binding of an isoform of the signaling protein 14-3-3 to GPIb-IX (144). In contrast, under lower stress conditions, such as exist in veins, interactions between integrins and their receptors become more important. Of particular significance in the platelet response cascade is aggregation of platelets by binding soluble fibrinogen to GPIIb-IIIa at the cell surfaces of activated platelets (145). Significantly, fibrinogen is absent from the walls of normal vessels, but present in atherosclerotic vessels (146). This binding of soluble fibrinogen occurs only after platelet activation has triggered the *inside-out* signaling mechanism to cause a conformational change in the integrin IIb-IIIa that allows it to bind soluble fibrinogen in the plasma (143). Thus, platelet adhesion occurs by two different mechanisms, both of which activate outside-in signaling and one of which requires inside-out signaling through an integrin conformational change.

Activation of platelets by the protease thrombin, an external agent, triggers sequential adhesion and aggregation, accompanied by microfilament rearrangements and a shape change (147). One aspect of understanding the relationship between signaling and the cytoskeleton in the platelet is determining which processes accompany adhesion and aggregation. Thrombin acts via a serpentine receptor by cleaving a receptor autoinhibitory peptide to permit its coupling to a trimeric G-protein (148) and to activate multiple signaling pathways (149). One of the early events after receptor activation is an increased affinity of IIb-IIIa for soluble fibrinogen (*inside-out* signaling) and a resultant *outside-in* signaling. The mechanism of the conformational activation of IIb-IIIa (*inside-out* signaling) has been partly delineated by crystallographic analyses (150). Transfection studies with other cell types suggest that it is a platelet-specific event (151), possibly involving a cytoplasmic repressor molecule (152) such as endonexin (153) or Rho A (154). Alternatively or additionally, the IIb-IIIa affinity may be controlled by PKC-mediated phosphorylation of its β_3 subunit (155). Thrombin-induced PI3K stimulation has also been implicated in the sustained activation of IIb-IIIa and platelet aggregation (156). The PI3K inhibitor wortmannin inhibits maintenance of integrin in an activated state and blocks antibody-induced aggregation without affecting secretion. Since PI3K appears not to be involved in actin polymerization, these results suggest that the activated integrin can trigger a set of pathways involved in aggregation without altering responses dependent on microfilament changes. PI3K can bind to

Table 4.5. Fractionation of platelet lysates

Platelet State	LSP	HSP	Supernatant
Unstimulated	Microfilaments	Microfilaments	Actin
	ABP	ABP	cortactin
	α-actinin	spectrin	PI3K
	tropomyosin	talin	PKC
		vinculin	FAK
		GPIb-IX	rap 1b
		GPIIb-IIIa	PECAM
		Src	
		Yes	
		GAP	
Activated	microfilaments	spectrin	rap 1b
	ABP	talin	PI3K
	cortactin	vinculin	PKC
	α-actinin	GPIb-IX	FAK
	tropomyosin	GPIIb-IIIa	PECAM
	myosin	Src	
		Yes	
		GAP	
Aggregating	microfilaments		
	ABP		
	cortactin		
	α-actinin		
	tropomyosin		
	myosin		
	spectrin		
	talin		
	vinculin		
	GPIb-IX		
	GPIIb-IIIa		
	Src		
	Yes		
	GAP		
	rap 1b		
	PI3K		
	PKC		
	FAK		
	PECAM		

LSP: low speed pelelt; HSP: high speed pellet
Modified from ref. f158 (Fox, JEB. Thromb Haemostasis 1993; 70:884-893), courtesy of FK Schattauer
Verlagsgesellschaft mblt.

signaling complexes through a phosphorylated Fcγ receptor ITAM motif, possibly mediated by the protein tyrosine kinase Syk (157).

Shape Change

One of the important platelet responses is the shape change which accompanies platelet activation (141). As indicated in chapters 1 and 3, cell shape is regulated by a combination of adhesion and the cytoskeleton. In the platelet, as in most cells, these elements appear to be tightly coordinated through interactions which occur at the cell membrane in response to signals propagated from the membrane to the cytoplasm. Circulating platelets are predominantly smooth, disc-shaped cells. However, they also contain cell surface invaginations leading to canalicular structures which present a less accessible cell surface and cytoplasmic granules, that can fuse with the plasma membrane during secretion (141). The canalicular system provides plasma membrane that can be utilized for rapid changes in cell shape while maintaining a smaller cell volume. Activation of platelets under non-aggregating conditions results in conversion to a spherical morphology and extension of pseudopodia and filopodia (141). This extension of processes suggests an important role for microfilaments in platelet shape changes, a possibility supported by observations of a rapid increase in actin polymerization during platelet activation (158). Platelets thus provide a useful model for the relationships between adhesion, signaling and microfilament behavior, because they undergo many though not all of the processes performed by motile cells.

Platelet Cytoskeleton and Membrane Skeleton

As mentioned in chapter 1, platelet actin is distributed among cytoplasmic microfilaments, the membrane cytoskeleton and soluble fractions (159, 160). In the unstimulated platelet, fractionation studies indicate that the cytoplasmic microfilaments are asssociated with 250 kDa actin binding protein (ABP), α-actinin and tropomyosin (158). In contrast, the membrane skeleton of the high speed pellet contains a variety of components (158) including microfilaments; actin crosslinking proteins ABP and spectrin; cell surface glycoproteins GPIb-IX, GPIa-IIa, and GPIIb-IIIa; focal adhesion proteins talin, vinculin and dystrophin-related protein (161); and signaling proteins of the Src tyrosine kinase and GAP families (Table 4.5). A simple interpretation of this composition is that the membrane skeleton is associated with membranes at *sites resembling focal adhesions* that serve as signal transduction complexes (158).

Upon activation of platelets under nonaggregating conditions, a rapid reorganization of the cytoskeleton occurs, including the rapid polymerization of actin. This actin reorganization is proposed to involve three stages (147):

1) The *first stage* results from changes in signaling elements which influence the cytoskeleton, such as phosphoinositides and calcium (see chapter 1), and is manifested by a calcium-dependent breakdown of cortical microfilaments (1).

2) The *second stage* involves an increase in F-actin from 30-40% to 60-70% of the total actin. Two types of polymerization predominate, one forming microfilament-containing filopodia and the other leading to the expansion of circumferential lamellae (162). This polymerization is inhibited by cytochalasin, indicating that it occurs at the barbed ends of filaments.

3) The *final stage* involves stabilizing the actin organization by crosslinking and membrane anchoring.

A key mechanism for increasing F-actin in platelets appears to be the generation of free barbed ends, predominantly by a Ca^{2+}-dependent mechanism involving gelsolin (162). Support for the importance of gelsolin in platelet function has come from gelsolin negative transgenic mice engineered by homologous recombination, whose platelet responses are diminished (35). However, the fact that these mice survive and have somewhat responsive platelets indicates a redundancy in the performance of the functions of gelsolin (1).

Severing by this mechanism leaves barbed ends which are capped by the gelsolin or a capping protein homolog of capZ (163). In either case, removal of the cap requires dissociation by polyphosphoinositide (164). Phosphoinositide changes may also facilitate participation of profilin in transferring G-actin from complexes with β-thymosin to microfilament barbed ends, as described in chapter 1. In a simplified scheme, hydrolysis of phosphatidylinositides promotes breakdown of microfilaments through the action of calcium-dependent severing proteins. Synthesis and clustering of these phospholipids would produce sites for uncapping microfilaments and additional actin polymerization. Thus, actin assembly would be both temporally and spatially regulated by phospholipid signals (1). The factors involved in inositide regulation of actin polymerization have been investigated using a new platelet permeabilization procedure and phosphoinositide-binding peptides from gelsolin (36). Phosphoinositide-mediated uncapping of F-actin via the thrombin receptor was demonstrated to involve the small G-protein Rac which induces the synthesis of phosphatidyl(4,5)diphosphate in the permeabilized platelets. The pathway between the thrombin receptor and Rac and between Rac and the phosphoinositide kinases involved in the synthesis is unclear. Although both 3- and 4-phosphorylated inositides can remove gelsolin from filament barbed ends, studies with inhibitors of PI3K, which is involved in synthesis of the 3-phosphorylated inositides (Fig. 2.12), suggest that it does not play a role in platelet actin assembly (156), even though it becomes associated with the cytoskeleton during activation. Tyrosine phosphorylation may be involved in platelet actin reorganization, because it is an early event of platelet activation and can activate PLCγ (165). Inositide hydrolysis by this enzyme generates IP_3, which can release Ca^{2+} from internal stores, and diacylglycerol, which may activate nucleating sites (166). Moreover, turnover of the phosphoinositides may facilitate their accumulation at sites favorable for promoting actin polymerization. A possible involvement of phospholipase isozymes is suggested by their integrin- and tyrosine kinase-dependent recruitment to the cytoskeleton (167).

In addition to actin polymerization, activation under nonaggregating conditions results in the relocalization of several cytoskeletal proteins (Table 4.5) (158). Tropomyosin and ABP- 280 are shifted to filopodia, and α-actinin becomes concentrated in filopodia and the submembrane cortex. Another protein which is phosphorylated and relocalized is cortactin, which is recruited to the actin filaments from the cytoplasm as a consequence of activation (168). Cortactin was originally described as a microfilament-associated Src substrate in the cortex of fibroblasts (169, 170), but little is known of its role in platelets. In contrast to the cytoplasmic microfilaments, *the membrane skeleton appears to undergo few compositional changes with activation under nonaggregating conditions.* However, substantial changes occur with aggregation, the most prominent of which is the association of the membrane skeleton with the platelet microfilaments in the low speed pellet from Triton lysates (158). Occupancy of the fibrinogen binding site on integrin IIb-IIIa is necessary, but not sufficient for the linkage of membrane skeletal IIb-IIIa to the cytoskeleton; platelet aggregation is also necessary. Interestingly, it is the fraction of IIb-IIIa linked to the cytoskeleton which preferentially binds ligand (171). Its linkage to the cytoskeleton results in a movement of integrin to the open canalicular system. Both ligand binding and redistribution of the receptor are inhibited by cytochalasin, suggesting a role for actin in controlling the behavior of integrin. Concomitant with linkage of IIb-IIIa to the microfilaments, the membrane skeletal signaling and "focal adhesion" components are linked to myosin-containing microfilaments. Thus, aggregation ties the platelets into a mass linked to an actomyosin motor which can effect contraction. This linkage must involve attachment of myosin to components of the surface membranes of the aggregated platelet masses. A key aspect of this process appears to be the thrombin-induced, aggregation-independent centralization of cell surface GPIb-IX and its movement to the open canalicular system (172). Two sequential steps are required: cytochalasin-sensitive attachment of the glycoprotein to microfilaments; and a slower step requiring myosin activation, implying a process by which cell surface GPIb-IX complexes are linked to contractile microfilaments of the platelet interior. It is thus of particular interest that the contractile motor protein myosin is phosphorylated and activated concomitant with its binding to the cytoplasmic microfilaments as an early step of platelet activation and the centralization of platelet granules (173).

The presence of several signaling components associated with the membrane skeleton raises questions about their role in platelet behavior. Particular attention has been accorded the protein tyrosine kinases, which catalyze phosphorylation of platelet proteins at several phases of the platelet responses. Three temporal waves of tyrosine phosphorylation have been described (174). Substrates for the *early tyrosine phosphorylation* include RasGAP, Vav (175) and cortactin. This phase is independent of fibrinogen binding to integrin IIb-IIIa and concomitant with activation of Src family kinases. Thus, it may play a role in the inside-out activation of IIb-IIIa, which is blocked by tyrosine kinase inhibitors (147), or in actin reorganization. Fibrinogen binding to IIb-IIIa (Fig. 4.8) activates the *second phase* which includes phosphorylation of 140 and 50-68 kDa proteins. This response is not accompanied by increased cytoplasmic Ca^{2+} or PKC activation or prevented by cAMP (142). Moreover, fibrinogen binding causes only a small increase in F-actin, suggesting that this activation mechanism is not involved in regulating the primary actin polymerization process. This phosphorylation may result in part from recruitment of the soluble kinase Syk to the membrane skeleton, since it is the only platelet tyrosine kinase identified to date whose activity is stimulated by the integrin ligation (174,176). Interestingly, Syk is recruited and activated during both the early and intermediate phases. The first is blocked by cytochalasin; the second is blocked by RGDS peptide, an inhibitor of integrin ligation (176). Vav, a GEF for Rho and cdc42Hs, is also phosphorylated at both phases (175). This phosphorylation may provide a mechanism to regulate transduction of thrombin- and integrin-mediated signals via a small G-protein to alter platelet cytoskeletal organization. Consistent with this model is the observation that cdc42Hs is translocated to the cytoskeleton by an integrin-mediated, cytochalasin-sensitive process that is blocked by the protein tyrosine kinase inhibitor genistein (177). The *third wave* is aggregation dependent and corresponds to the phosphorylation of focal adhesion kinase (FAK) (see Fig. 4.8), suggesting a possible involvement in linking the focal adhesion-like complexes to the microfilaments. This phase also results in the tyrosine phosphorylation of the β_3 subunit of integrin at sites which can link to SHC and GRB2 of the mitogenic pathway (178). In the presence of mM extracellular

Ca^{2+} rapid dephosphorylation occurs, a result of calpain-mediated PTP-1B cleavage and association with the plasma membrane (179). IgG stimulation of platelets via the FcRII receptor results in phosphorylation of both Syk and FAK. The latter is inhibited by PKC inhibitors, suggesting that PKC is an intermediary in the activation cascade under these conditions (180).

The results cited above demonstrate similarities in the adhesion of platelets and that of other cell types, particularly in the components involved. However, there are noteworthy disparities, e.g., the temporal difference in recruitment of FAK to adhesion complexes, which will undoubtedly prove important in understanding the specific mechanisms.

Summary

Cell motility is achieved by a cycle involving membrane protrusion, attachment, traction and release. Cell crawling over a substratum has been described by two mechanisms: amoeboid and fibroblast-like. The two have many common features and probably differ primarily in the strength of the forces involved in the different steps of progression. Cell motility absolutely requires formation of membrane protrusions, but the mechanisms involved are unclear. Actin polymerization, osmotic forces, filament association with the membrane, myosin motors and crosslinking proteins have all been implicated. Protrusions develop in response to cell surface stimuli, including activation of both serpentine (heterotrimeric G-protein-linked) and tyrosine kinase receptors. The small G-proteins Rac and cdc42 are involved, but the pathways linked to these signaling switches still need to be defined. Directed migration involves attachment of protrusions to the substratum by focal adhesions. Focal adhesions contain transmembrane integrins associated with extracellular matrix components and attached to microfilaments at the plasma membrane cytoplasmic surface. They are extremely complex structures with multiple cytoplasmic components and redundant interactions. Formation of focal adhesions has been described as a hierarchical process dependent on integrin ligation, integrin clustering, tyrosine phosphorylation and microfilament assembly. Over 30 differing signaling and cytoskeletal molecules have been shown to be recruited during formation. The small G-protein Rho is required and can be activated by multiple mechanisms. Although a number of tyrosine kinases are present in focal adhesions, none has been demonstrated to be required specifically for adhesion formation. FAK, a complex multidomain tyrosine kinase localized to focal adhesions, has been implicated in cell migration but not in focal adhesion formation. Traction processes to move the cell body and detachment of the cell tail are necessary for continued movement. Both contractile processes (actomyosin motors) and tyrosine phosphorylation have been implicated in these events. Growth cone motility is a special case of cell motility in which the contractile and retraction phases have been replaced by growth and stabilization to permit neurite extension, processes primarily involving microtubules. Finally, directed cell movements can be stimulated by external factors such as chemoattractants. One mechanism for explaining the directionality involves weakening of the cell cortical actin meshwork to enhance localization of protrusions.

Platelets provide a useful model because they undergo the first three of the motility steps in a model which is essentially a cell fragment. Platelets have also been an important system for studying inside-out and outside-in signaling via integrins. Platelet activation by a number of mechanisms induces rapid shape changes with concomitant changes in adhesions and actin polymerization, and finally platelet aggregation. Platelet actin is found in soluble form in cytoplasmic microfilaments and in the membrane skeleton. The membrane skeleton comprises components similar to those of focal adhesions and contains numerous signal transduction elements. Platelet actin polymerization is regulated by a barbed end capping process controlled by inositides, calcium and the small G-protein Rac. Other factors, including tyrosine kinases, have been implicated. Association of the membrane skeleton with microfilaments occurs concomitant with platelet aggregation, thus linking an actomyosin motor to the plasma membrane for clot retraction. Tyrosine phosphorylation has been implicated at several phases of platelet activation, including inside-out signaling, outside-in signaling and linkage of the membrane skeleton to microfilaments. However, the molecular pathways involved in these signaling steps still remain largely undefined.

References

1. Stossel TP. The machinery of blood cell movements. Blood 1994; 84:367-379.
2. Mitchison TJ, Cramer LP. Actin-based cell motility and cell locomotion. Cell 1996; 84:371-379.
3. Stossel TP. On the crawling of animal cells. Science 1993; 260:1086-1094.
4. Lauffenburger DA, Horwitz AF. Cell migration: A physically integrated molecular process. Cell 1996; 84: 359-369.
5. Huttenlocher A, Ginsberg MH, Horwitz, AF. Modulation of cell migration by integrin-mediated cytoskeletal linkages and ligand-binding affinity. J Cell Biol 1996; 134:1551-1562.
6. Condeelis J. Life at the leading edge: the formation of cell protrusions. Annu Rev Cell Biol 1993; 9:411-444.
7. Oliver T, Lee J, Jacobson K. Forces exerted by locomoting cells. Sem Cell Biol 1994; 5:139-147.
8. Sheetz MP. Cell migration by graded attachment to substrates and contraction. Sem Cell Biol 1994; 5:149-155.
9. Sheetz MP, Wayne DB, Pearlman AL. Extension of filopodia by motor-dependent actin assembly. Cell Motil Cytoskel 1992; 22:160-169.
10. Gingell D. Contact signalling and cell motility. Symp Soc Exper Biol 1993; 47: 1-33.
11. Ostap EM, Pollard TD. Overlapping functions of myosin-I isoforms? J Cell Biol 1996; 133:221-224.
12. Fath KR, Burgess DR. Membrane motility mediated by unconventional myosin. Curr Opin Cell Biol 1994; 6:131-135.
13. Novak KD, Peterson MD, Reedy MC, Titus MA. *Dictyostelium* myosin I double mutants exhibit conditional defects in pinocytosis. J Cell Biol 1995; 131: 1205-1221.
14. Wang F-S, Wolenski JS, Cheney RE, Mooseker MS, Jay DG. Function of myosin-V in filopodial extension of neuronal growth cones. Science 1996; 273:660-663.
15. Small JV. Lamellipodia architecture: actin filament turnover and the lateral flow of actin filaments during motility. Sem Cell Biol 1994; 5:157-163.
16. Theriot JA, Mitchison TJ. Actin microfilament dynamics in locomoting cells. Nature 1991; 352:126-131.
17. Zigmond SH. Signal transduction and actin filament organization. Curr Opin Cell Biol 1996; 8:66-73.
18. Luna EJ, Hitt A. Cytoskeleton-plasma membrane interactions. Science 1992; 258:955-964.
19. Cunningham CC, Stossel TP, Kwiatkowski DJ. Enhanced motility in NIH 3T3 fibroblasts that overexpress gelsolin. Science 1991; 251:1233-1236.
20. Cox D, Condeelis J, Wessels D, Soll D, Kern H, Knecht D. Targeted disruption of the ABP-120 gene leads to cells with altered motility. J Cell Biol 1992; 116: 943-955.

21. Tsukita S, Yonemura S, Tsukita S. ERM (ezrin/radixin/moesin) family: from cyto-skeleton to signal transduction. Curr Opin Cell Biol 1997; 9:70-75.

22. Martin M, Andreoli C, Sahuquet A, Montcourrier P, Algrain M, Mangeat P. Ezrin NH2-terminal domain inhibits the cell extension activity of the COOH-terminal domain. J Cell Biol 1995; 128: 1081-1093.

23. Bassell G, Singer RH. mRNA and cytoskeletal filaments. Curr Opin Cell Biol 1997; 9:109-115.

24. Bretscher MS. Moving membrane up to the front of migrating cells. Cell 1996; 85: 465-467.

25. Bergmann JE, Kupfer A, Singer SJ. Membrane insertion at the leading edge of motile fibroblasts. Proc Natl Acad Sci USA 1983; 80:1367-1371.

26. Lee J, Ishihara A, Jacobson K. How do cells move along surfaces? Trends Cell Biol 1993; 3:366-370.

27. Theriot JA. Regulation of the actin cytoskeleton in living cells. Sem Cell Biol 1994; 5:193-199.

28. Nobes CD, Hall A. Rho, Rac, and Cdc42 GTPases regulate the assembly of multi-molecular focal complexes associated with actin stress fibers, lamellipodia, and filopodia. Cell 1995; 81:53-62.

29. Lim L, Hall C, Monfries C. Regulation of actin cytoskeleton by Rho-family GTPases and their associated proteins. Sem. Cell Devel Biol 1996; 7:699-706.

30. Rodriguez-Viciana P, Warne PH, Khwaja A, Marte BM, Pappin D, Das P, Waterfield MD, Ridley A, Downward J. Role of phosphoinositide 3-OH kinase in cell transformation and control of the actin cytoskeleton by Ras. Cell 1997; 89: 457-467.

31. Erickson JW, Zhang C-j, Kahn RA, Evans T, Cerione RA. Mammalian cdc42 is a brefeldin A-sensitive component of the Golgi apparatus. J Biol Chem 1996; 271: 26850-26854.

32. Ridley AJ. Membrane ruffling and signal transduction. BioEssays 1994; 16:321-327.

33. Ridley AJ, Hall A. Signal transduction pathways regulating rho-mediated stress fibre formation: requirement for a tyrosine kinase. EMBO J 1994; 13:2600-2610.

34. Chen P, Murphy-Ullrich JE, Wells A. A role for gelsolin in actuating EGF recep-tor-mediated cell motility. J Cell Biol 1996; 134:689-698.

35. Witke W., Sharpe AH, Hartwig JH, Azuma T, Stossel TP, Kwiatkowski DJ. Hemo-static, inflammatory, and fibroblast responses are blunted in mice lacking gelsolin. Cell 1995; 81:41-51.

36. Hartwig JH, Bokoch GM, Carpenter CL, Janmey PA, Taylor LA, Toker A, Stossel TP. Thrombin receptor ligation and activated rac uncap actin filament barbed ends through phosphoinositide synthesis in permeabilized human platelets. Cell 1995; 82:643-653.

37. Symons M, Derry JMJ, Kartak B, Jiang S, Lemahieu V, McCormick F, Francke U, Abo A. Wiskott-Aldrich Syndrome protein, a novel effector for the GTPase CDC42Hs, is implicated in actin polymerization. Cell 1996; 84:723-734.

38. Miki H, Miura K, Takenawa T. N-WASP, a novel actin-depolymerizing protein, regulates the cortical cytoskeletal rearrangement in PIP₂-dependent manner down-stream of tyrosine kinases. EMBO J 1996; 15:5326-5335.

39. Abercrombie M, Heaysman J, Pegrum SM. The locomotion of fibroblasts in cul-ture. Exp Cell Res 1971; 67:359-367.

40. Turner CE, Burridge K. Transmembrane molecular assemblies in cell-extracellu-lar matrix interactions. Curr Opin Cell Biol 1991; 3:849-853.

41. Burridge K, Fath K, Kelly T, Nuckolls G, Turner, C. Focal adhesions: Transmem-brane junctions between the extracellular matrix and the cytoskeleton. Ann Rev Cell Biol 1988; 4:487-525.

42. Lo SH, Chen LB. Focal adhesion as a signal transduction organelle. Cancer Metas Rev 1994; 13:9-24.
43. Hynes RO. Integrins: versatility, modulation, and signaling in cell adhesion. Cell 1992; 69:11-25.
44. Schwartz MA, Schaller MD, Ginsberg MH. Integrins: emerging paradigms of signal transduction. Annu Rev Cell Dev Biol 1995; 11:549-599.
45. Sastry SK, Horwitz AF. Integrin cytoplasmic domains: mediators of cytoskeletal linkages and extra- and intracellular initiated transmembrane signaling. Curr Opin Cell Biol 1993; 5:819-831.
46. Dedhar S, Hannigan GE. Integrin cytoplasmic interactions and bidirectional transmembrane signalling. Curr Opin Cell Biol 1996; 8:657-669.
47. Wu C, Keivens VM, O'Toole TE, McDonald JA, Ginsberg MH. Integrin activation and cytoskeletal interaction are essential for the assembly of a fibronectin matrix. Cell 1995; 83:715-724.
48. Burridge K, Chrzanowska-Wodnicka M. Focal adhesions, contractility, and signaling. Annu Rev Cell Dev Biol 1996; 12: 463-519.
49. Plopper G, Ingber DE. Rapid induction and isolation of focal adhesion complexes. Biochem Biophys Res Commun 1993; 193:571-578.
50. Plopper GE, McNamee HP, Dike LE, Bojanowski K, Ingber DE. Convergence of integrin and growth factor receptor signaling pathways within the focal adhesion complex. Mol Biol Cell 1995; 6: 1349-1365.
51. Johnson RP, Craig SW. F-actin binding site masked by the intramolecular association of vinculin head and tail domains. Nature 1995; 373:261-264.
52. Soderling TR. Protein kinases. Regulation by autoinhibitory domains. J Biol Chem 1990; 265:1823-1826.
53. Isenberg G. New concepts for signal perception and transduction by the actin skeleton at cell boundaries. Sem Cell Devel Biol 1996; 7: 707-715.
54. Gilmore AP, Burridge K. Regulation of vinculin binding to talin and actin by phosphatidylinositol-4,5-bisphosphate. Nature 1996; 381: 531-535.
55. Craig SW, Johnson RP. Assembly of focal adhesions: progress, paradigms, and portents. Curr Opin Cell Biol 1996; 8: 74-85.
56. Yamada K, Miyamoto S. Integrin transmembrane signaling and cytoskeletal control. Curr Opin Cell Biol 1995; 7: 681-689.
57. Miyamoto S, Teramoto H, Coso OA, Gutkind JS, Burbelo PD, Akiyama SK, Yamada KM. Integrin function: Molecular hierarchies of cytoskeletal and signaling molecules. J Cell Biol 1995; 131: 791-805.
58. Ridley AJ. Signal transduction through the GTP-binding proteins Rac and Rho. J Cell Sci 1994; (Suppl)18:127-131.
59. Symons M. Rho family GTPases: the cytoskeleton and beyond. Trends Biochem Sci 1996; 21:178-181.
60. Hotchin NA, Hall A. The assembly of integrin adhesion complexes requires both extracellular matrix and intracellular rho/rac GTPases. J Cell Biol 1995; 131:1857-1865.
61. Reif K, Nobes CD, Thomas G, Hall A, Cantrell DA. Phosphatidylinositol 3-kinase signals activate a selective subset of Rac/Rho-dependent effector pathways. Curr Biol 1996; 6:1445-1455.
62. Lopez-Ilasaca M, Crespo P, Pillici PG, Gutkind JS, Wetzker R. Linkage of G protein-coupled receptors to the MAPK signaling pathway through PI 3-kinase gamma. Science 1997; 275:394-397.
63. Peppelenbosch MP, Qiu R-G, De Vries-Smits AMM, Tertoolen LGJ, de Laat SW, McCormick F, Hall A, Symons MH, Bos JL. Rac mediates growth factor-induced arachidonic acid release. Cell 1995; 81: 849-856.

64. Maher PA, Pasquale EB, Wang JY, Singer SJ. Phosphotyrosine containing proteins are concentrated in focal adhesions and intercellular junctions in normal cells. Proc Nat Acad Sci USA 1985; 82: 6576-6580.

65. Nobes CD, Hawkins P, Stephens L, Hall A. Activation of the small GTP-binding proteins rho and rac by growth factor receptors. J Cell Sci 1994; 108:225-233.

66. Rosales C, O'Brien V, Kornberg L, Juliano R. Signal transduction by cell adhesion receptors. Biochim Biophys Acta 1995; 1242:77-98.

67. Parsons JT, Schaller MD, Hildebrand J, Leu T-H, Richardson A, Otey C. Focal adhesion kinase: structure and signalling. J Cell Sci 1994; (Suppl)18:109-113.

68. Clark EA, Brugge JS. Integrins and signal transduction pathways: the road taken. Science 1995; 268:233-239.

69. Richardson A, Parsons JT. Signal transduction through integrins: a central role for focal adhesion kinase? BioEssays 1995; 17:229-236.

70. Gilmore AP, Romer LH. Inhibition of focal adhesion kinase (FAK) signaling in focal adhesions decreases cell motility and proliferation. Mol Biol Cell 1996; 7:1209-1224.

71. Ilic D, Kanazawa S, Furuta Y, Yamamoto T, Aizawa S. Impairment of mobility in endodermal cells by FAK deficiency. Exp Cell Res 1996; 222:298-303.

72. Richardson A, Parsons JT. A mechanism for regulation of the adhesion-associated protein tyrosine kinase pp125^{FAK}. Nature 1996; 380:538-540.

73. Brown MC, Perrotta JA, Turner CE. Identification of LIM3 as the principal determinant of paxillin focal adhesion localization and characterization of a novel motif on paxillin directing vinculin and focal adhesion kinase binding. J Cell Biol 1996; 135:1109-1123.

74. Turner CE. Paxillin, a cytoskeletal target for tyrosine kinases. BioEssays 1994; 16: 47-52.

75. Han J-D, Rubin CS. Regulation of cytoskeleton organization and paxillin dephosphorylation by cAMP. Studies on murine Y1 adrenal cells. J Biol Chem 1996; 271:29211-29215.

76. Hannigan GE, Leung-Hagesteijn C, Fitz-Gibbon L, Coppolino M, Redeva G, Filmus J, Bell J, Dedhar S. Regulation of cell adhesion and anchorage-dependent growth by a new β1-integrin-linked protein kinase. Nature 1996; 379:91-96.

77. Leung T, Chen X-Q, Manser E, Lim L. The p160 RhoA-binding kinase ROK is a member of a kinase family and is involved in the reorganization of the cytoskeleton. Mol Cell Biol 1996; 16: 5313-5327.

78. LaFlamme SE, Auer KL. Integrin signaling. Sem. Cancer Biol. 1996; 7: 111-118.

79. Chong LD, Traynor-Kaplan A, Bokoch GM, Schwartz MA. The small GTP-binding protein rho regulates a phosphatidylinositol 4-phosphate 5-kinase in mammalian cells. Cell 1994; 79: 507-513.

80. Tapon N, Hall A. Rho, Rac and Cdc42 GTPases regulate the organization of the actin cytoskeleton. Curr Opin Cell Biol 1997; 9:86-92.

81. Auer KL, Jacobson BS. β₁ integrins signal lipid second messengers required during cell adhesion. Mol Biol Cell 1995; 6:1305-1313.

82. Woods A, Couchman JR. Protein kinase C involvement in focal adhesion formation. J Cell Sci 1992; 101:277-290.

83. Lewis JM, Cheresh DA, Schwartz MA. Protein kinase C regulates αvβ5-dependent cytoskeletal associations and focal adhesion kinase phosphorylation. J Cell Biol 1996; 134:1323-1332.

84. Chrzanowska-Wodnicka M, Burridge K. Rho-stimulated contractility drives the formation of stress fibers and focal adhesions. J Cell Biol 1996; 133:1403-1415.

85. Reinhard J, Scheel AA, Diekmann D, Hall A, Ruppert C, Bahler M. A novel type of myosin implicated in signaling by rho family GTPases. EMBO J 1995; 14: 697-704.

86. Bahler M. Myosins on the move to signal transduction. Crit Opin Cell Biol 1996; 8:18-22.

87. Kimura K, Ito M, Amano M, Chihara K, Fukata Y, Nakafuku M, Yamamori B, Feng J, Nakano T, Okawa K, Iwamatsu A, Kaibuchi K. Regulation of myosin phosphatase by rho and rho-associated kinase (rho-kinase). Science 1996; 273: 245-248.

88. Nagata K, Hall A. The rho GTPase regulates protein kinase activity. BioEssays 1996; 18:529-531.

89. Amano M, Chihara K, Kimura K, Fukata Y, Nakamura N, Matsuura Y, Kaibuchi K. Formation of actin stress fibers and focal adhesions enhanced by Rho-kinase. Science 1997; 275:1308-1311.

90. Kolodney MS, Elson EL. Contraction due to microtubule disruption is associated with increased phosphorylation of myosin regulatory light chain. Proc Natl Acad Sci USA 1995; 92:10252-10256.

91. Crowley E, Horwitz AF. Tyrosine phosphorylation and cytoskeletal tension regulate the release of fibroblast adhesions. J Cell Biol 1995; 131:525-537.

92. Small JV. Microfilament-based motility in non-muscle cells. Curr Opin Cell Biol 1989; 1:75-79.

93. Brundage RA, Fogarty KE, Tuft RA, Fay FS. Calcium gradients underlying polarization and chemotaxis of eosinophils. Science 1991; 254:703-706.

94. Morris CE. Mechanosensitive ion channels. J Membr Biol 1990; 113:93-107.

95. Hartwig JH, Shevin P. The architecture of actin filaments and the ultrastructural location of actin-binding protein in the periphery of lung macrophages. J Cell Biol 1986; 103:1007-1020.

96. Stossel TP. From signal to pseudopod. How cells control cytoplasmic actin assembly. J Biol Chem 1989; 264:18261-18264.

97. Anderson KI, Wang YL, Small JV. Coordination of protrusion and translocation of the keratocyte involves rolling of the cell body. J Cell Biol 1996; 134:1209-1218.

98. Jay PY, Pham PA, Wong SA, Elson EL. A mechanical function of myosin II in cell motility. J Cell Sci 1995; 108:387-393.

99. Lawson MA, Maxfield FR. Ca^{2+}- and calcineurin-dependent recycling of an integrin to the front of migrating neutrophils. Nature 1994; 377:75-79.

100. Diaz-Nido J, Ulloa L, Sanchez C, Avila J. The role of the cytoskeleton in the morphological changes occurring during neuronal differentiation. Sem Cell Devel Biol 1996; 7:733-739.

101. Tanaka E, Sabry J. Making the connection: cytoskeletal rearrangements during growth cone guidance. Cell 1995; 83: 171-176.

102. Williams EJ, Furness J, Walsh FS, Doherty P. Activation of the FGF receptor underlies neurite outgrowth stimulated by L1, N-CAM, and N-cadherin. Neuron 1994; 13:583-594.

103. Doherty P, Williams EJ, Walsh FS. A soluble chimeric form of the L1 glycoprotein stimulates neurite outgrowth. Neuron 1995; 14:57-66.

104. Garrity PA, Rao Y, Salecker I, McGladie J, Pawson T, Zipursky SL. Drosophila photoreceptor axon guidance and targeting requires the dreadlocks SH2/SH3 adapter protein. Cell 1996; 85:639-650.

105. Bixby JL, Jhabvala P. Tyrosine phosphorylation in early embryonic growth cones. J Neurosci 1993; 13:3421-3432.

106. Ruoslahti E, Obrink B. Common principles in cell adhesion. Exp Cell Res 1996; 227:1-11.

107. Gertler FB, Hill KK, Clark MJ, Hoffman FM. Dosage-sensitive modifiers of *Drosophila abl* tyrosine kinase function: *prospero*, a regulator of axonal outgrowth, and *disabled*, a novel tyrosine kinase substrate. Genes Dev 1993; 7: 441-453.

108. Desai CJ, Gindhart JG Jr, Goldstein LSB, Zinn K. Receptor tyrosine phosphatases are required for motor axon guidance in the Drosophila embryo. Cell 1996; 84: 599-609.

109. Krueger NX, Vactor DV, Wan HI, Gelbart WM, Goodman CS, Saito H. The transmembrane tyrosine phosphatase DLAR controls motor axon guidance in Drosophila. Cell 1996; 84:611-622.

110. Tigyi G, Fischer DJ, Sebok A, Yang C, Dyer DL, Miledi R. Lysophosphatidic acid-induced neurite retraction in PC12 cells: control by phosphoinositide-Ca^{2+} signaling and Rho. J Neurochem 1996; 66:537-548.

111. Lin C-H, Forscher P. Growth cone advance is inversely proportional to retrograde F-actin flow. Neuron 1995; 14: 763-771.

112. Fan J, Mansfield SG, Redmond T, Gordon-Weeks PR, Raper JA. The organization of F-actin and microtubules in growth cones exposed to a brain-derived collapsing factor. J Cell Biol 1993; 121: 867-878.

113. Sydor AM, Su AL, Wang F-S, Xu A, Jay DG. Talin and vinculin play distinct roles in filopodial motility in the neuronal growth cone. J Cell Biol 1996; 134: 1197-1207.

114. Challacombe JF, Snow DM, Letourneau PC. Actin filament bundles are required for microtubule reorientation during growth cone turning to avoid an inhibitory guidance cue. J Cell Science 1995; 109:2031-2040.

115. Lin C-H, Forscher P. Cytoskeletal remodeling during growth cone-target interactions. J Cell Biol 1993; 121: 1369-1383.

116. Goslin K, Birgbauer E, Banker G, Solomon F. The role of the cytoskeleton in organizing growth cones: a microfilament-associated growth cone component depends upon microtubules for its localization. J Cell Biol 1989; 109:1621-1631.

117. Devreotes PN, Zigmond SH. Chemotaxis in eukaryotic cells: a focus on leukocytes and *Dictyostelium*. Ann Rev Cell Biol 1988; 4:649-686.

118. Zigmond SH. Consequences of chemotactic peptide receptor modulation for leukocyte orientation. J Cell Biol 1981; 88:644-647.

119. Parent CA, Devreotes PN. Molecular genetics of signal transduction in Dictyostelium. Annu Rev Biochem 1996; 65: 411-440.

120. van Haastert PJM. Transduction of the chemotactic cAMP signal across the plasma membrane of *Dictyostelium* cells. Experientia 1995; 51:1144-1154.

121. Newell PC, Liu G. Streamer F mutants and chemotaxis of *Dictyostelium*. BioEssays 1992; 14:473-479.

122. Browning DD, The T, O'Day DH. Comparative analysis of chemotaxis in Dictyostelium using a radial bioassay method: protein tyrosine kinase activity is required for chemotaxis to folate but not to cAMP. Cell Signal 1995; 7: 481-489.

123. Gerisch G, Albrecht R, De Hostos E, Wallraff E, Heizer C, Kreitmeier M, Muller-Taubenberger A. Actin-associated proteins in motility and chemotaxis. Symp Soc Exp Biol 1993; 47:297-315.

124. Noegel AA, Luna JE. The *Dictyostelium* cytoskeleton. Experientia 1995; 51: 1135-1143.

125. Abu-Elneel K, Karchi M, Ravid S. *Dictyostelium* myosin II is regulated during chemotaxis by a novel protein kinase C. J Biol Chem 1996; 271:977-984.

126. Dembinsky A, Rubin H, Ravid S. Chemoattractant-mediated increases in cGMP induce changes in *Dictyostelium* myosin II heavy chain-specific protein kinase C activities. J Cell Biol 1996; 134: 911-921.

127. Edwards SW. *Biochemistry and Physiology of the Neutrophil*. Cambridge, UK: Cambridge University Press.

128. Bokoch GM. Chemoattractant signaling and leukocyte activation. Blood 1995; 86: 1649-1660.

129. Downey GP, Fukushima T, Fialkow L, Waddell TK. Intracellular signaling in neutrophil priming and activation. Sem Cell Biol 1995; 6:345-356.

130. Edwards SW. Cell signalling by integrins and immunoglobulin receptors in primed neutrophils. Trends Biochem Sci. 1995; 20:362-367.

131. Bokoch GM, Knaus UG. The role of small GTP-binding proteins in leukocyte function. Curr Opin Immunol 1994; 6: 98-105.

132. Klotz K-N, Jesaitis AJ. Neutrophil chemoattractant receptors and the membrane skeleton. BioEssays 1994; 16: 193-198.

133. Springer TA. Adhesion receptors of the immune system. Nature 1990; 346: 425-434.

134. Butcher EC, Picker LJ. Lymphocyte homing and homeostasis. Science 1996; 272: 60-66.

135. Ager A. Lymphocyte recirculation and homing: roles of adhesion molecules and chemoattractants. Trends Cell Biol 1994; 4:326-333.

136. McEver RP. Selectins. Curr Opin Immunol 1994; 6:75-84.

137. Kansas GS. Selectins and their ligands: current concepts and controversies. Blood 1996; 88:3259-3287.

138. Pavalko FM, Walker DM, Graham L, Goheen M, Doerschuk CM, Kansas GS. The cytoplasmic domain of L-selectin interacts with cytoskeletal proteins via α-actinin: receptor positioning in microvilli does not require interaction with α-actinin. J Cell Biol 1995; 129:1155-1164.

139. Brenner B, Gulbins E, Schlottmann K, Koppenhoefer U, Busch GL, Walzog B, Steinhausen M, Coggeshall KM, Linderkamp O, Lang F. L-selectin activates the Ras pathway via the tyrosine kinase p56[lck]. Proc Natl Acad Sci USA 1996; 93:15376-15381.

140. Yoshida M, Westlin WF, Wang N, Ingber DE, Rosenzweig A, Resnick N, Gimbrone MA Jr. Leukocyte adhesion to vascular endothelium induces E-selectin linkage to the actin cytoskeleton. J Cell Biol 1996; 133:445-455.

141. Frojmovic MM, Milton JG. Human platelet size, shape, and related functions in health and disease. Physiol Rev 1982; 62:185-261.

142. Shattil SJ, Ginsberg MH, Brugge JS. Adhesive signaling in platelets. Curr Opin Cell Biol 1994; 6:695-704.

143. Williams MJ, Du X, Loftus JC, Ginsberg MH. Platelet adhesion receptors. Sem Cell Biol 1995; 6:305-314.

144. Du X, Harris SJ, Tetaz TA, Ginsberg MH, Berndt MC. Association of a phospholipase A2 (14-3-3 protein) with the platelet glycoprotein Ib-IX complex. J Biol Chem 1994; 269:18287-18290.

145. Shattil SJ. Regulation of platelet anchorage and signaling by integrin $\alpha_{IIb}\beta_{IIIa}$. Thromb Haemostas 1993; 70:224-228.

146. van Santen GH, de Graaf S, Heijnen HFG, Connolly TM, de Groot PG, Sixma JJ. Increased platelet deposition on atherosclerotic coronary arteries. J Clin Invest 1994; 93:615-632.

147. Furman MI, Gardner TM, Goldschmidt-Clermont PJ. Mechanisms of cytoskeletal reorganization during platelet activation. Thromb Haemostas 1993; 70:229-232.

148. Vu TK, Hung DT, Wheaton VI, Coughlin SR. Molecular cloning of a functional thrombin receptor reveals a novel proteolytic mechanism of receptor activation. Cell 1991; 64:1057-1068.

149. Hung DT, Wong YH, Vu TKH, Coughlin SR. The cloned platelet thrombin receptor couples to at least two distinct effectors to stimulate phosphoinositide hydrolysis and inhibit adenylyl cyclase. J Biol Chem 1992; 267:20831-20834.

150. Gumbiner BM. Cell adhesion: the molecular basis of tissue architecture and morphogenesis. Cell 1996; 84:345-357.

151. Calvete JJ. Clues for understanding the structure and function of a prototypic human integrin: the platelet glycoprotein IIb/IIIa complex. Thromb Haemostas 1994; 72:1-15.

152. O'Toole TE, Mandelman D, Forsyth J, Shattil SJ, Plow EF, Ginsberg MH. Modulation of the affinity of integrin αIIbβ3 (GPIIb-IIIa) by the cytoplasmic domain of IIb. Science 1991; 254:845-857.

153. Shattil S, O'Toole T, Eigenthaler M, Thon V, Williams M, Babior BM, GinsbergMH. β_3-Endonexin, a novel polypeptide that interacts specifically with the cytoplasmic tail of the integrin β_3 subunit. J Cell Biol 1995; 131:807-816.

154. Morii N, Teru-uchi T, Tominaga T, Kumagai N, Kozaki S, Ushibubi F, Narimiya S. A rho gene product in human blood platelets. II. Effects of the ADP-ribosylation by botulinum C3 ADP-ribosyltransferase on platelet aggregation. J Biol Chem 1992; 267:20921-20926.

155. van Willigen G, Hers I, Gorter G, Akkerman, J-WN. Exposure of ligand-binding sites on platelet integrin α_{IIB}/β_3 subunit. Biochem J 1996; 314:769-779.

156. Kovacsovics TJ, Bachelot C, Toker A, Vlahos, CJ. Duckworth B, Cantley LC, Hartwig JH. Phosphoinositide 3-kinase inhibition spares actin assembly in activating platelets but reverses platelet aggregation. J Biol Chem 1995; 270: 11358-11366.

157. Chacko GW, Brand JT, Coggeshall KM, Anderson CL. Phosphoinositide 3-kinase and p72^{syk} noncovalently associated with the low affinity FC receptor on human platelets through an immunoreceptor tyrosine-based activation motif. Reconstitution with synthetic phosphopeptides. J Biol Chem 1996; 271:10775-10781.

158. Fox JEB. The platelet cytoskeleton. Thromb. Haemostas. 1993; 70: 884-893.

159. Fox JEB, Boyles JK, Berndt MC, Steffen PK, Anderson LK. Identification of a membrane skeleton in platelets. J Cell Biol 1988; 106:1525-1538.

160. Hartwig JH, DeSisto M. The cytoskeleton of the resting human blood platelet: structure of the membrane skeleton and its attachment to actin filaments. J Cell Biol 1991; 112:407-425.

161. Earnest JP, Santos GF, Zuerbig S, Fox JEB. Dystrophin-related protein in the platelet membrane skeleton. Integrin-induced change in detergent-insolubility and cleavage by calpain in aggregating platelets. J Biol Chem 1995; 270:27259-27265.

162. Hartwig JH. Mechanism of actin rearrangements mediating platelet activation. J Cell Biol 1992; 118:1421-1442.

163. Barkalow K, Witke W, Kwiatkowski DJ, Hartwig JH. Coordinated regulation of platelet actin filament barbed ends by gelsolin and capping protein. J Cell Biol 1996; 134:389-399.

164. Janmey PA, Stossel TP. Modulation of gelsolin function by phosphatidylinositol 4,5-bisphosphate. Nature 1987; 325: 362-364.

165. Wahl M, Carpenter G. Selective phospholipase C activation. BioEssays 1991; 13:107-113.

166. Shariff A, Luna EJ. Diacylglycerol-stimulated formation of actin nucleation sites at plasma membranes. Science 1992; 256: 245-247.

167. Banno Y, Nakashima S, Ohzawa M, Nozawa Y. Differential translocation of phospholipase C isozymes to integrin-mediated cytoskeletal complexes in thrombin-stimulated human platelets. J Biol Chem 1996; 271:14989-14994.

168. Fox JEB, Lipfert L, Clark EA, Reynolds CC, Austin CD, Brugge JS. On the role of the platelet membrane skeleton in mediating signal transduction. Association of GP IIb-IIIa, pp60^{c-src}, pp60^{c-yes}, and the p21^{ras} GTPase-activating protein with the membrane skeleton. J Biol Chem 1993; 268:25973-2984.

169. Wu H, Reynolds AB, Danner SB, Vines RR, Parsons JT. Identification and characterization of a novel cytoskeleton-associated pp60^{src} substrate. Mol Cell Biol 1991; 11:5113-5124.

170. Wu H, RR, Parsons JT. Cortactin, an 80/85-kilodalton pp60^{src} substrate, is a filamentous actin-binding protein enriched in the cell cortex. J Cell Biol 1993; 120: 1417-1426.

171. Fox JEB, Shattil SJ, Kinlough-Rathbone RL, Richardson M, Packham MA, Sanan DA. The platelet cytoskeleton stabilizes the interaction between $\alpha_{IIb}\beta_3$ and its ligand and induces selective movements of ligand-occupied integrin. J Biol Chem 1996; 271:7004-7011.

172. Kovacsovics TJ, Hartwig JH. Thrombin-induced GPIb-IX centralization on the platelet surface requires actin assembly and myosin II activation. Blood 1996; 87: 618-629.

173. Fox JEB, Phillips DR. Role of phosphorylation in mediating the association of myosin with the cytoskeletal structures of human platelets. J Biol Chem 1982; 257: 4120-4126.

174. Clark EA, Shattil SJ, Ginsberg MH, Bolen J, Brugge JS. Regulation of the protein tyrosine kinase pp72^{syk} by platelet agonists and the integrin $\alpha_{IIb}\beta_3$. J Biol Chem 1994; 269:28859-28864.

175. Cichowski K, Brugge JS, Brass LF. Thrombin receptor activation and integrin engagement stimulate tyrosine phosphorylation of the proto-oncogene product, p95^{vav}, in platelets. J Biol Chem 1996; 271:7544-7550.

176. Tohyama Y, Yanagi S, Sada K, Yamamura H. Translocation of p72^{syk} to the cytoskeleton in thrombin-stimulated platelets. J Biol Chem 1994; 269:32796-32799.

177. Dash D, Aepfelbacher M, Siess W. Integrin α_{IIb}-β_3-mediated translocation of CDC42Hs to the cytoskeleton in stimulated human platelets. J Biol Chem 1995; 270:17321-17326.

178. Law DA, Nannizzi-Alaimo L, Phillips DR. Outside-in integrin signal transduction. $\alpha_{IIb}\beta_3$-(GPIIb-IIIa) tyrosine phosphorylation induced by platelet aggregation. J Biol Chem 1996; 271:10811-10815.

179. Frangioni JV, Oda A, Smith M, Salzman EW, Neel BG. Calpain-catalyzed cleavage and subcellular relocation of protein phosphotyrosine phosphatase 1B (PTP-1B) in human platelets. EMBO J 1993; 12:4843-4856.

180. Haimovich B, Regan C, DiFazio L, Ginalis E, Ji P, Purohit U, Rowley RB, Bolen J, Greco R. The FcγRII receptor triggers pp125^{FAK} phosphorylation in platelets. J Biol Chem 1996; 271:16332-16337.

CHAPTER 5

Cell Cycle and Cell Division

Cell Cycle

Cell division is the most complex process undertaken by a single cell and must be stringently regulated to maintain fidelity of reproduction. Proliferative cells are in a continuous loop or cycle composed of four stages: S, during which DNA synthesis occurs, M (mitosis), G1 and G2. The gap phases G1 and G2 are growth and regulatory periods which are required to assure fidelity of the synthetic and division processes, as illustrated in Figure 5.1 (1). G1 is required for most cell types to complete cell growth. A critical checkpoint in G1, called *START* in yeast or the *restriction point* in mammalian cell, is the point in the cycle at which the cell commits to DNA replication (2). Both positive and negative signals operate at this point to determine whether the cell continues through the cycle (3). G2 is necessary to assure that DNA synthesis is complete to initiate mitosis and can be extended if DNA replication is incomplete or DNA has been damaged. In the rapidly dividing cells of early embryogenesis following fertilization, the cell cycle and G1 are short because no cell growth is required (4). Much of the research on the cell cycle has been done in unicellular eukaryotic organisms, particularly yeast. Control of the cell cycle in the unicellular organism is relatively simple because their cellular requirements are simple. They respond primarily to cues concerning nutrition and proliferation. In contrast, cells of multicellular organisms must be regulated by the overall needs of the organism. Thus, many cell types are required to be in a quiescent state (G0) most of their life. Other cell types are transiently proliferative and undergo cycles of cell growth and cell death according to the needs of the organism. Multicellular organisms have therefore evolved a highly complex set of mechanisms for regulating the cell cycle, which must then be integrated with all of the processes involved in cell growth and division.

Cyclins and Cyclin-Dependent Kinases (CDKs)

In eukaryotes the cell cycle is regulated by the activities of serine/threonine protein kinase complexes, *cyclin dependent kinases (cdk)*. In yeast, a single kinase, cdk1, called cdc2 kinase in fission yeast and cdc28 kinase in budding yeast, serves as the master control element of the cell cycle. It plays a regulatory role at both the G1/S and G2/M transition points and is activated by forming complexes with *cyclins* (5), cdk binding proteins expressed at different stages of the cell cycle. There are two primary classes of cyclins, G1 and G2/M, based on their synthesis and degradation at different points of the cell cycle. In general, G1 cyclins have short half-lives and are regulated by transcription, while G2/M cyclins are regulated by both transcription and rapid degradation at mitosis (6). Passage through and out of the

Signaling and the Cytoskeleton, by Kermit L. Carraway, Coralie A. Carothers Carraway and Kermit L. Carraway III. © 1998 Springer-Verlag and R.G. Landes Company.

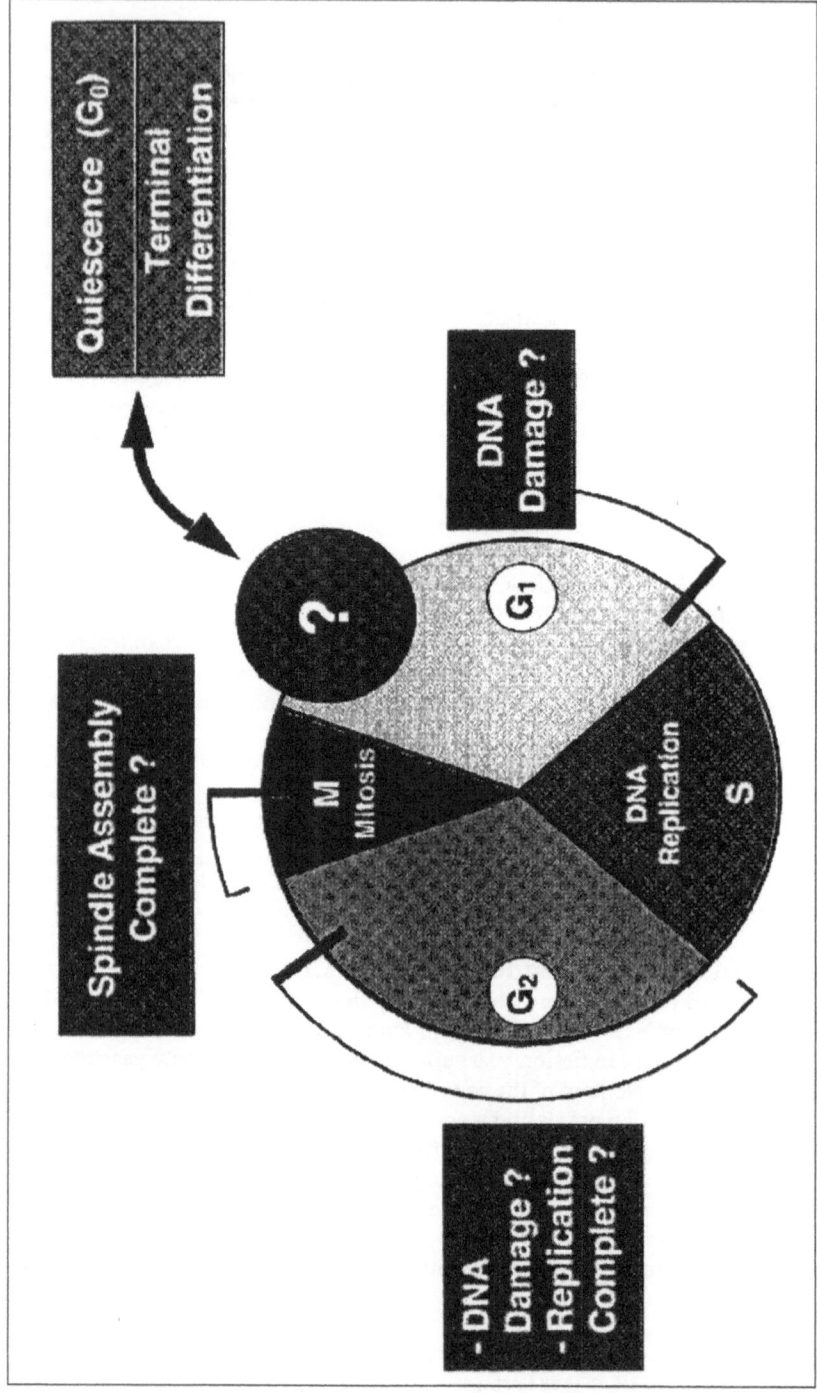

Fig. 5.1. Schematic view of the cell cycle. Figure from ref. 1 (Nigg EA. BioEssays 1995; 17: 471-480), courtesy of ICSU Press.

G1 phase is controlled by cdk association with G1 cyclins. For example, the G1/S transition in budding yeast is controlled by G1 cyclins CLN1, CLN2 and CLN3 in association with cdc28. Cdk complexes with G2/M cyclins control passage into and through mitosis.

Cdk-cyclin complex formation and degradation provide only a part of the cdk regulation story. The activities are also regulated by phosphorylation and dephosphorylation. Both activating (MO15, cdk7) and inactivating (Wee1) kinases have been described; a primary regulating event is the reversal of Wee1 inactivation of cdk/cyclin kinases by the phosphatase *cdc25* (6), as indicated in Figure 5.2 (7). Another level of regulation of the activities of cdk/cyclin complexes is provided by inhibitors which repress kinases at specific stages and thus block cell cycle progression (8). For example, mating pheromones in budding yeast can specifically block passage through G1 prior to START by activating an inhibitor (Far1) via a pheromone receptor-G protein-MAPK pathway (6).

Because of the complexity of extracellular requirements on the cells of multicellular organisms, their cell cycle control system is more complex than in yeast (5). Control mechanisms in the G1 phase of the cycle specify cell fate into one of four pathways: *proliferation, quiescence (Go), differentiation or apoptosis*. These decisions occur in response to a variety of extracellular agents, including growth factors and the extracellular matrix, but only when the cells are in the G1 phase of their cycle prior to the restriction point. Control of these decisions and progression through the rest of the cell cycle depends on the cyclins and cdks of these cell types. In higher eukaryotes, different cyclins (A-H, see Table 5.1) (1) and cyclin-dependent kinases (cdk 1-7, see Table 5.2) (1) have been identified and partially characterized (1). Assuming combinatorial interactions, these could create an enormous number of kinase/cyclin complexes; however, only a few of those are likely to form and be functional. At least five different types of parameters are important for controlling the cell cycle via the kinases: expression level of cyclin, expression level of kinase, interactions between cyclins and kinases, kinase inhibitors and posttranslational modifications of kinases. The key to understanding cell division is the relationship between these kinases and the processes involved in the cell cycle (9).

As in the case of yeast, vertebrate cyclins can be classified as G1 or G2/M. A simplified version of the involvement of specific cyclin/cdk complexes in the vertebrate cell cycle is shown in Figure 5.3 (1). Cyclin B/cdk1 regulates entry into mitosis. Cdk2 is required for the G1-S transition and for progression through S phase as complexes with cyclin E and cyclin A, respectively. D-type cyclins, as complexes with cdk4/6 are regulators of G1 progression. Examples of other complexes of these kinases and cyclins have been observed and the roles of additional cyclins and cdks and their complexes remain to be defined. Undoubtedly, some of the latter will prove to be cell specific while others provide functional redundancies. In considering the relationship of cell cycle control, signaling and the cytoskeleton, the progression through G1 and the progression into and through mitosis are of primary interest and will be the focus of the following discussion. In each case two questions arise. What are the regulators of the cdk activities important for driving their functions? What are the targets that contribute to signaling and cytoskeletal changes in the cells?

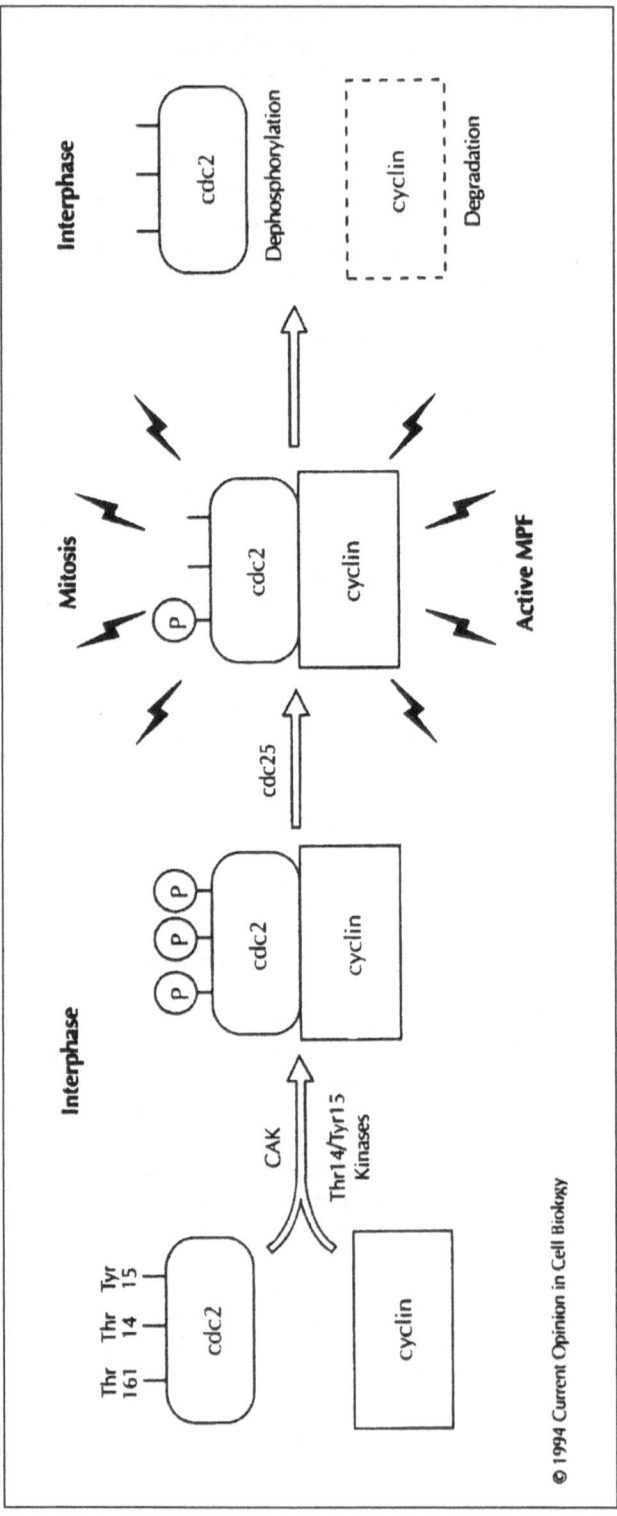

Fig. 5.2. Model for regulation of cdkα(cdc2)-cyclin B complex (MPF) by phosphorylation and dephosphorylation to control entry into mitosis. Figure from ref. 7 (Coleman TR, Dunphy WG. Curr Opin Cell Biol 1994; 6: 877-882), courtesy of Current Biology Ltd.

Table 5.1. Cyclins in vertebrates

Name	Function	cdk partner(s)	Expression	Localization
A	S and G2/M	cdc2/cdk2	Peak at G2/M	Nuclear
B1	G2/M	cdc2	Peak at G2/M	Cytoplasmic
B2	G2/M	cdc2	Peak at G2/M	Cytoplasmic
B3	G2/M (and S ?)	cdc2/cdk2	Peak at G2/M	Nuclear
C	?	?	?	?
D1	G1	cdk4/6	Predominantly G1	Nuclear
D2	G1	?	?	Nuclear (?)
D3	G1	cdk4/6	Constant	Nuclear
E	G1/S	cdk2	Peak at G1/S	Nuclear
F	?	?	Peak at G2/M	Nuclear/ Cytoplasmic
G	?	?	?	?
H	Multiple phases	cdk7	?	?

Modified from ref. 1 (Nigg EA. BioEssays 1995; 17: 471-480), courtesy of ICSU Press.

Table 5.2. cdk subunits in vertebrates and in yeasts

Name	Regulatory subunit(s)	Proposed function
Vertebrates		
cdc2 (= cdk1)	Cyclins A, B1, B2, B3	G2/M transition
cdk2	Cyclins A, E, D1, (D2, D3)	G1/S transition and S phase
cdk3	?	G1/S transition
cdk4	Cyclins D1 (D2 ?), D3	G0/G1 (and G1/S ?) transitions
cdk5	p35, Cyclins D1, (D2 ?), D3	Neurofilament phosphorylation
cdk6	Cyclins D1, (D2, D3 ?)	G0/G1 (and G1/S ?) transitions
cdk7	Cyclin H, p36[c]	Required to activate cdk1-6
Yeasts		
cdc28 (*S. cerevisiae*)	CLN1-3, CLB1-6	Multiple transitions
PHO85 (*S. cerevisiae*)	PHO80, PCL 1 (HCS26), PCL 2 (ORFD)	Nutritional control of G1 (?)[d]
cdc2 (*S. pombe*)	cdc13, cig 1,2, (puc1, cyc 17 ?)	Multiple transitions

Modified from ref. 1 (Nigg EA. BioEssays 1995; 17: 471-480), courtesy of ICSU Press.

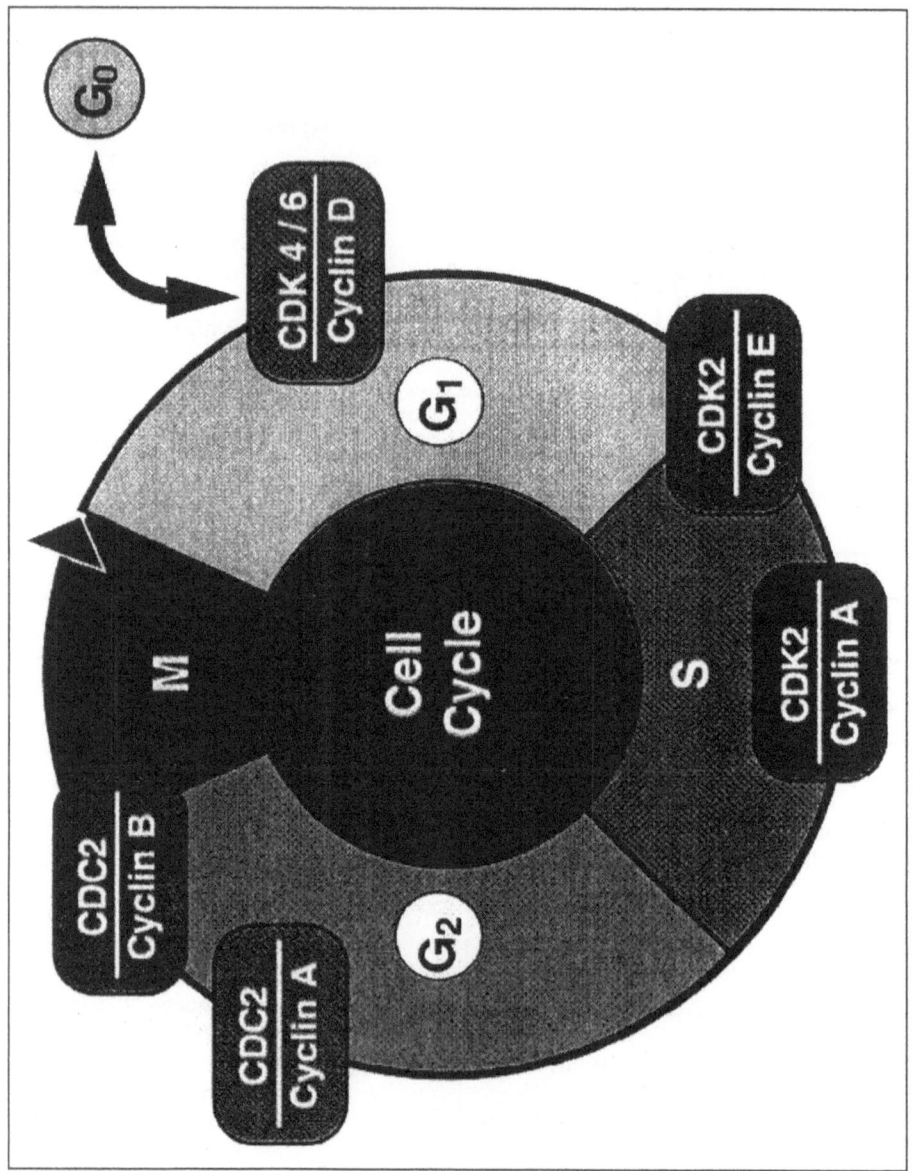

Fig. 5.3. Schematic representation of involvement of cyclin-cdk complexes at various stages of mammalian cell cycle. Figure from ref.1 (Nigg EA. BioEssays 1995; 17: 471-480), courtesy of ICSU Press.

G1 Progression

Signal Integration in Yeast

One of the key aspects of the cell cycle is the response to extracellular cues. In budding yeast, as noted above, mating pheromones can block progression through the cell cycle in G1. A critical question in cell cycle control is how the cell integrates various events which must be regulated at specific phases of the cycle. The mating response is coupled to the cell polarization that permits cell mating (see chapter 3). The integration of these events is regulated by the small G protein cdc42 (Fig. 5.4) (10), which has been previously noted as a regulator of filopodial protrusion in cultured animal cells (chapters 1 and 4). In budding yeast, cdc42 is required for bud formation, accumulating at the incipient bud site and localizing there as the bud grows (11). It regulates this morphogenetic event through its associations with cdc24, a GEF, and Bem1, an actin binding protein. These relationships are described more completely in chapter 3 on cell polarity. The pheromone signal for cell cycle arrest is also dependent on cdc42, which is coupled through the kinase Ste20, an analog of mammalian PAK65, to the yeast MAPK cascade (Fig. 5.4) (10). Fus3, the final component of this cascade, phosphorylates and activates Far1, the inhibitor of the cdc28/cyclin kinase complex to block cell cycle progression. Meanwhile, the yeast cell grows in a polarized fashion following the pheromone gradient released by the cells of the opposite mating type. *The key to the integration of the cell cycle block and polarization appears to be the formation of a multimeric complex containing elements of both pathways.* The scaffolding component in the complex is Ste5, which apparently can interact with multiple components of the MAPK cascade, cdc42, Bem1 and Ste4/Ste18, two components of the heterotrimeric G protein linked to the pheromone receptor (see chapter 3, section on yeast polarization) (10). Thus, integration is achieved by localizing both pathways in a single signaling particle with a common intermediate regulatory step. Interestingly, a similar microfilament-linked, multimeric complex containing many of the components implicated in mitogenesis and regulation of the cytoskeleton, including all of the proteins of the MAPK cascade, Ras, ErbB-2 receptor tyrosine kinase, PI3K, PKC, Src and Abl has been isolated from microvilli of an ascites tumor cell (see chapter 3), suggesting that this type of mechanism for integration may be common.

Extracellular Signals in Animal Cells

To integrate functions in multicellular organisms, individual cells must often be able to respond to extracellular cues provided by other cells, including proliferation signals. G1 is noteworthy as the phase of the cell cycle at which mitogens act to stimulate cell proliferation. In the absence of growth stimulation, most cells can withdraw from the cell cycle into G0 or undergo apoptosis. Their *transition from G0 to G1 and progression through the cycle requires not only mitogens, but also an appropriate cellular environment including cell adhesions and an intact cytoskeleton* (12). Likewise, cell adhesion plays a role in determining whether some cells undergo apoptosis (13). In each case these factors must act before the cycle reaches the restriction point late in G1. Once a cell passes that point, it is committed to continue through cell division unless artificially interrupted. As noted above, two different cyclins are involved in progression through and out of G1: D-type and E. Mitogen-stimulated transitions from G0 to G1 activate synthesis of both D-type cyclins and cyclin E. However, the expression of E and D-type cyclins are

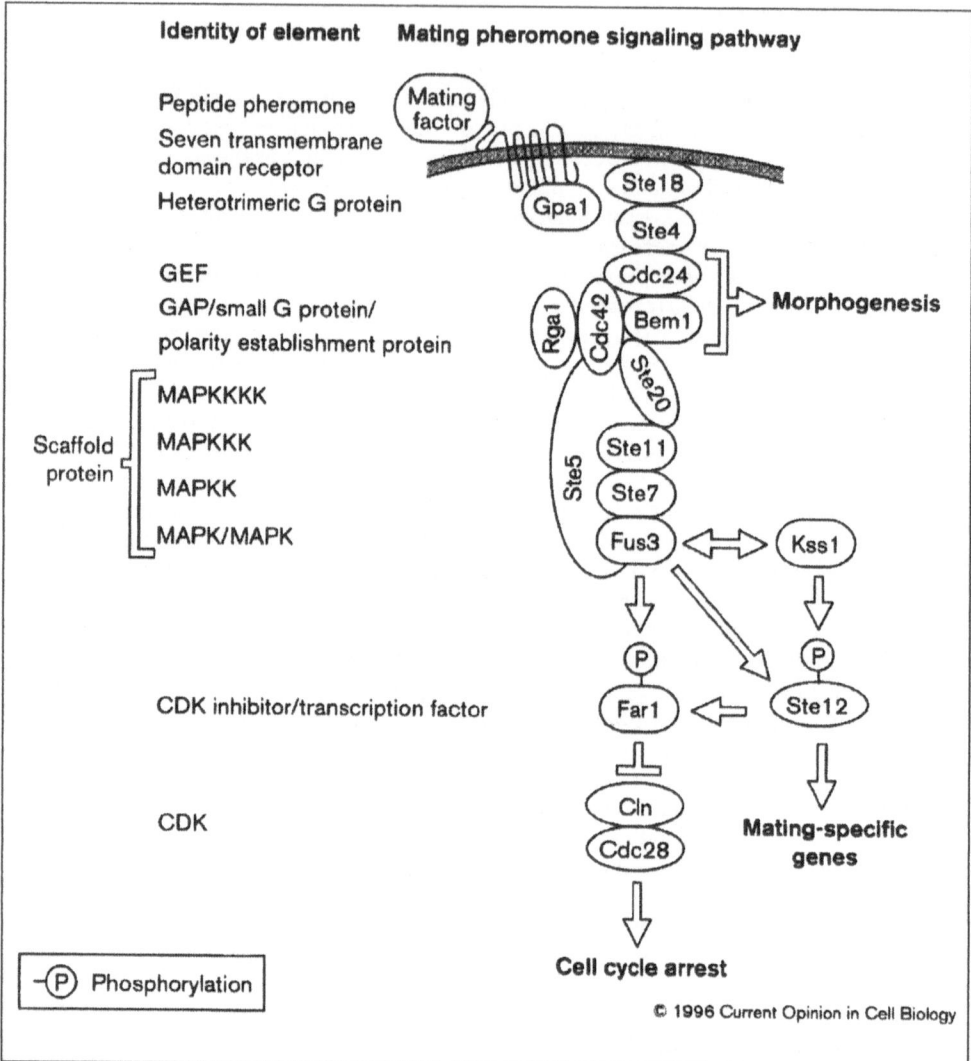

Fig. 5.4. Schematic representation showing integration of mating and cell cycle arrest pathways. Figure from ref. 10 (Wittenberg C, Reed SI. Curr Opin Cell Biol 1996; 8: 223-230), courtesy of Current Biology Ltd.

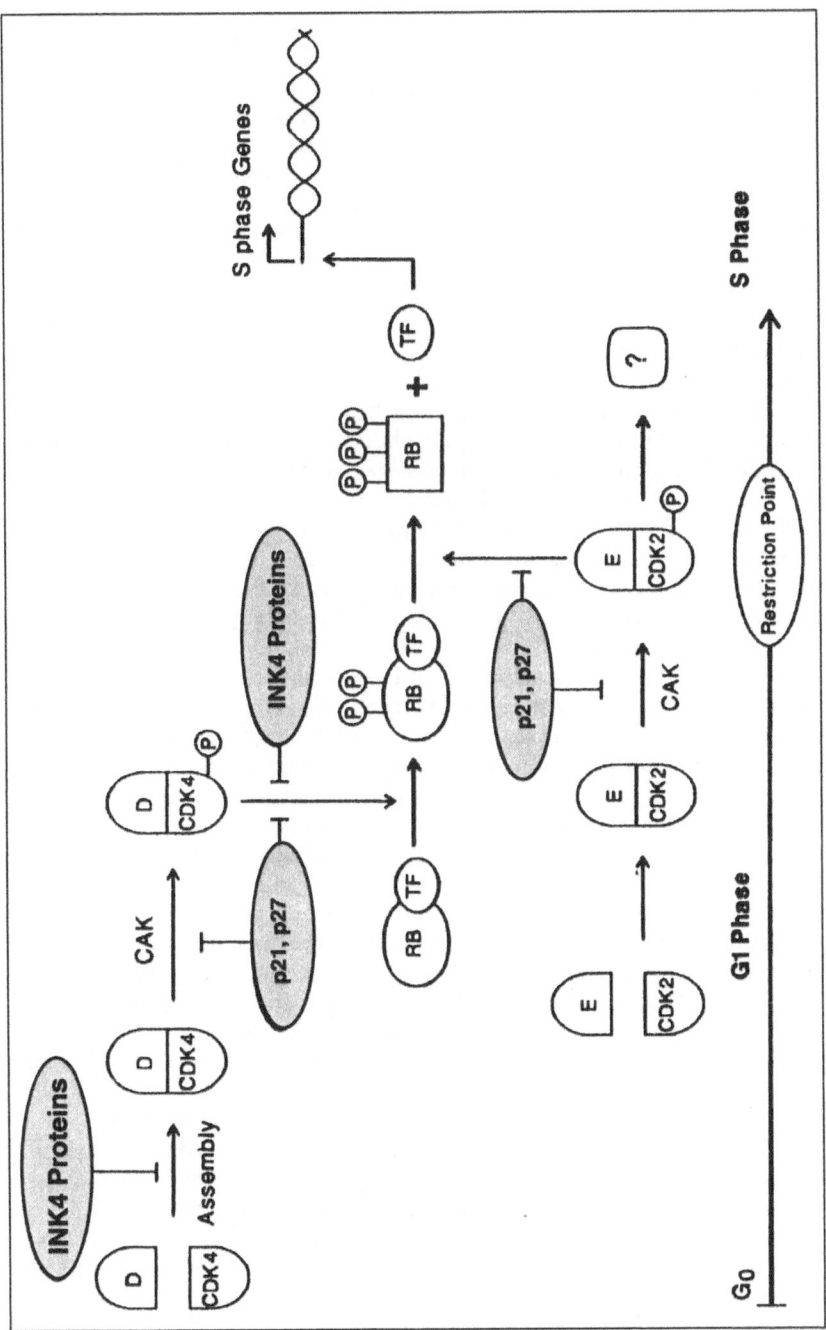

Fig. 5.5. Schematic representation of positive and negative regulators of G1 progression. Figure from ref. 8 (Sherr CJ, Roberts JM. Genes Devel 1995; 9: 1149-1163), courtesy of Cold Spring Harbor Laboratory Press.

regulated differently. E cyclin undergoes a cycle dependent regulation of synthesis and degradation, also observed for cyclins A and B. In contrast, levels of the D-type cyclins are regulated primarily by the extracellular signals (14). They are synthesized in response to mitogen stimulation and rapidly degraded with withdrawal of the mitogens (5). D-type cyclins act through formation of complexes with cdk4 or cdk6, but other cdk complexes with D cyclins have been observed. The primary cdk partner for cyclin E is cdk2. As previously noted, cyclin E/cdk2 is important for passage through the G1/S boundary, and it is degraded during S phase (14).

Regulation of Cyclin Dependent Kinases

The activities of G1 cdks are controlled by cyclin expression, phosphorylation and inhibitors, as illustrated in Figure 5.5 (8). Active complexes must first be assembled from cyclins and kinases, then activated by cyclin activating kinase (CAK), a cdk7/cyclin H complex. Assembly can be blocked by the INK4 family of proteins which can also act as kinase inhibitors. Phosphorylation by CAK and the activities of the cdk4/cyclin D and cdk2/cyclin E complexes can be inhibited by p21 and p27. The INK proteins p15, p16 and p18 contain four ankyrin-like repeats and compete with cyclin D for binding to cdk4 and cdk6 (3). TGFβ can induce overexpression of p15, implicating it in the growth inhibitory response of this growth factor. The p21 inhibitor binds directly to cdk1, cdk2 and cdk4 in vitro and can be observed in their complexes with cyclins A, B, D and E (8). Surprisingly, in nontransformed cells these complexes exhibit cdk activity. The fact that they can be inhibited by addition of further p21 suggests that the ratio of p21 to other components in the complex is important rather than just the presence of p21 in the complex. Some of these multimeric complexes also contain PCNA, a subunit of DNA polymerase δγ, suggesting that p21 may play a direct role in regulating the initiation of DNA synthesis (3). Expression of p21 is regulated in part by the tumor suppressor p53, a transcription factor activated in response to DNA damage which can halt G1 progression and promote apoptosis, depending on the cellular context. The inhibitor p27 is related to p21 and blocks activation of cdk2 and cdk4 complexes (3). It was identified as the factor responsible for TGFβ and cell-cell contact inhibition of cell growth (15). Surprisingly, p27 gene expression appears to be independent of TGFβ and to be regulated post-translationally by release from a sequestered state (16). The TGFβ growth inhibitor effects could result from p27 and/or INK proteins binding to cdk/cyclin complexes. p27 has also been implicated in the growth arrest of some cell types by cAMP since its expression can be induced by the nucleotide (17).

The discovery of the G1 inhibitors has provided insights into the mechanisms by which mitogens and cell adhesion regulate G1 progression. Two different mechanisms are involved in this regulation: stimulation of cyclin D expression and downregulation of the expression of cdk inhibitors. Both of these effects depend jointly on the presence of mitogens and adhesion to substratum, including cytoskeleton integrity (12). Adhesion regulates both transcription and translation of cyclin D1 (18). Adhesion also appears to be required for the degradation of the inhibitor p27 (19) and for repression of the expression of p21 (18, 20). These results indicate that there is a redundancy in the factors involved in control of G1 progression. One of the important elements of cell proliferation, particularly for fibroblasts, is anchorage dependence, whose loss is closely correlated with tumorigenicity (18). This redundancy in the control of G1 progression may explain the variations in the ability of different external agents to induce anchorage independence in different cell lines which may have lost some of their G1 control mechanisms during immortalization and cell passage (18).

Adhesion, the Cytoskeleton and G1 Progression

The mechanisms by which adhesion and the cytoskeleton contribute to cell cycle progression are still unclear. In addition to their effects on cell proliferation, many mitogens also stimulate changes in cell shape, cytoskeleton organization and cell motility. For example, the serum lipid mitogen LPA can trigger changes in both gene expression and the cytoskeleton through its effects on G-protein linked receptors (21). One of the earliest effects of growth factors is on cell shape and cell surface activity, particularly ruffling (22). Both of these types of agents causes changes in cell adhesion through focal contacts and microfilament reorganization (23). Previous studies, described in chapter 4, have shown a hierarchy of focal contact changes mediated through integrins (24). Studies on hepatocyte adhesion and spreading suggested differential regulation of effects on specific gene expression and progression to S phase, respectively (25). Cytochalasin can also block progression to S phase, indicating a requirement for microfilament integrity (26). These results suggest that the cell cycle effects not only require adhesion and recruitment of focal contact and signaling components, but also stabilization of the adhesion sites by the recruitment and organization of microfilaments (24). Consistent with this hypothesis is the observation that cytochalasin D inhibits DNA synthesis more readily than the microtubule-perturbing agent nocodozole, but that nocodozole acts synergistically with suboptimal cytochalasin to inhibit both cell spreading and DNA synthesis (27). Similarly, expression of tropomyosin 1 (28), a microfilament stabilizing protein associated with stress fibers, and vinculin (29, 30), an actin-binding protein of focal adhesions, have been implicated in anchorage-dependent growth.

Members of the Rho family of small G proteins participate in the control of cell adhesion and microfilament organization. Microinjection studies have shown that all three, Rho, Rac and cdc42 are also important in the progression through G1 (31). Unlike Ras which activates MAPK (ERK) through Raf, the Rho family members stimulate the stress-activated MAPK (JNK/SAPK) pathways (Fig. 2.7). Site-directed mutants of Rac and cdc42 were used to investigate the downstream targets involved in G1 progression and cytoskeletal changes. Specific mutants of Rac and cdc42 did not bind or activate PAK and were unable to activate JNK MAPK (32, 33). However, they still induced cytoskeletal changes and G1 progression, indicating that PAK and JNK are not involved in these phenomena. In contrast, a mutant which blocked Rac interaction with ROK failed to induce either lamellipodia or G1 progression. As noted in chapter 4, ROK has been implicated in focal adhesion formation and stress fiber assembly (34). Thus, ROK may be on the pathway leading to the cytoskeletal changes. The parallel between the cytoskeletal changes and G1 progression is also consistent with a cytoskeletal involvement in G1 progression. Recent studies (described in chapter 4) have demonstrated a Ser/Thr kinase ILK which binds integrins and is implicated in anchorage-independent, but not serum-independent growth, of epithelial cells (35). ILK overexpression in intestinal epithelial cells results in an increase of cyclin D expression and activation of cdk4 and cyclin E-associated kinases, suggesting a role in the pathway for the regulation of the cycle by adhesion.

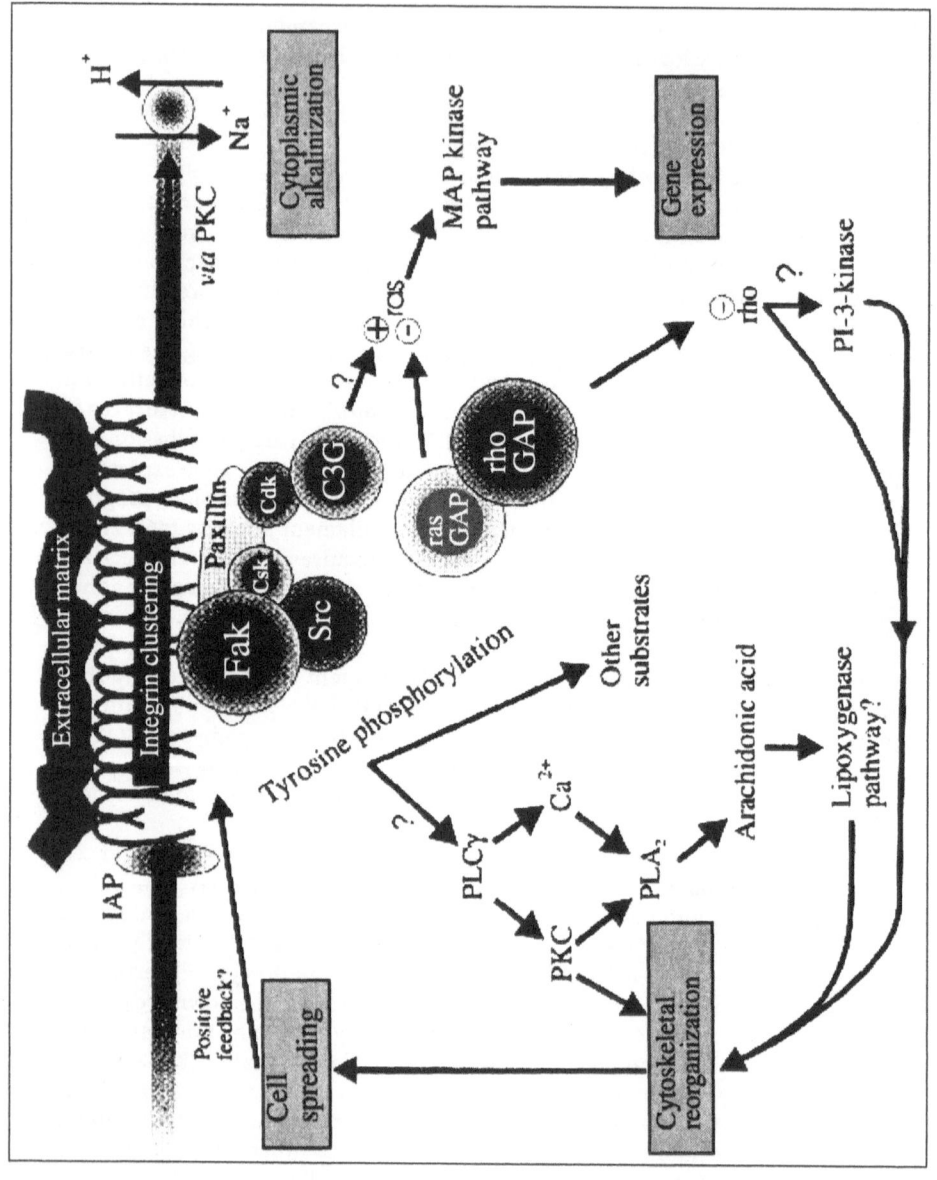

Fig. 5.6. Model for integrin-mediated signal transduction showing potential integration of gene expression and cytoskeleton-modulating pathways. Figure from ref. 37 (Richardson A, Parsons JT. BioEssays 1995; 17: 229-236), courtesy of ICSU Press.

The results above indicate that passage through G1 and its control via D-type cyclins and their inhibitors require signals from both integrin- and mitogen-mediated pathways (35). Thus, an important question is how these pathways are related (36). Integrin dependent signaling for both microfilament organization and proliferation is proposed to proceed via recruitment and phosphorylation of FAK (Fig. 5.6) (37, 38), as described in chapter 4. Integrin β 1 cytoplasmic domain, which links to FAK, has been demonstrated to stimulate cell proliferation (39). How FAK activation is linked to the transcriptional events involved in regulation of gene expression is not clear, but activation of a MAPK cascade has been implicated (40, 41). However, mitogen stimulation also proceeds through activation of MAPK, raising the question of why both events should be needed if they are mediated through a common pathway. A direct comparison of EGF and integrin activation of MAPK in 3T3 cells provides some insights (42). EGF and integrin stimulate strong tyrosine phosphorylation of the EGF receptor and FAK, respectively, but not reciprocally. Moreover, patterns of other tyrosine-phosphorylated proteins are different. Significantly, Ras may not always be required for MAPK stimulation via integrin. Since Ras is coupled to MAPK (Erk) activation via Raf recruitment to the plasma membrane and cytoskeleton (43), one explanation for the differences between integrin and mitogen effects is that integrin-stimulated FAK or a downstream element in its pathway recruits Raf or an analogous kinase to a different site from that of Raf of the mitogenic pathway. Reminiscent of the integration of the budding and G1 control in yeast, this integrin-stimulated Raf (or relative) could even be associated with the same signal transduction complex as the mitogen-activated Raf/Erk, but associated differently. Again, the paradigm is that integration of signaling pathways may be achieved by multifunctional, multimeric signaling complexes organized at membrane-cytoskeleton sites.

A second important difference between integrin and growth factor signaling is that integrins can activate alternative pathways to the nucleus: 1) the stress-activated pathway involving JNK-type MAPK; 2) the Src/NF-κB pathway (44, 45); and 3) a Shc- and mitogen-dependent, FAK-independent stimulation of G1 progression (46). Thus, combinatorial effects of these pathways may determine the expression levels of the D-type cyclins and the cdk/cyclin D inhibitors, providing regulatory mechanisms which control differentiation, proliferation and apoptosis. However, it is important to remember that the same external agent in different cell types may act through alternative pathways, and different external agents which induce similar phenotypic changes may act through different pathways. Several alternative mechanisms have already been described. FAK can be activated by a number of different factors, including some mitogens and oncogenes, as well as integrin (47). LPA can activate both the Ras and FAK pathways. FAK autophosphorylation creates binding sites for Grb2, PI3K and Src kinases (37, 48), all of which have been implicated in proliferation pathways. Insulin stimulates DNA synthesis, cell shape changes and microfilament reorganization through receptor-mediated phosphorylation of IRS-1. Adhesion of insulin-stimulated cells to a vitronectin substratum stimulates MAPK and increased proliferation via association of αvβ3 with IRS-1, which can also bind to Grb2 when phosphorylated (49), linking integrins to MAPK through a FAK-independent, Ras-dependent route. Moreover, insulin stimulation leads to dephosphorylation instead of phosphorylation of FAK and paxillin (50). The dephosphorylation is catalyzed by PTP1D through a pathway that is Ras independent. Thus, there appear to be many alternatives for regulating cell cycle progression, mitogenesis, adhesion and the cytoskeleton which remain to be explored. What is still unclear is how the combined information from the adhesion and mitogenic pathways is transduced into signals for the transcriptional machinery of the nucleus.

Two possibilities for the passage of information to the nucleus have been advanced. The *first* and more widely known uses a biochemical pathway in which kinases such as MAPK or Rsk migrate into the nucleus after activation. The tyrosine kinase Abl may be particularly important in transmitting signals through integrin (51). c-Abl has both a nuclear localization signal and microfilament-binding domain and is found in both the nucleus and cytoplasm (52). Its nuclear kinase activity is regulated during cell cycle progression, kept in an inactive state by association with Rb in an Rb-E2F complex (53). In contrast, cytoplasmic Abl is not regulated during cell cycle progression. Adhesion of fibroblasts to the integrin ligand fibronectin induces transient recruitment of a subset of the Abl to focal adhesions, followed by export to the nucleus (51). Concomitantly, cytoplasmic and nuclear Abl are activated, suggesting that the active nuclear Abl comes from the cytoplasm. These results provide a model in which c-Abl is a component of an integrin-mediated pathway regulating gene expression in which cell adhesion modulates both Abl localization and activity.

The *second* pathway for transmission of information from the cell surface to the nucleus involves mechanical transmission of the signal and may be particularly important for adhesion-generated signals. Sites of adhesion between the cell membrane and the extracellular matrix are known to be sensitive areas for the transduction of mechanical stimuli to the cytoplasm and nucleus, a fact demonstrated particularly well by the requirement of many cells in culture that they be attached to a substratum before they can divide. Investigations of mechanotransduction at the cell surface have led to the development of two types of mechanisms. In the first type, the mechanical stimulus acts upon the cell surface to activate/inactivate signaling components (mechanotransducers) in the plasma membrane (54), which in turn generate cytoplasmic signals via biochemical second messengers and other pathways described in chapter 2. One of the first mechanotransducers to be described was the stretch-activated and -inactivated ion channel (55). The alternative mechanism involves a direct connection from the cell surface to the nucleus through the cytoskeleton. Ingber (56) has proposed a model for explaining cytoskeletal involvement in cellular mechanical changes called the cellular tensegrity model. In this model, cytoskeletal elements interconnect through the plasma membrane to the extracellular matrix (57) to provide a continuous deformable, cytoplasmic network responsive to mechanical forces at the cell surface. Moreover, the nuclear matrix can also be constructed according to similar principles and connected to the cytoplasmic network at the nuclear membrane (58). The cytoskeletal elements involved in transmitting force to the nucleus were investigated using integrin-binding microbeads or micropipettes which could be micromanipulated at the cell surface (59). Microfilaments were implicated at low strain levels, while intermediate filaments mediated the force transfer at both low and higher strain levels. Mechanical perturbations of the cell surface can thus be transmitted to the nucleus and can potentially alter chromosome structure and gene transcription.

Apoptosis

The regulation of apoptosis is equally complex and cell-type dependent. In some cell types, apoptosis appears to be a default pathway which must be overridden by survival factors (60). For example, neuron survival requires the presence of nerve growth factor whose absence leads to apoptosis (61). Investigations on PC12 pheochromocytoma cells have shown that apoptosis triggered by NGF withdrawal involves activation of the JNK and p38 MAPK isoforms and inhibition of Erk, suggesting a need for a balance between the stress-activated and growth factor-activated MAPKs for apoptosis (62). In other cells, a triggering mechanism is needed. Cytotoxic T cell killing proceeds via induced apoptosis, in which the trigger is ligand binding by the cell surface receptor Fas, a member of the tumor necrosis factor receptor family (see chapter 2) (63). In epithelial and endothelial cells apoptosis is promoted by the loss of adhesion to the basement membrane, a process sometimes called anoikis (13). This process can be suppressed by integrin expression, apparently by a mechanism involving FAK phosphorylation that does not activate MAPK or proliferation (64). Myc is a proto-oncogene and transcription factor (see chapter 2) whose over-expression triggered by mitogens can produce uncontrolled cell proliferation. However, in the absence of growth stimulus, Myc over-expression can lead to apoptosis (60). Recently, Myc has been implicated in the expression of cdc25 (65), the cdk phosphatase, and in cdk activation (66). Myc also appears to act on an inhibitor of cdk2/cyclin E (66). Thus, Myc may be able to act at multiple levels of G1 as part of the regulation of the cell cycle control process and of apoptosis. Finally, microtubule perturbation by taxol can trigger apoptosis via a pathway leading to Raf-1 activation (67). How the cytoskeletal effect is linked to the signal pathway remains to be investigated.

Role of Rb

One of the most important questions in cell cycle control is how the cdk/cyclin complexes regulate transcriptional events linked to passage through transition from G1 to S phase. A key substrate for these kinase complexes is the retinoblastoma protein pRb (2) (Fig. 5.5) (8), a tumor suppressor whose hyperphosphorylation correlates with passage through the restriction point (8, 12, 68). pRb is phosphorylated by Cdk/cyclin D complexes, which bind pRb through an N-terminal cyclin D motif (3). Cyclin D transcription is stimulated by hypophosphorylated pRb, providing a feedback control mechanism over progression through G1. The pRb and cyclin D connection also appears to be important in the G0 to G1 transition. The ability of pRb to repress cell proliferation results from the association of hypophosphorylated pRb with the transcription factor E2F to inhibit E2F-mediated transactivation (69). A number of cellular promoters which control genes involved in growth control have E2F sites. These growth regulating components include c-myc, DNA polymerase-α, thymidine kinase, thymidylate synthase, cdk1 and cyclin A. This last observation provides evidence for a pathway connecting cdk and cyclin A via pRb and E2F. Since cdk activity can be regulated by cell adhesion, this pathway may provide an explanation for the adhesion dependence of cyclin A expression (70) for the next stage of the cell cycle, progression through S phase to mitosis.

Mitosis

Regulation of CDK1

Historically, control of entry into mitosis was the first cell cycle event to be studied at the molecular level (4). The recognition of the periodic expression of cyclin B and its formation of a complex with cdk1 to form MPF (M-phase or maturation-promoting factor) provided the initial insights into the role of cyclins and their modulators in the regulation of cell division (71). *Activation of MPF, the cdk1/cyclin B complex, is the key to G2/M passage* and is dependent on the cyclic expression of cyclin B. However, the accumulation of cyclin is gradual compared to the abrupt onset of mitosis. The explanation is that MPF activity, as mentioned previously, is tightly regulated by phosphorylation/dephosphorylation events (Fig. 5.2) (7). Although the formation of the cyclin/kinase complex is necessary for kinase activation, it is not sufficient to achieve full activity (72). Further activation is achieved by phosphorylation of cdk1 by a second cdk, called CAK, which is cdk7 in complex with cyclin H. However, CAK activity is also not tightly regulated by the cell cycle. Instead, cycle-dependent control of the cdk1/cyclin B activity is attained by negative modulation, phosphorylation at two different sites by the inactivating dual function kinase Wee1. Wee1 can be inhibited by phosphorylation by another kinase Nim1, whose responsiveness to nutritional conditions may provide a link between external conditions and mitosis (72). A more important determinant of cdk1 activity is dephosphorylation of the tyrosine phosphate introduced by Wee1. This phosphate can be specifically removed by the tyrosine phosphatase cdc25, whose activity fluctuates dramatically with the mitotic cycle. Cdc25 activity is also regulated by phosphorylation/dephosphorylation through a positive feedback loop with cdk1 and other cellular protein phosphatases. One of the kinases involved in its regulation is a member of the polo kinase family, a conserved family of Ser/Thr kinases found in most organisms (73). The message from these considerations is that regulation of cdk1/cyclin B kinase is multifactorial and highly complex. The necessity for such complexity probably derives from two sources. First, the fidelity of the steps before entrance into mitosis is essential to cell survival. Second, as demonstrated below, cdk1 (cdc2) is a critical determinant in many of the processes in mitosis and cell division.

Mechanics of Mitosis

Before embarking on an analysis of the molecular controls, it is necessary to provide a brief description of the mechanics of mitosis. Mitotic events are generally divided into five phases, which lead into cytokinesis: prophase, prometaphase, metaphase, anaphase and telophase (74). *Prophase* is noteworthy for two events:

1) Interphase cytoplasmic microtubules disassemble, and the mitotic spindle, composed of microtubules and associated proteins organized around two centrosomes, begins to form outside the nucleus. Separation of the centrosomes forms the two spindle poles.

2) The diffuse chromatin of the interphase slowly condenses into chromosomes. Each chromosome consists of two sister chromatids and contains a specific DNA sequence, the centromere, attached to a protein complex, the kinetochore.

In *prometaphase* the nuclear envelope is disrupted and disperses into membrane vesicles which remain around the spindle. The kinetochores undergo maturation and attach to a set of microtubules, the kinetochore microtubules. Other microtubules in the spindle are called polar microtubules; those outside the spindle are astral microtubules. In *metaphase* chromosomes attached to the spindle, microtubules are aligned in a plane, each sister kinetochore attached to microtubules linked to opposite poles of the spindle. *Anaphase* begins abruptly with the separation of the paired kinetochores. The sister chromosomes are pulled toward the spindle poles, separating them first by shortening of the kinetochore microtubules,

Table 5.3. Candidate substrates for the p34^{cdc2} protein kinase in animal cells

Substrate	Same site(s) in vivo as in vitro?	Possible role
Nuclear lamins	yes	Nuclear lamina disassembly
Vimentin	yes	Intermediate filament disassembly
Caldesmon	yes	Microfilament contraction
Histone H1	yes	Chromosome condensation
pp60$^{c\text{-}src}$	yes	Cytoskeletal rearrangements
NO38, nucleolin	yes	Nucleolar reorganization
SV40 T antigen	yes	DNA replication
c-Abl	yes	Unknown
p105^{Rb}	yes	Unknown
P53	yes	Unknown
RNA polymerase II	unknown	Transcription inhibition
EF-1γ	unknown	Translation inhibition
Cyclin B	unknown	Regulation of p34^{cdc2}
Myosin light chain	unknown	Contractile ring activation
Casein kinase II	unknown	unknown

Modified from ref. 71 (Norbury C, Nurse P. Annu Rev Biochem 1992; 61: 441-470. with permission, from the Annual Review of Biochemistry, Volume 61, 1992, by Annual Reviews Inc.

then by elongating the polar microtubules to further separate the spindle poles. With the arrival of the daughter chromosomes at the poles in *telophase,* the kinetochore microtubules disappear, the polar microtubules further elongate and the nuclear envelope reforms.

Almost every element of the cell is drastically remodeled at mitosis (4). These events include chromosome condensation, disassembly of the nuclear lamina and other intermediate filament proteins, nuclear membrane disruption, blockage of membrane trafficking, reorganization of microtubules for spindle formation and reorganization of microfilaments in preparation for cytokinesis. *All of these events are triggered by MPF protein kinase (cyclin B/cdk1).* Thus, it is critical to know which proteins are specifically phosphorylated by MPF at appropriate times in the mitosis process. A list of substrates is shown in Table 5.3. Two complementary approaches have been taken: in vitro assays of MPF candidate substrates (75) and analysis of proteins phosphorylated at appropriate times in mitosis (71). By comparing the positions of residues phosphorylated in these proteins in vitro and in vivo, one can make educated assessments of which proteins are regulated by MPF. Combining these results with information about functional roles of these proteins provides a picture of how events in mitosis unfold. That picture still resembles an incompletely assembled jigsaw puzzle, but the outline of the major steps has been accomplished. Two types of MPF regulation have emerged from these studies: direct and indirect. In the first MPF directly phosphorylates components active in the mechanical processes driving mitosis. In the second MPF phosphorylates other regulatory components which modify the agents involved in mechanical changes. Obviously, the direct effects are most easily observed. Many processes probably involve both direct and indirect mechanisms, but the latter are less well understood.

Role of Microtubules

Much of what happens in mitosis is driven by changes in microtubule organization (76). The microtubules which radiate from the centrosome in interphase shorten at the onset of mitosis and reorganize to form the mitotic spindle (77). As described in chapter 1, changes in the dynamic instability of microtubules, regulated by their binding proteins, is fundamental to that organization. The second determinant is the action of the microtubule motor proteins. Thus, a key element in regulating mitotic events must be the effect of MPF and its secondary regulatory effectors on microtubules. Such an effect was demonstrated by Verde et al (78) in cell free extracts of *Xenopus* eggs. Shortening of centrosome-nucleated microtubules, a key event in spindle formation, was shown to be cdk1 dependent and a result of a decreased stability of the plus ends and an increased catastrophe rate modulated by a cdk1 substrate. Recent studies have described a candidate protein, a small protein called stathmin or oncoprotein 18 which is a cdk target and increases the catastrophe rate of microtubules (79, 80). Since MAPs are important regulators of microtubule dynamics, MAP phosphorylation by cdk1 could be critical in the organization of the spindle. Ookata et al. (81) have shown that cdk1/cyclin B is associated with the spindle. In HeLa cells the kinase was demonstrated biochemically to associate with microtubules via MAP-4 (82), the major MAP present in mammalian non-neuronal cells. Moreover, recombinant fragments of MAP-4 formed a complex with the kinase and were phosphorylated. This phosphorylation did not block MAP-4 binding to microtubules, but did abolish its microtubule stabilizing activity. Interestingly, association with microtubules appears to be a unique characteristic of the cyclin B/cdk1 complex (82), suggesting that it is the only cdk complex involved in spindle formation, though not necessarily the only regulatory mechanism, since indirect mechanisms may also contribute (83).

Other protein kinases may also contribute to spindle formation. Polo kinases have been shown to be critical for spindle bipolarity in most organisms (73). Both kinesins and dyneins have also been implicated in spindle formation (84). Antibody injection studies have demonstrated that cytoplasmic dynein (85) and the kinesin-related motor Eg5 (84) are involved in centrosome migration during spindle assembly and are essential for bipolar spindle formation. Thus, centrosome separation appears to depend on cooperation between plus- and minus-end-directed microtubule motors. Eg5 is phosphorylated in a cycle-dependent manner, exclusively on Ser in S phase, but on both Ser and Thr during mitosis (84). Site-directed mutations and phosphopeptide analyses indicate, respectively, that the conserved Thr residue is involved in binding Eg5 to the centrosome and that the Thr phosphorylated during mitosis is a substrate site for cdk1/cyclin B in vitro. These results provide one example of the general mechanism of regulating the functions of the microtubule motor proteins by phosphorylation, either by regulating their associations or their motor activities (86) as described further in chapter 6.

Chromosome Condensation

Most of the cdk1/cyclin B complex translocates into the nucleus before chromosome condensation and nuclear envelope breakdown (87), suggesting a cycle dependent loss of affinity for microtubules/MAPs and an increased affinity for chromosome and envelope proteins. Chromosome condensation is accompanied by extensive phosphorylation of histone H1 and other proteins (88), but in vitro experiments suggest that H1 is not required for condensation (89). In the mitotic chromosome approximately one-third of the mass is DNA, one-third histones and one-third other proteins. Fractionation of the chromosomes yields an insoluble protein mass which has been termed the *chromosome scaffold* (90). The most abundant protein in the scaffold is *topoisomerase II* (topo II) (91), which controls DNA topology by transiently breaking and rejoining DNA double helical strands to prevent aberrant structures from forming. Topo II is bound to specific sequences called scaffold attachment regions, that are spaced about 50-200 kb apart along chromosomal DNA. The stoichiometry of topo II suggests that it has both enzymatic and structural roles, though the latter is controversial (90). Topo II enzyme activity and scaffold binding are regulated by phosphorylation. Several kinases have been shown to phosphorylate topo II and increase its activity, including cdk1, casein kinase II (CK II) and PKC (75, 92, 93). Phosphorylation of topo II increases as cells pass through G2 and enter mitosis (92). In yeast, CK II is the major enzyme involved in phosphorylating topo II in vivo. CK II is phosphorylated by cdk1/cyclin B, thus providing a link to the master mitotic regulator. However, it is also possible that a phosphatase plays a role in the specific regulation of topo II activity at mitosis (92). A second group of proteins implicated in the mitotic chromosome scaffold and chromosome condensation are the SMC (structural maintenance of chromosomes) family (94). These proteins have similar structures to microtubule motor proteins. Two types of models have been proposed for their function: a self-assembly model to form a matrix to support chromosome structure via their central coiled-coil domains and a chromatin motor model based on their similarities to other motor proteins (95). Additional studies are needed to test these models and to determine how the participation of these proteins in chromosome changes is regulated.

Lamin Phosphorylation

Prometaphase begins with breakdown of the nuclear envelope which is composed of a membrane stabilized by a network of lamin-type intermediate filaments. Disassembly of this network to release membrane vesicles appears to primarily be the result of direct phosphorylation of the lamins by cdk1/cyclin B, since disassembly can be blocked by mutation of specific cdk1 phosphorylation sites (96). However, since this phosphorylation in vitro does not cause disassembly of the filaments, other factors are probably involved (97). Both A- and B-lamins become hyperphosphorylated during mitosis. The phosphorylated A-lamins are soluble, but the B-lamins which have prenylated C-termini remain associated with nuclear membrane vesicles (97, 98). Phosphorylation and breakdown of cytoplasmic intermediate filaments also occurs during mitosis. Studies using phosphopeptide-specific monoclonal antibodies indicate that vimentin is phosphorylated by cdk1/cyclin B during early mitotic phases and by PKC during reorganization of intracellular membranes (99). One explanation for the mitosis-specific vimentin phosphorylation by PKC is that the active form of the enzyme is bound to organelles or the plasma membrane during interphase and not accessible to IFs. Vesiculation of the organelles provides a greater mobility for the active enzyme to associate with and phosphorylate vimentin. Alternatively, the IFs may be breaking down and being phosphorylated as multimers. The IF crosslinking protein plectin is also phosphorylated by cdk1 during mitosis, decreasing its ability to associate with IFs and converting it to a more soluble form (100). Since plectin can crosslink IFs with other cytoskeletal elements (chapter 1), the phosphorylation may play a role in mitosis-dependent cytoskeletal rearrangements.

Kinetochores and Motors

The key to precise chromosome separation during mitosis is the attachment of the chromosomes to spindle microtubules at the centromere. Considering the significance and conservation of their function, centromeres are surprisingly variegated (101). Budding yeast point centromeres are contained within a ≈250 bp DNA sequence, while fission yeast and mammalian regional centromeres encompass kb of DNA. Kinetochores assembled on *point centromeres* bind a single microtubule, while those on *regional centromeres* bind multiple microtubules. The two types of kinetochores differ in both composition (102) and structure (101). A model for the budding yeast kinetochore shows the structure assembled around a modified nucleosome containing a histone H3 variant (101). The adjacent DNA is bound to a multimeric complex which links to microtubules. Other DNA and microtubule-associated components, including a kinesin-related protein (KRP) are implicated in the complex. The mammalian centromere-kinetochore complex has been characterized by high-voltage, electron-microscopic tomography as a trilaminar structure containing a well-defined outer plate loosely connected to a less well-defined inner plate associated with heterochromatin (103). The outer plate is a 35-40 nm structure composed of 10-20 nm fibers. It is also composed of three different domains and connected to the inner plate by 10-20 nm fibers. The heterochromatin domain contains α-satellite DNA and its binding proteins and a KRP (101). The inner plate contains both DNA and proteins necessary for assembly. Microtubules bind to the outer plate which contains microtubule binding proteins and a KRP. Moreover, a fibrous corona associated with the outer surface of the outer plate contains cytoplasmic dynein. Thus, *the mammalian kinetochore has motors for moving chromosomes in either direction along microtubules.*

During prometaphase the disruption of the nuclear envelope allows access of the dynamic centrosome-associated microtubules to the chomosomes. The microtubules become randomly attached to kinetochores and stabilize (104), first one sister, then the other to microtubules of opposite poles. During this and subsequent processes, the chromosomes undergo oscillatory movements via a mechanism known as directional instability (76). The movements are driven by a combination of microtubule assembly/disassembly, primarily at the kinetochore, and kinetochore motors to move the chromosomes along the microtubules until they are aligned at the metaphase plate. How this movement is controlled is unknown, but it may involve microtubule motors plus a repulsive effect at the spindle poles resulting from microtubule density (105). One possible participant in these chromosome movements is the KRP CENP-E. CENP-E is a member of the class of mitotic apparatus proteins known as *passenger proteins* (90). They concentrate at centromeres during prometaphase and transfer to the spindle midzone during anaphase. Although their functions are obscure, several roles are possible, including involvement in chromatid pairing, spindle structure and stabilization and localization of the cleavage furrow for cytokinesis. CENP-E is transiently bound to kinetochores during early mitosis, but becomes redistributed to the microtubules of the spindle midzone at anaphase (106). It is degraded after cytokinesis. CENP-E contains two microtubule binding sites which allows it to crosslink microtubules.

However, one of those can be blocked by phosphorylation with MPF, effectively restricting its binding before dephosphorylation at anaphase. Thus, CENP-E may be able to participate in kinetochore-driven chromosome movements before anaphase because of the kinase regulation of its binding activity. Other avenues for regulating kinetochore-based motility mechanisms need to be investigated. Kinetochores contain cdk1 and a number of phosphorylated proteins (107), but whether these phosphorylations contribute to the control of chromosome movements still needs to be clarified.

Intracellular Organelles

During mitosis, intracellular organelles such as the Golgi apparatus must be partitioned between the daughter cells. The Golgi is fragmented into vesicles which disperse in the cytoplasm during prophase and metaphase, concomitant with re-organization of interphase microtubules into the mitotic spindle (108). Similarly, intracellular vesicular transport for exocytosis and endocytosis (chapter 6) ceases. Surprisingly, treatment of interphase cells with okadaic acid, an inhibitor of protein phosphatase 1 and 2A, can induce the same type of Golgi fragmentation and cessation of membrane traffic (109), as well as partial microtubule disassembly (110). How this protein dephosphorylation is related to the cdk1 phosphorylating processes which trigger many of the other events during mitosis is unclear. A reasonable model for Golgi disruption is that dephosphorylation of specific microtubule-associated proteins causes their dissociation from Golgi sites and reorganization of the microtubules into the spindle apparatus.

Anaphase Chromosome Separations

Anaphase onset is indicated by the separation of the sister kinetochores. Commitment to this step requires passage through the spindle assembly checkpoint (111) which assures proper formation of the mitotic spindle and attachment of the chromosomes to kinetochores. Three features of abnormal spindles have been suggested to lead to mitotic arrest: unoccupied microtubule binding sites at the kinetochore, absence of tension at the kinetochore and aberrant dynamics of the kinetochore microtubules (112). Studies in budding yeast have identified six nonessential genes involved at this checkpoint *MAD1-3* and *BUB1-3*. MAD2 localization during the cycle suggests that it monitors kinetochore attachment to microtubules (113, 114). The functions of the other gene products are less certain, but hyperphosphorylation of MAD1 appears to be involved in monitoring microtubule integrity in the spindle (115). This phosphorylation can be catalyzed by the essential kinase Mps1p and under physiological conditions, requires Bub3p and Bub1p, which is also a kinase (116). Exactly how these kinases are connected to the cycle and the master regulator cdk1 is still uncertain (117). A second aspect of this checkpoint was discovered from studies on an antibody against a phosphopeptide epitope which stains kinetochores which are not under tension (101). Microinjection of this antibody can block anaphase without disrupting chromosome movements. Thus, this phosphorylation may provide a signal for the state of the kinetochore to prevent premature kinetochore disjunction.

Separation of the sister kinetochores at initiation of anaphase allows the chromosomes to be pulled toward their respective poles. Sister chromatid disjunction requires calcium and can be blocked by inhibitors of ubiquitin-mediated proteolysis, even in the presence of cdk1/cyclin B (118). Thus the degradation of the cyclin complex is not necessary to signal anaphase. However, it is necessary for the exit from mitosis (119). These observations suggest that a similar proteolytic step may be involved in both cyclin B degradation and kinetochore disjunction. The targets of the proteolysis for sister chromatid disjunction may be linking proteins within the centromere. Topo II is also required for chromatid disjunction, possibly for unraveling DNA tangles (90). Two different operations are involved in separating the chromosomes in anaphase, known as anaphase A and B. Anaphase A presumably involves kinetochore activities similar to those in prometaphase. Thus, the kinetochore acts as a motor to pull the attached chromatid along the kinetochore fiber (120), using either dynein or a minus-end-directed KRP. Moreover, shortening of the kinetochore fibers results from depolymerization at the kinetochore, not at the spindle poles. Phosphorylation of a 62 kDa protein has been implicated in microtubule depolymerization in the mitotic apparatus of sea urchin eggs (121). In anaphase B, chromosome movement results from the spindle poles becoming further separated, suggesting a microtubule sliding mechanism requiring crosslinking motor proteins such as the CENP-E described above.

Chromosome movements during metaphase and anaphase are powered in part by dynein and KRPs. The specifics of these mechanisms are still under investigation. Members of four of the eight kinesin subfamilies are involved in cell division (122). A minus-end directed KRP is involved in microtubule attachment to centrosomes and modulation of the dynamics of microtubules. One plus-end-directed KRP plays a role in spindle formation. A second may mediate anaphase B spindle elongation, and a third is associated with chromosome arms and may function during prometaphase and metaphase movements. Unfortunately, little is known about how these are regulated.

Reassembly

During anaphase and cytokinesis cellular structural elements begin to reassemble in the daughter cells in response to the decreased activity of MPF. Presumably the formation of cytoskeletal networks results from a phosphatase-dependent reversal of mitosis-induced phosphorylations. For example, protein phosphatase 2A is present on microtubules throughout the cell cycle. Its activity is higher in S phase than in mitosis, and it is proposed to reverse mitotic phosphorylation of MAP-4 by cdk1 (123). Unfortunately, the molecular aspects of the specific reassemby processes are still poorly understood, though lamins and nuclear vesicle binding to chromatin are clearly important (124). The reformation of ER networks has been studied in vitro. Such studies suggest the importance of dynein in the mitotic breakdown and reconstitution of the ER (124). Reformation of the Golgi occurs as interphase microtubules reassemble. A model for this process suggests that Golgi vesicles are transported to the region of the centrosome where they assemble into the Golgi stacks (125). One question which arises is how the Golgi reassembles its structure. Golgi consists of a *cis* network, three types of flattened cisternae and a *trans* network. Furthermore, the ends of the Golgi stacks are dilated relative to the centers and appear to be the sites for vesicle budding and fusion (126). One hypothesis for the organization of the cisternae is that they contain a membrane skeleton related to that of the erythrocyte (chapter 1) (126). In fact, isoforms of both spectrin and ankyrin have been localized in the Golgi, as has an actin binding protein called comitin which is a mannose-binding lectin (127). From these, a matrix could be assembled which would not only provide structure to the organelle, but also provide binding sites for specific localization of proteins, such as the glycosyltransferases (128).

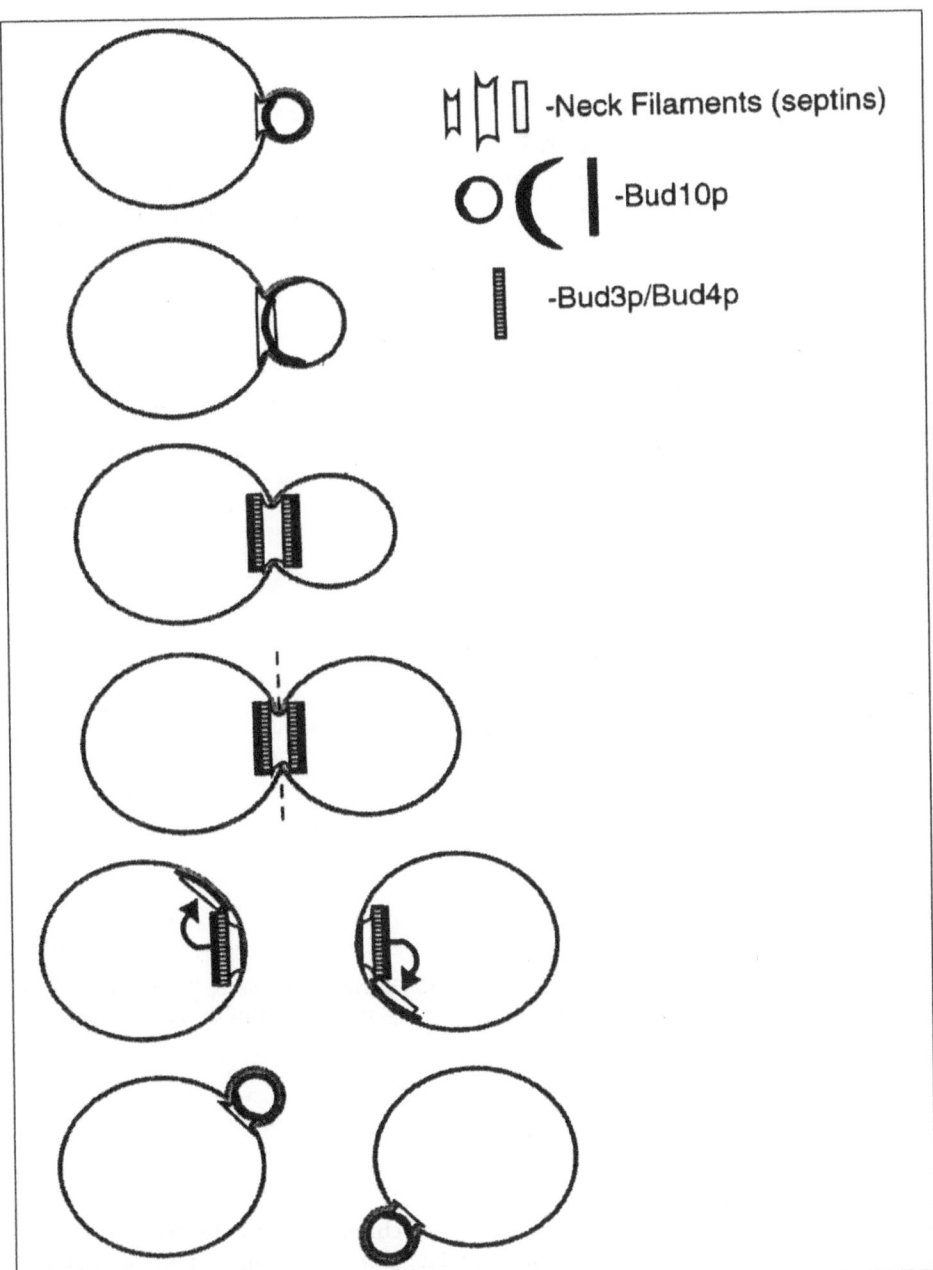

Fig. 5.7. Orientation of septins during cleavage of budding yeast. Heavy line is septin; dashed line is Bud3p for comparison. Figure from ref. 135 (Chant J Curr Opin Cell Biol 1996; 8: 557-565), courtesy of Current Biology Ltd.

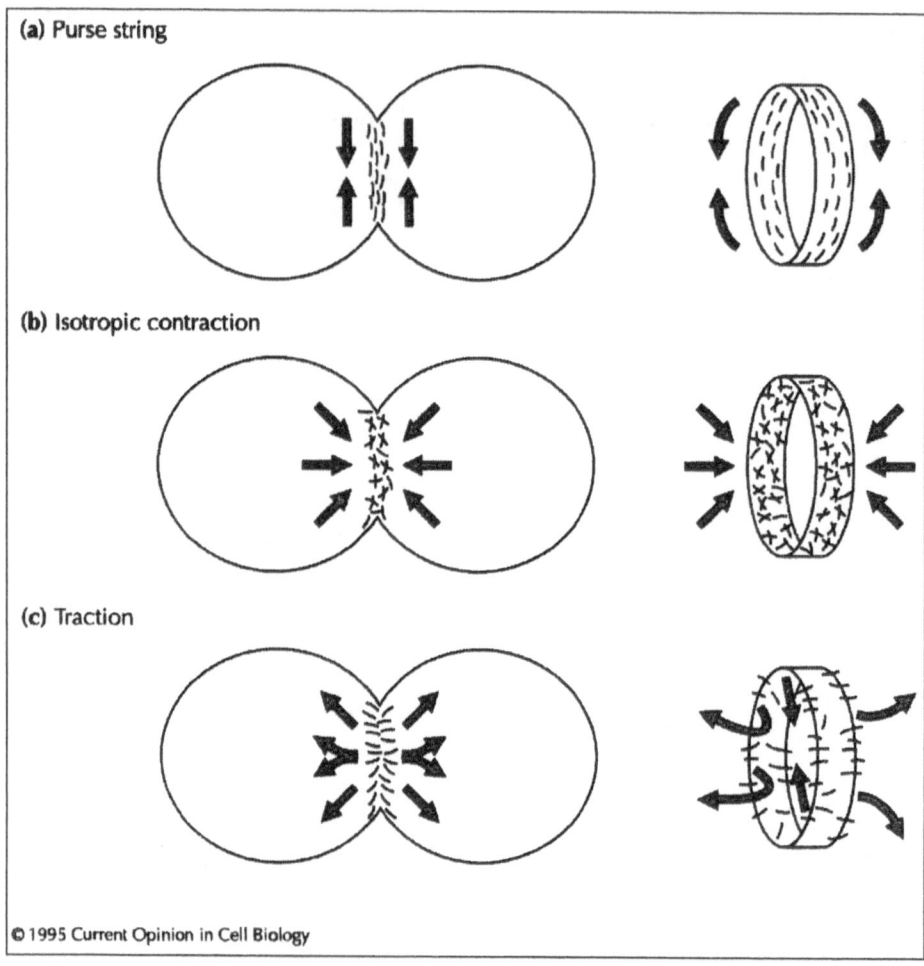

Fig. 5.8. Models for force generation and transduction during cytokinesis. Figure from ref. 137 (Fishkind DJ, Wang Y-l. Curr. Opin. Cell Biol. 1995; 7: 23-31), courtesy of Current Biology Ltd.

Cytokinesis

Furrow Localization

The events of anaphase not only move the chromosomes to the two poles, but also prepare the cell for division into two daughter cells. An important question is how the cleavage mechanism is initiated. Different types of organisms have evolved different cleavage mechanisms; we will focus on the mechanism in mammalian cells first. In these cells a cleavage furrow is developed which contains actin and myosin. According to the contractile ring hypothesis, contractile forces in the furrow act like a purse string. Obviously, the localization of this furrow is critical to the precise division of the cell. The mitotic spindle determines the site of formation of the furrow (129), using signals emanating from the spindle midzone (130).

Recent studies suggest that this localization is determined by passenger proteins which move from the kinetochore to the spindle midzone during anaphase. Margolis and Andreassen (131) have suggested that a cytoplasmic structure called the telophase disc, is the critical element in establishing the furrow and plane of cleavage. This structure, which can be identified and localized by antibodies to the specific antigen TD-60, bisects the anaphase cell and prefigures the site of telophase cleavage.

TD-60 is present on centromeres during prophase and migrates to the equator in the unfurrowed anaphase cell. During telophase, TD-60 is aligned with the furrow and becomes concentrated near the furrow membrane. Immediately after cleavage TD-60 is located at the position of maximum constriction. This is the behavior expected of a component involved in localizing the cleavage furrow, but does not establish whether the behavior is a cause or a consequence of that localization. Regardless, the hypothesis is that the telophase disc signals the location and timing of cleavage by organizing actin and myosin at the membrane for contraction of the furrow (131). Movement of the TD-60 and other passenger proteins to the spindle midzone may involve a plus-end-directed motor such as the KRP CENP-E (132). Perhaps the KRP moves complexes of these passenger proteins along the spindle microtubules to the midzone. However, a link to the membrane must also be established to initiate cleavage furrow formation. One model suggests that the telophase disc serves as a recruitment site for myosin, which then aligns microfilaments between the cytoskeletal cortex and the disc (131). This model would use the prior membrane linkages to microfilaments and would not require actin or myosin recruitment to the membrane.

In budding yeast the location of the cleavage plane is established by a different mechanism. Committment of the cell to division occurs with activation of cdk1/cyclin G1 complex and results in activation of the cell polarization machinery (see chapter 3). Three types of structural elements are polarized virtually simultaneously to the bud site: microfilaments, microtubules and a septin scaffold. Septins comprise a family of proteins reported to form the 10 nm filaments observed in the neck regions of the buds of budding yeast (133). They are noteworthy for the presence of a P-loop consensus region suggested to be a guanine nucleotide binding site (134). It is the septin scaffold which establishes the cleavage plane (Fig. 5.7) (135). This scaffold then functions for recruitment of the components involved in cytokinesis. Interestingly, the septin ring does not contract, but is itself divided during cleavage and partitioned between the daughter cells. Since septin analogs have been found in *Drosophila* and mice (134), their functions cannot be limited to structures formed at bud sites. In these animal cells the septins are concentrated during furrow contraction to the midbody region between the dividing cells (136), but their role is unclear. How formation of the septin ring is regulated is yet uncertain, but a member of the polo kinase family has been identified as a septum promoting factor in fission yeast (73).

Contractile Models

Although the purse string model is widely cited as the mechanism for contraction in the cleavage furrow, recent studies have raised questions about its generality. An organized contractile ring may not be necessary for division in all cell types. Fishkind and Wang (137) have presented two additional alternatives for force generation: isotropic contraction and traction (Fig. 5.8) (137). In the former cleavage is achieved through a random network of sliding actin filaments. In the traction model cleavage involves actin filaments associated end-on with the plasma membrane. The three mechanisms are not mutually exclusive and may coexist and cooperate to achieve cleavage in some cell types. Myosin II is considered to be an essential component not only for cytokinesis, but also for cell shape changes throughout the cell cycle, from the initial rounding at prophase through cytokinesis and the spreading of the daughter cells (138). However, recent studies of myosin mutants of budding yeast suggest that the myosin is not required for cytokinesis; instead it is implicated in cell separation after division (139).

Role of the Cytoskeleton

A number of cytoskeleton and regulatory components have been implicated in cytokinesis. *Profilin* (140) and *cofilin* (141) are essential, almost certainly for the establishment of the microfilament structures involved in the contraction of the furrow. Knockouts of cofilin are lethal; injection of inhibitory anti-cofilin antibodies into two-cell *Xenopus* blastomeres blocked cytokinesis (142). *Tropomyosin* is present at the site of septation and believed to stabilize microfilaments during formation of the contractile apparatus (143). The microfilament association of the microfilament-binding protein *caldesmon* varies with the cell cycle, decreasing in prophase as a consequence of cdk1-dependent phosphorylation (144). Caldesmon inhibits actomyosin ATPase and contraction aids in blocking filament severing and may stimulate actin polymerization (145). All of these activities are diminished by phosphorylation. During the early stages of mitosis, caldesmon has a diffuse distribution. During cytokinesis caldesmon becomes increasingly dephosphorylated and associated with microfilaments. These results suggest that the absence of caldesmon may be required for the formation of the contractile apparatus in cleavage furrows, and that it plays a role in events subsequent to cleavage, such as spreading of the daughter cells (145). Phosphorylation of caldesmon may also be important in the loss of microfilaments, such as stress fibers, which occurs with the disruption of the cytoskeleton during prometaphase. Loss of caldesmon from microfilaments can increase their susceptibility to severing by gelsolin. Moreover, the caldesmon affinity for microfilaments is decreased by calcium-mediated binding of calmodulin to a specific site on caldesmon.

The action of calcium on caldesmon and microfilaments is just one indication of its possible importance in cytokinesis. A localized transient rise in calcium at anaphase onset in early embryos is correlated with cyclic increases in inositol trisphosphate (146). As mentioned previously, IP_3 changes are usually regulated by controlling the hydrolytic activity of phospholipase C enzymes together with the levels of phosphoinositides. A second potential role for calcium is in the direct regulation of contractility. Calmodulin has been implicated both in cytokinesis (147) and in the regulation of the activity of myosin II (148). However, myosin II may also be directly regulated by phosphorylation of its regulatory light chains by cdk1 (149). This phosphorylation has been proposed to regulate timing of cytokinesis by inhibition of the myosin during prophase and metaphase, though other studies suggest that light chains are not essential for regulation of cell division (137). Instead, or in addition, phosphorylation of the myosin heavy chain has been implicated as a negative regulator of the assembly of myosin in the cleavage furrows of dividing sea urchin eggs (150) and *Dictyostelium* (151). Interestingly, the heavy chain kinase is a member of the PAK family which can be modulated by Rac and Cdc42 small G proteins (chapter 1) (152). Recently, a new member of the Rac family has been isolated from *Dictyostelium*. It is one of eight *rac* genes in this organism and appears to be essential solely for cytokinesis (150). Although a number of studies have implicated small G proteins in cytokinesis (153), little is known about either their activation or targets.

Membrane Fusion

One unanswered question concerning cytokinesis is the mechanism for fusing the adjacent plasma membranes to complete the separation of the daughter cells, a process which must involve membrane lipid reorganization. Phosphatidylethanolamine (PE) is unique in its ability to participate in membrane fusion events including myoblast fusion. However, it is normally found at the inner surface of plasma membrane bilayers. Using a fluorescent PE-specific cyclic peptide, PE was shown to appear at the cleavage furrow on the cell surface during late telophase (154). A peptide-strepavidin complex blocked cytokinesis and inhibited both actin filament disassembly and subsequent membrane fusion, but had no effect on furrowing or microtubule rearrangements. Thus, PE may play an important role in the final stages of cytokinesis.

Summary

The cell cycle consists of four phases: M, G1, S and G2. Regulation of progression through these phases determines whether the cell undergoes division. This regulation is controlled by protein kinases which are activated by cyclin-type proteins. The cyclin/kinase complexes are further regulated by proteolysis, phosphorylation/dephosphorylation and specific inhibitors. In eukaryotic cells the progression through G1 is significant for its susceptibility to external factors. During this phase, prior to the restriction point, the cell can be removed from the cycle into quiescent, differentiated or apoptotic states. Progression through G1 in most cells requires external factors, cell adhesion and cytoskeleton integrity. Adhesion and cytoskeleton integrity act primarily on the expression and inhibitors of cyclin D and its cdk complexes and must be integrated with effects of mitogens. Progression into M phase requires passage of checkpoints for DNA and cytoskeletal integrity. Simplistically, cell division requires two elements of the cytoskeleton: microtubules for chromosome separation during mitosis and microfilaments for cytokinesis. Mitosis is a complex process in which virtually every structural element in the cell is modified. Breakdown of the nuclear envelope is driven by phosphorylation of lamin. Spindle formation and chromosome separations require both microtubule dynamic instability and microtubule motors. How these are regulated is unclear. Precise division of the cell requires positioning of the cleavage furrow. The mechanism for accomplishing this varies in different organisms. Cytokinesis is driven by actomyosin motors, but how the cleavage furrow complexes are organized and regulated remain unclear.

References

1. Nigg EA. Cyclin-dependent protein kinases: key regulators of the eukaryotic cell cycle. BioEssays 1995; 17:471-480.
2. Bartek J, Bartkova J, Lukas J. The retinoblastoma protein pathway and the restriction point. Curr Opin Cell Biol 1996; 8:805-814.
3. Hunter T, Pines J. Cyclins and cancer: cyclin D and cdk inhibitors come of age. Cell 1994; 79:573-582.
4. Kirschner M. The cell cycle then and now. Trends Biochem Sci 1992; 17: 281-285.
5. Sherr CJ. G1 phase progression: cycling on cue. Cell 1994; 79:551-555.

6. Pines J. Protein kinases and cell cycle control. Sem Cell Biol 1994; 5:399-408.
7. Coleman TR, Dunphy WG. Cdc2 regulatory factors. Curr Opin Cell Biol 1994; 6: 877-882.
8. Sherr CJ, Roberts JM. Inhibitors of mammalian G, cyclin-dependent kinases. Genes Devel 1995; 9:1149-1163.
9. Hartwell LH, Kastan MB. Cell cycle control and cancer. Science 1994; 266: 1821-1828.
10. Wittenberg C, Reed SI. Plugging it in: signaling circuits and the yeast cell cycle. Curr Opin Cell Biol 1996; 8:223-230.
11. Chant J. Cell polarity in yeast. Trends in Genetics 1994; 10:328-333.
12. Assoian RK, Zhu X. Cell anchorage and the cytoskeleton as partners in growth factor dependent cell cycle progression. Curr Opin Cell Biol 1997; 9:93-98.
13. Ruoslahti E, Reed JC. Anchorage dependence, integrins, and apoptosis. Cell 1994; 77:477-478.
14. Peters G. The D-type cyclins and their role in tumorigenesis. J Cell Sci 1994; (Suppl)18:89-96.
15. Polyak K, Lee M, Erdjument-Bromage H, Koff A, Roberts JM, Tempst P, Massague J. Cloning of p27^{Kip1}, a cyclin-dependent kinase inhibitor and a potential mediator of extracellular antimitogenic signals. Cell 1994; 78:59-66.
16. Polyak K, Kato JY, Solomon MJ, Sherr CJ, Massague J, Roberts JM, Koff A. p27^{Kip1}, a cyclin-cdk inhibitor, links transforming growth factor-β and contact inhibition to cell cycle arrest. Genes Devel 1994; 8:9-22.
17. Graves LM, Lawrence JC Jr. Insulin, growth factors and cAMP: antagonism in the signal transduction pathways. Trends Endocrin Metab 1996; 7:43-50.
18. Zhu X, Ohtsubo M, Bohmer RM, Roberts JM, Assoian RK. Adhesion-dependent cell cycle progression linked to the expression of cyclin D1, activation of cyclin E-cdk2, and phosphorylation of the retinoblastoma protein. J Cell Biol 1996; 133:391-403.
19. Schulze A, Thome-Zerfass K, Berges J, Middendorp S, Jansen-Durr P, Henglein B. Anchorage-dependent transcription of the cyclin A gene. Mol Cell Biol 1996; 16:4632-4638.
20. Fang F, Orend G, Watanabe N, Hunter T, Ruoslahti E. Dependence of cyclin E-cdk2 kinase activity on cell anchorage. Science 1996; 271:499-502.
21. Jalink K, Hordijk PL, Moolenaar WH. Growth factor-like effects of lysophosphatidic acid, a novel lipid mediator. Biochim Biophys Acta 1994; 1198: 185-196.
22. Bretscher A. Microfilament structure and function in the cortical cytoskeleton. Annu Rev Cell Biol 1991; 7:337-374.
23. Craig SW, Johnson RP. Assembly of focal adhesions: progress, paradigms, and portents. Curr Opin Cell Biol 1996; 8: 74-85.
24. Yamada K, Miyamoto S. Integrin transmembrane signaling and cytoskeletal control. Curr Opin Cell Biol 1995; 7: 681-689.
25. Hansen LK, Mooney DJ, Vacanti JP, Ingber DE. Integrin binding and cell spreading on extracellular matrix act at different points in the cell cycle to promote hepatocyte growth. Mol Biol Cell 1994; 5:967-975.
26. Bohmer RM, Scharf E, Assoian RK. Cytoskeletal integrity is required throughout the mitogen stimulation phase of the cell cycle and mediates the anchorage-dependent expression of cyclin D1. Mol Biol Cell 1996; 7: 101-111.
27. Ingber DE, Prusty D, Sun Z, Betensky H, Wang N. Cell shape, cytoskeletal mechanics, and cell cycle control in angiogenesis. J Biomechanics 1996; 28: 1471-1480.
28. Boyd J, Risinger JI, Wiseman RW, Merrick BA, Selkirk JK, Barrett JC. Regulation of microfilament organization and anchorage-independent growth by tropomyosin 1. Proc Natl Acad Sci USA 1995; 92:11534-11538.
29. Rodriguez Fernandez JL, Geiger B, Salomon D, Ben-Ze'ev A. Suppression of vinculin expression by antisense transfection confers changes in cell morphology,

motility, and anchorage-dependent growth of 3T3 cells. J Cell Biol 1993; 122: 1285-1294.

30. Rodriguez Fernandez JL, Geiger B, Salomon D, Sabanay I, Zoller M, Ben-Ze'ev A. Suppression of tumorigenicity in transformed cells after transfection with vinculin cDNA. J Cell Biol 1992; 119:427-438.

31. Olson MF, Ashworth A, Hall A. An essential role for Rho, Rac and cdc42 GTPases in cell cycle progression through G1. Science 1995; 269:1270-1272.

32. Lamarche N, Tapon N, Stowers L, Burbelo PD, Aspenstrom P, Bridges T, Chant J, Hall A. Rac and cdc42 induce actin polymerization and G1 cell cycle progression independently of p65PAK and the JNK/SAPK MAP kinase cascade. Cell 1996; 87:519-529.

33. Joneson T, McDonough M, Bar-Sagi D, Van Aelst L. RAC regulation of actin polymerization and proliferation by a pathway distinct from Jun kinase. Science 1996; 274:1374-1376.

34. Leung T, Chen X-Q, Manser E, Lim L. The p160 RhoA-binding kinase ROK is a member of a kinase family and is involved in the reorganization of the cytoskeleton. Mol Cell Biol 1996; 16: 5313-5327.

35. Dedhar S, Hannigan GE. Integrin cytoplasmic interactions and bidirectional transmembrane signalling. Curr Opin Cell Biol 1996; 8:657-669.

36. Juliano R. Cooperation between soluble factors and integrin-mediated cell anchorage in the control of cell growth and differentiation. BioEssays 1996; 18: 911-917.

37. Richardson A, Parsons JT. Signal transduction through integrins: a central role for focal adhesion kinase? BioEssays 1995; 17:229-236.

38. LaFlamme SE, Auer KL. Integrin signaling. Sem. Cancer Biol. 1996; 7: 111-118.

39. Pasqualini R, Hemler ME. Contrasting roles for integrins β1 and β5 cytoplasmic domains in subcellular localization, cell proliferation, and cell migration. J Cell Biol 1994; 125:447-460.

40. Clark EA, Brugge JS. Integrins and signal transduction pathways: the road taken. Science 1995; 268:233-239.

41. Rosales C, O'Brien V, Kornberg L, Juliano R. Signal transduction by cell adhesion receptors. Biochim Biophys Acta 1995; 1242:77-98.

42. Chen Q, Lin TH, Der CJ, Juliano RL. Integrin-mediated activation of MEK and mitogen-activated protein kinase is independent of Ras. J Biol Chem 1996; 271: 18122-18127.

43. Carraway KL, Carraway CAC. Signaling, mitogenesis and the cytoskeleton: Where the action is. BioEssays 1995; 17:171-175.

44. Devary Y, Rosette C, DiDonato JA, Karin M. NF-κB activation by ultraviolet light not dependent on a nuclear signal. Science 1993; 261:1442-1445.

45. Qwarnstrom EE, Ostberg CO, Turk GL, Richardson CA, Bomsztyk, K. Fibronectin attachment activates the NF-κB p50/p65 heterodimer in fibroblasts and smooth muscle cells. J Biol Chem 1994; 269:30765-30768.

46. Wary KK, Mainiero F, Isakoff SJ, Marcantonio EE, Giancotti FG. The adaptor protein Shc couples a class of integrins to the control of cell cycle progression. Cell 1996; 87:733-748.

47. Zachary I, Rozengurt E. Focal adhesion kinase (p125FAK): a point of convergence in the action of neuropeptides, integrins, and oncogenes. Cell 1992; 71:891-894.

48. Schaller MD, Parsons JT. Focal adhesion kinase and associated proteins. Curr Opin Cell Biol 1994; 6:705-710.

49. Vuori K, Ruoslahti E. Association of insulin receptor substrate-1 with integrins. Science 1994; 266:1576-1578.

50. Ouwens DM, Mikkers HMM, van der Zon GC, Stein-Gerlach M, Ullrich A, Maassen JA. Insulin-induced tyrosine dephosphorylation of paxillin and focal adhesion kinase requires active phosphotyrosine phosphatase 1D. Biochem J 1996; 318:609-614.

51. Lewis JM, Baskaran R, Taagepera S, Schwartz MA, Wang JYJ. Integrin regulation of c-Abl tyrosine kinase activity and cytoplasmic-nuclear transport. Proc Natl Acad Sci USA 1996; 93: 15174-15179.

52. Wang JYJ. Abl tyrosine kinase in signal transduction and cell-cycle regulation. Curr Opin Genet Dev 1993; 3:35-43.

53. Welch LJ, Wang JYJ. A C-terminal protein-binding domain in the retinoblastoma protein regulates nuclear c-Abl tyrosine kinase in the cell cycle. Cell 1993; 75:779-790.

54. Watson, PA. Function follows form: generation of intracellular signals by cell deformation. FASEB J 1991; 5:2013-2019.

55. Morris CE. Mechanosensitive ion channels. J Membr Biol 1990; 113: 93-107.

56. Ingber DE. Cellular tensegrity: defining new rules of biological design that govern the cytoskeleton. J Cell Science 1993; 104:613-627.

57. Ingber DE. Integrins as mechanochemical transducers. Curr Opin Cell Biol 1991; 3:841-848.

58. Ingber DE, Dike L, Hansen L, Karp S, Liley H, Maniotis A, McNamee H, Mooney D, Plopper G, Sims S, Wang N. Cellular tensegrity: exploring how mechanical changes in the cytoskeleton regulate cell growth, migration, and tissue pattern during morphogenesis. Int Rev Cytol 1994; 150:173-224.

59. Maniotis AJ, Chen CS, Ingber DE. Demonstration of mechanical connections between integrins, cytoskeletal filaments, and nucleoplasm that stabilize nuclear structure. Proc Natl Acad Sci USA 1997; 94:849-854.

60. Thompson CB. Apoptosis in the pathogenesis and treatment of disease. Science 1995; 267:1456-1462.

61. Raff MC. Social controls on cell survival and cell death. Nature 1992; 356:397-400.

62. Xia Z, Dickens M, Raingeaud J, Davis RJ, Greenberg ME. Opposing effects of ERK and JNK-p38 MAP kinases on apoptosis. Science 1995; 270:1326-1331.

63. Nagata S, Golstein P. The Fas death factor. Science 1995; 267:1449-1456.

64. Frisch SM, Vuori K, Ruoslahti E, Chan-Hui P-Y. Control of adhesion-dependent cell survival by focal adhesion kinase. J Cell Biol 1996; 134:793-799.

65. Galaktionov K, Chen X, Beach D. Cdc25 cell-cycle phosphatase as a target of c-*myc*. Nature 1996; 382:511-517.

66. Steiner P, Philipp A, Lukas J, Godden-Kent D, Pagano M, Mittnacht S, Bartek J, Eilers M. Identification of a Myc-dependent step during the formation of active G, cyclin-cdk complexes. EMBO J 1995; 14:4814-4826.

67. Blagosklonny MV, Giannakakou P, El-Deiry WS, Kingston DGI, Higgs PI, Neckers L, Fojo T. Raf-1/bcl-2 phosphorylation: a step from microtubule damage to cell death. Cancer Res 1997; 57: 130-135.

68. Weinberg RA. The retinoblastoma protein and cell cycle control. Cell 1995; 81: 323-330.

69. Dyson N. pRB, p107 and the regulation of the E2F transcription factor. J Cell Science 1994; (Suppl)18:81-87.

70. Guadagno TM, Ohtsubo M, Roberts JM, Assoian RK. A link between cyclin A expression and adhesion-dependent cell cycle progression. Science 1993; 262: 1572-1575.

71. Norbury C, Nurse P. Animal cell cycles and their control. Annu Rev Biochem 1992; 61:441-470.

72. King RW, Jackson PK, Kirschner MW. Mitosis in transition. Cell 1994; 79: 563-571.

73. Glover DM, Ohkura H, Tavares A. Polo kinase: the choreographer of the mitotic stage. J Cell Biol 1996; 135:1681-1684.

74. Alberts B, Bray D, Lewis J, Raff M, Roberts K, Watson JD. Molecular Biology of the Cell, 3rd ed. Chap. 18, New York: Garland Publishing.

75. Nigg EA. Targets of cyclin-dependent protein kinases. Curr Opin Cell Biol 1993; 5:187-193.
76. Inoue S, Salmon ED. Force generation by microtubule assembly/disassembly in mitosis and related movements. Mol Biol Cell 1995; 6:1619-1640.
77. Sanger JW, Sanger JM. The cytoskeleton and cell division. Meth Achiev Exp Pathol 1979; 8:110-142.
78. Verde F, Labbe J-c, Doree M, Karsenti E. Regulation of microtubule dynamics by cdc2 protein kinase in cell-free extracts of *Xenopus* eggs. Nature 1990; 343: 233-238.
79. Belmont LD, Mitchison TJ. Identification of a protein that interacts with tubulin dimers and increases the catastrophe rate of microtubules. Cell 1996; 84:623-631.
80. Marklund U, Larsson N, Gradin HM, Brattsand G, Gullberg M. Oncoprotein 18 is a phosphorylation-responsive regulator of microtubule dynamics. EMBO J 1996; 15:5290-5298.
81. Ookata K, Hisanaga S, Okumura E, Kishimoto T. Association of p34^{cdc2}/cyclin B complex with microtubules in starfish oocytes. J Cell Sci 1993; 105:873-881.
82. Ookata K, Hisanaga S, Bulinski JC, Murofushi H, Aizawa H, Itoh TJ, Hotani H, Okumura E, Tachibana K, Kishimoto T. Cyclin B interaction with microtubule-associated protein 4 (MAP4) targets p34^{cdc2} kinase to microtubules and is a potential regulator of M-phase microtubule dynamics. J Cell Biol 1995; 128: 849-862.
83. Mori A, Aizawa H, Saido T, Kawasaki H, Mizumo K, Murofushi H, Suzuki K, Sakai H. Site-specific phosphorylation by protein kinase C inhibits assembly-promoting activity of microtubule-associated protein 4. Biochem 1991; 30:9341-9346.
84. Blangy A, Lane HA, d'Herin P, Harper M, Kress M, Nigg EA. Phosphorylation by p34^{cdc2} regulates spindle association of human Eg5, a kinesin-related motor essential for bipolar spindle formation in vivo. Cell 1995; 83:1159-1169.
85. Vaisberg EA, Koonce MP, McIntosh JR. Cytoplasmic dynein plays a role in mammalian mitotic spindle formation. J Cell Biol 1993; 123:849-858.
86. Holzbauer ELF, Vallee RB. Dyneins: molecular structure and cellular function. Annu Rev Cell Biol 1994; 10:339-372.
87. Gallant P, Fry AM, Nigg EA. Protein kinases in the control of mitosis: focus on nucleocytoplasmic trafficking. J Cell Sci 1995; (Suppl)19:21-28.
88. Bradbury EM. Reversible histone modification and the chromosome cell cycle. BioEssays 1992; 14:9-16.
89. Ohsumi K, Katagiri C, Kishimoto T. Chromosome condensation in *Xenopus* mitotic extracts without histone H1. Science 1993; 262:2033-2035.
90. Earnshaw WC, Mackay AM. Role of nonhistone proteins in the chromosomal events of mitosis. FASEB J 1994; 8: 947-956.
91. Earnshaw WC, Pluta AF. Mitosis. BioEssays 1994; 16:639-643.
92. Cardenas ME, Gasser SM. Regulation of topoisomerase II by phosphorylation: a role for casein kinase II. J Cell Sci 1993; 194:219-225.
93. Wells NJ, Fry AM, Guano F, Norbury C, Hickson ID. Cell cycle phase-specific phosphorylation of human topoisomerase II. Evidence of a role for protein kinase C J Biol Chem 1995; 270: 28357-28363.
94. Saitoh N, Goldberg I, Earnshaw WC. The SMC proteins and the coming of age of the chromosome scaffold hypothesis. BioEssays 1995; 17:759-766.
95. Hirano T, Mitchison TJ, Swedlow JR. The SMC family: from chromosome condensation to dosage compensation. Curr Opin Cell Biol 1995; 7:329-336.
96. Inagaki M, Matsuoka Y, Tsujimura K, Ando S, Tokui T, Takahashi T, Inagaki N. Dynamic property of intermediate filaments: regulation by phosphorylation. BioEssays 1996; 18:481-487.
97. Foisner R. Dynamic organisation of intermediate flaments and associated proteins during the cell cycle. BioEssays 1997; 19:297-305.

98. Gerace L, Blobel G. The nuclear envelope lamina is reversibly depolymerized during mitosis. Cell 1980; 19:277-287.

99. Takai Y, Ogawara M, Tomono Y, Moritoh C, Imajoh-Ohmi S, Tsutsumi W, Taketani Y, Inagaki M. Mitosis-specific phosphorylation of vimentin by protein kinase C coupled with reorganization of intracellular membranes. J Cell Biol 1996; 133:141-149.

100. Foisner R, Malecz N, Dressel N, Stadler C, Wiche G. M-phase-specific phosphorylation and structural rearrangement of the cytoplasmic cross-linking protein plectin involve p34^{cdc2} kinase. Mol Biol Cell 1996; 7:273-288.

101. Pluta AF, Mackay AM, Ainsztein AM, Goldberg IG, Earnshaw WC. The centromere: hub of chromosomal activities. Science 1995; 270:1591-1594.

102. Bloom K. The centromere frontier: kinetochore components, microtubule-based motility, and the CEN-value paradox. Cell 1993; 73:621-624.

103. Ault JG, Rieder CL. Centrosome and kinetochore movement during mitosis. Curr Opin Cell Biol 1994; 6:41-49.

104. Hyman AA, Karsenti E. Morphogenetic properties of microtubules and mitotic spindle assembly. Cell 1996; 84:401-410.

105. Rieder CL, Salmon ED. Motile kinetochores and polar ejection forces dictate chromosome position on the vertebrate mitotic spindle. J Cell Biol 1994; 124: 223-233.

106. Liao H, Li G, Yen TJ. Mitotic regulation of microtubule cross-linking activity of CENP-E kinetochore protein. Science 1994; 265:394-398.

107. Brinkley BR, Ouspenski I, Zinkowski RP. Structure and molecular organization of hte centromere-kinetochore complex. Trends Cell Biol 1992; 2:15-21.

108. Warren G. Membrane partitioning during cell division. Annu Rev Biochem 1993; 62:323-348.

109. Lucocq J, Warren G, Pryde J. Okadaic acid induces Golgi apparatus fragmentation and arrest of intracellular transport. J Cell Science 1991; 100:753-759.

110. Lucocq J. Mimicking mitotic Golgi disassembly using okadaic acid. J Cell Science 1992; 103:875-880.

111. Wells WAE. The spindle-assembly checkpoint: aiming for a perfect mitosis, every time. Trends Cell Biol 1996; 6: 228-234.

112. Murray A. Cell cycle checkpoints. Curr Opin Cell Biol 1994; 6:872-876.

113. Li Y, Benezra R. Identification of a human mitotic checkpoint gene: *hsMAD2*. Science 1996; 274:246-248.

114. Chen R-H, Waters JC, Salmon ED, Murray AW. Association of spindle assembly XMAD2 with unattached kinetochores. Science 1996; 274:242-246.

115. Hardwick KG, Murray, AW. Mad1p, a phosphoprotein component of the spindle assembly checkpoint in budding yeast. J Cell Biol 1995; 131:709-720.

116. Hardwick KG, Weiss E, Luca FC, Winey M, Murray, AW. Activation of the budding yeast spindle assembly checkpoint without mitotic spindle disruption. Science 1996; 273:953-956.

117. Elledge SJ. Cell cycle checkpoints: preventing an identity crisis. Science 1996; 274:1664-1671.

118. Holloway SL, Glotzer, King RW, Murray AW. Anaphase is initiated by proteolysis rather than by the inactivation of maturation-promoting factor. Cell 1993; 73: 1393-1402.

119. King RW, Deshaies RJ, Peters J-M, Kirschner MW. How proteolysis drives the cell cycle. Science 1996; 274: 1652-1659.

120. Gorbsky GJ. Chromosome motion in mitosis. BioEssays 1992; 14:73-80.

121. Dinsmore JH, Sloboda RD Calcium and calmodulin-dependent phosphorylation of a 62 kD protein induces microtubule depolymerization in sea urchin mitotic apparatuses. Cell 1988; 53:769-783.

122. Moore JD, Endow SA. Kinesin proteins: a phylum of motors for microtubule-based motility. BioEssays 1996; 18:207-219.

123. Thaler CD, Haimo LT. Microtubules and microtubule motors: mechanisms of regulation. Int Rev Cytol 1996; 164: 269-327.

124. Wilson KL, Wiese C. Reconstituting the nuclear envelope and endoplasmic reticulum in vitro. Sem Cell Devel Biol 1996; 7:487-496.

125. Saraste J, Thyberg J. (1996). In: Treatise on the Cytoskeleton. Hesketh HE, Pryme IF, eds. Greenwich, CT: JAI Press, 239-273.

126. Barr FA, Warren G. Disassembly and reassembly of the Golgi apparatus. Sem Cell Devel Biol 1996; 7:505-510.

127. Jung E, Fucini, P, Stewart M, Noegel AA, Schleicher M. Linking microfilaments to intracellular membranes: the actin-binding and vesicle-associated protein comitin exhibits a mannose-specific lectin activity. EMBO J 1996; 15: 1238-1246.

128. Slusarewicz P, Nilsson T, Hui N, Watson R, Warren G. Isolation of a matrix that binds medial Golgi enzymes. J Cell Biol 1994; 124:405-413.

129. Rappaport R. Establishment of the mechanism of cytokinesis in animal cells. Int Rev Cytol 1986; 105:245-281.

130. Cao L-g, Wang Y-l. Signals from the spindle midzone are required for the stimulation of cytokinesis in cultured epithelial cells. Mol Biol Cell 1996; 7: 225-232.

131. Margolis RL, Andreassen PR. The telophase disc: its possible role in mammalian cell cleavage. BioEssays 1993; 15: 201-207.

132. Yen TJ, Compton DA, Wise D, Zinkowski RP, Brinkley BR, Earnshaw WC, Cleveland DW. CENP-E, a novel human centromere-associated protein required for progression from metaphase to anaphase. EMBO J 1991; 10: 1245-1254.

133. Chant J. Septin scaffolds and cleavage planes in Saccharomyces. Cell 1996; 84: 187-190.

134. Longtine MS, DeMarini DJ, Valencik ML, Al-Awar OS, Fares H, De Virgilio C, Pringle JR. The septins: roles in cytokinesis and other processes. Curr Opin Cell Biol 1996; 8:106-119.

135. Chant J. Generation of cell polarity in yeast. Curr Opin Cell Biol 1996; 8: 557-565.

136. Sanders SL, Field CM. Cell division. Septins in common. Curr Biol 1994; 4: 907-910.

137. Fishkind DJ, Wang Y-l. New horizons for cytokinesis. Curr Opin Cell Biol 1995; 7: 23-31.

138. Maciver SK. Myosin II function in non-muscle cells. BioEssays 1996; 18: 179-182.

139. Brown SS. Myosins in yeast. Curr Opin Cell Biol 1997; 9:44-48.

140. Balasubramanian MK, Hirani BR, Burke JD, Gould KL. The *Schizosaccharomyces pombe* cdc3+ gene encodes a profilin essential for cytokinesis. J Cell Biol 1994; 125:1289-1302.

141. Moon AL, Janmey PA, Louie KA, Drubin DG. Cofilin is an essential component of the yeast cortical cytoskeleton. J Cell Biol 1993; 120:421-435.

142. Welch MD, Nallavarapu A, Rosenblatt J, Mitchison TJ. Actin dynamics in vivo. Curr Opin Cell Biol 1997; 9: 54-61.

143. Balasubramanian MK, Helfman DM, Hemmingsen SM. A new tropomyosin essential for cytokinesis in the fission yeast *S. pombe*. Nature 1992; 360:84-87.

144. Hosoya N, Hosoya H, Yamashiro S, Mohri H, Matsumura F. Localization of caldesmon and its dephosphorylation during cell division. J Cell Biol 1993; 121: 1075-1082.

145. Yamashiro S, Yoshida K, Yamakita Y, Matsumura F. Caldesmon: possible functions in microfilament reorganization during mitosis and cell transformation. Actin: Biophysics, Biochemistry, and Cell Biology. In: Estes, JE and Higgins PJ, eds. New York: Plenum Press, 1994:113-122.

146. Ciapa B, Pesando D, Wilding M, Whitaker M. Cell-cycle calcium transients driven by cyclic changes in inositol triphosphate levels. Nature 1994; 368: 875-878.

147. Liu T, Williams JG, Clarke M. Inducible expression of calmodulin antisense RNA in *Dictyostelium* cells inhibits the completion of cytokinesis. Mol Biol Cell 1992; 3: 1403-1413.
148. Mabuchi I. Regulation of cytokinesis in animal cells-possible involvement of protein phosphorylation. Biomed Res 1993; 14:155-159.
149. Satterwhite LL, Lohka MJ, Wilson KL, Scherson TY, Cisek LJ, Corden JL, Pollard TD. Phosphorylation of myosin-II regulatory light chain by cyclin-p34^{cdc2}: a mechanism for the timing of cytokinesis. J Cell Biol 1992; 118:595-605.
150. Larochelle DA, Epel D. Myosin heavy chain dephosphorylation during cytokinesis in dividing sea urchin embryos. Cell Motil Cytskel 1993; 25:369-380.
151. Egelhoff TT, Lee RJ, Spudich JA. Dictyostelium myosin heavy chain phosphorylation sites regulate myosin filament assembly and localization in vivo. Cell 1993; 75:363-371.
152. Brzeska H, Korn ED. Regulation of class I and clsass II myosins by heavy chain phosphorylation. J Biol Chem 1996; 271:16983-16986.
153. Song K, Mach KE, Chen C-Y, Reynolds T, Albright CF. A novel suppressor of *ras1* in fission yeast, *byr4*, is a dosage-dependent inhibitor of cytokinesis. J Cell Biol 1996; 133:1307-1319.
154. Emoto K, Kobayashi T, Yamaji A, Yahara I, Inoue K, Umeda M. Redistribution of phosphatidylethanolamine at the cleavage furrow of dividing cells during cytokinesis. Proc Natl Acad Sci USA 1996; 93: 12867-12872.

CHAPTER 6

Intracellular Membrane Trafficking, Secretion/Exocytosis and Endocytosis

Cell Organization and Intracellular Trafficking

Organization of cells into compartments (cellular organelles) is necessary to separate synthetic and degradative activities and prevent the cell from degenerating into a series of futile cycles. As indicated in chapter 1, the location and form of many organelles, at least in interphase cells, is regulated by their interactions with microtubules (1, 2). During mitosis, most of the organellar structures are converted to vesicles, then reformed during anaphase; these processes also involve microtubules, as described in chapter 5. During the interphase period of the cell cycle, the intracellular organelles are highly dynamic. A schematic view of the major pathways involved in intracellular membrane trafficking through the organelles is shown in Figure 6.1. The primary pathways are: *secretion/exocytosis*, in which material produced in the endoplasmic reticulum (ER) is transported to the cell surface for export or incorporation into the plasma membrane; *uptake/endocytosis*, in which materials are taken from outside the cell, passaged through early (sorting) and/or late (prelysosomal) endosomal compartments and often shunted to the lysosomes for degradation; *recycling*, endocytosis of plasma membrane into an early endosomal compartment in the cell and back to the plasma membrane for reutilization, often as part of the endocytosis/degradation process; and *transcytosis*, a variant of recycling in polarized cells in which material and membrane transits from one side of the cell to the other. The cytoskeleton appears to contribute to membrane trafficking in three ways:

1) Cytoskeleton attachments to the plasma membrane, usually microfilaments, can alter membrane properties to facilitate or reduce endocytosis or exocytosis (3).
2) Microtubules can act as struts for maintaining organelle localization and geometries.
3) Microtubules or microfilaments may act as tracks for directed membrane movements (1). These membrane movements depend largely on the activities of a large number of cytoskeleton-associated motor proteins, some of which are indicated in Table 6.1 (4).

Movement of membrane between organelles, including the plasma membrane, is proposed to occur by two mechanisms:

Signaling and the Cytoskeleton, by Kermit L. Carraway, Coralie A. Carothers Carraway and Kermit L. Carraway III. © 1998 Springer-Verlag and R.G. Landes Company.

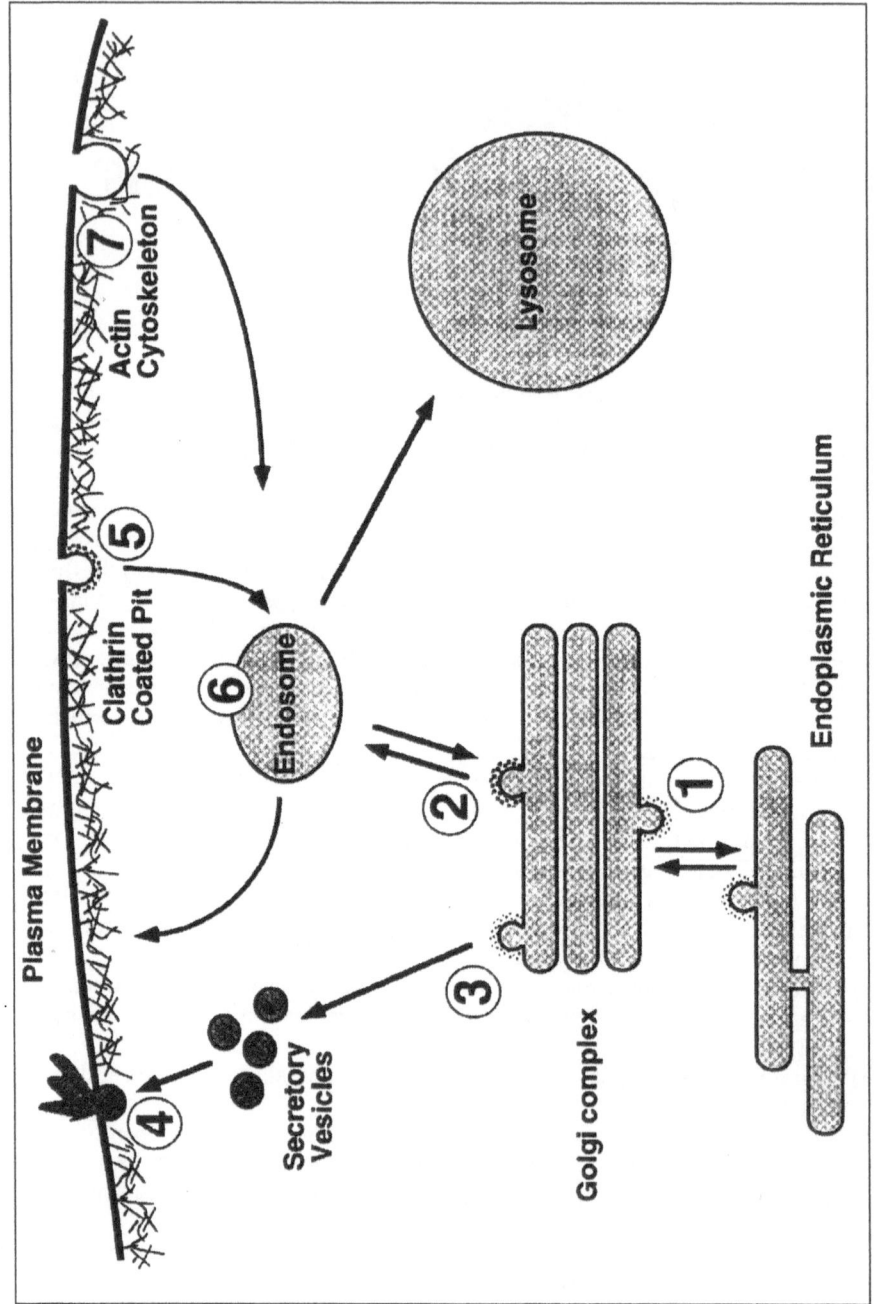

Fig. 6.1. Schematic diagram showing membrane traffic in eukaryotic cells. Figure from ref. 94, reprinted with permission (De Camilli P, Emr SD, McPherson PS, Novick P, Science 1996; 271:1533-1539). © 1996 American Association for the Advancement of Science.

Table 6.1. Motors implicated in membrane traffic

Membrane/ membrane traffic	Motor	Observation	System
ER	KHC/CD	Colocalization	Chick brain
	KHC	Δ morphology	Rat astrocytes
	CD	↓ motility	Xenopus egg extracts
	nd (actin-based)	↓ motility	Locust photoreceptor cells
Golgi complex	KHC	Colocalization	Rat hepatocytes
	CD (DHC2)	Colocalization	NRK, COS, HeLa
	KHC	Δ morphology	Rat astrocytes
	CD, MI	Colocalization	Chicken epithelial cells
	MV (p190)	Colocalization	Rat brain sections
Golgi complex-ER	KHC	Colocalization, ↓ transport	NRK
PM	KHC	↓ motility	Sea urchin eggs, 3T3
Apical membrane	KHC, CD	↓ transport	MDCK, HeLa cytosol
Basolateral membrane	KHC	↓ transport	MDCK, HeLa cytosol
Endosomes	KHC	↓ transport	Rat astrocytes
Endosomes and lysosomes	CD	Colocalization	NRK
	MI (myr4)	Colocalization	MDCK
	KHC	Δ morphology	Rat astrocytes, mouse macrophages NRK, L6E9
Endocytosis	KHC	↓ transport	Mouse fibroblasts
	CD	Colocalization	Rat hepatocytes
Pinocytosis	MI (MYO3, MYO5)	↓ internalization	S. cerevisiae
Phagocytosis	MI (myo A,B,C)	↓ internalization	Dictyostelium
	MI (myo B,C,D)	↓ internalization	Dictyostelium

Table continued next page

Table 6.1 (continued)

Membrane/Membrane traffic	Motor	Observation	System
Early—late endosomes	KHC, CD	↓ interaction	MDCK
	CD	↓ interaction	BHK
Axonal transport			
brain			
Synaptic vesicles	CD	Colocalization	Bovine
	Unc-104 (unc-104)	Δ morphology	Neurons, C. elegans
	KHC	↓ transport	Rabbit retinal nerve
Synaptic vesicle precursors	Unc-104 (KIF1A)	Proximal acc	Rat neurons

Specific genes or protein names are in parentheses. BHK, baby hamster kidney; CD, cytoplasmic dynein; colocalization, with membrane in question; injection, microinjection; IP, immunoprecipitation; KHC, kinesin heavy chain; L6E9, Rat myoblasts; mAb, monoclonal antibodies; MDCK, Madin Darby canine kidney cells; MI, myosin I; MT, microtubules; MV, myosin V; MVI, myosin VI; nd, not defined; NRK, normal rat kidney cells; NZ, nocodazole; pAb, polyclonal antibodies; proximal acc., proximal accumulation; TGN, trans-Golgi network; UV-van, UV vanadate; VM, video microscopy; ↓, decrease; Δ, change. Modified from ref. 4 (Goodson HV, Valetti C, Kreis TE. Curr Opin Cell Biol 1997; 9: 18–28), courtesy of Current Biology Ltd.

1) *vesicle transit* and
2) *organellar membrane fusion mediated by tubular extensions* connecting the organelles (5, 6).

In the former vesicles bud from the original organelle, then migrate to and are fused with the destination organelle. Much of the research in membrane trafficking in recent years has been concerned with the mechanisms and components defining the mechanics and specificity of these vesicular processes (7, 8). Budding is a complex process involving the GTP-mediated assembly of coat proteins on the membrane which stabilizes the vesicle for budding (Fig. 6.2A). Specificity of budding and cargo selection are achieved by the use of different coat proteins for different organelles (8). *Coat formation is initiated by binding of the small G-protein ARF* in its GTP form to the donor membrane, followed by recruitment of the coat proteins and budding. GTP hydrolysis destabilizes the coat to release the uncoated vesicle, which is ready for docking with its acceptor organelle. The targeting and fusion mechanisms proposed for the destination step of vesicle trafficking are shown in Figure 6.2B and 6.2C. This model depicts vesicle targeting as the interaction between vesicle and target membrane *SNARE* proteins. Targeting specificity is achieved with different SNAREs on different organelles. Binding of SNAP and the ATPase NSF are necessary to initiate the fusion process. However, in addition to the specificity of vesicle interactions, membrane vesicle trafficking is a directed process, dependent on cellular cytoskeletal elements. In general, traffic near the cytoplasmic surface of the plasma membrane involves microfilaments (9), while that deeper in the cell uses microtubules. Material transfer by organellar fusion also involves directed movement using microtubules. In this case, membrane projections are extended along microtubule tracks to provide a pathway for the migration of luminal contents between compartments, as illustrated in Figure 6.3.

Microtubule Involvement in Intracellular Trafficking

Transfer of secretory or membrane components between the ER and Golgi is a temperature-sensitive process which occurs in the presence of some microtubule-polymerizing drugs and involves a poorly defined intermediate compartment (6). Interestingly, during reduced temperature experiments, elements of this intermediate compartment codistribute with cold-resistant microtubules. Moreover, the Golgi is associated with stabilized microtubules containing detyrosinated tubulin (chapter 1) which have been implicated in the ER-Golgi transfer (10). Oligosaccharide analyses indicate that ER-Golgi transit also transfers ER resident proteins (11). The compositional integrity of the ER is maintained by a recycling process which brings these proteins back to the ER. Morphological studies of this redistribution induced by the drug brefeldin A suggest that the recycling occurs through organellar fusion via membrane projections along microtubules (11, 12), though the physiological significance of this mechanism has been questioned (6). Microinjection of anti-kinesin will block the redistribution from centriole-associated (minus end) Golgi to the ER (13), as predicted if kinesin is involved as a motor for the membrane projections. Surprisingly, kinesin is also found associated with membranes moving from the ER to Golgi. One explanation is that the motor is inactivated on these membranes, and the kinesin is just part of the cargo.

Protein secretion is inhibited by microtubule-active drugs, apparently at the stage between the Golgi apparatus and the plasma membrane (14). Thus, microtubules are thought to play important roles in the transit of secretory products to the cell surface. In many cell types an array of microtubules extends from the

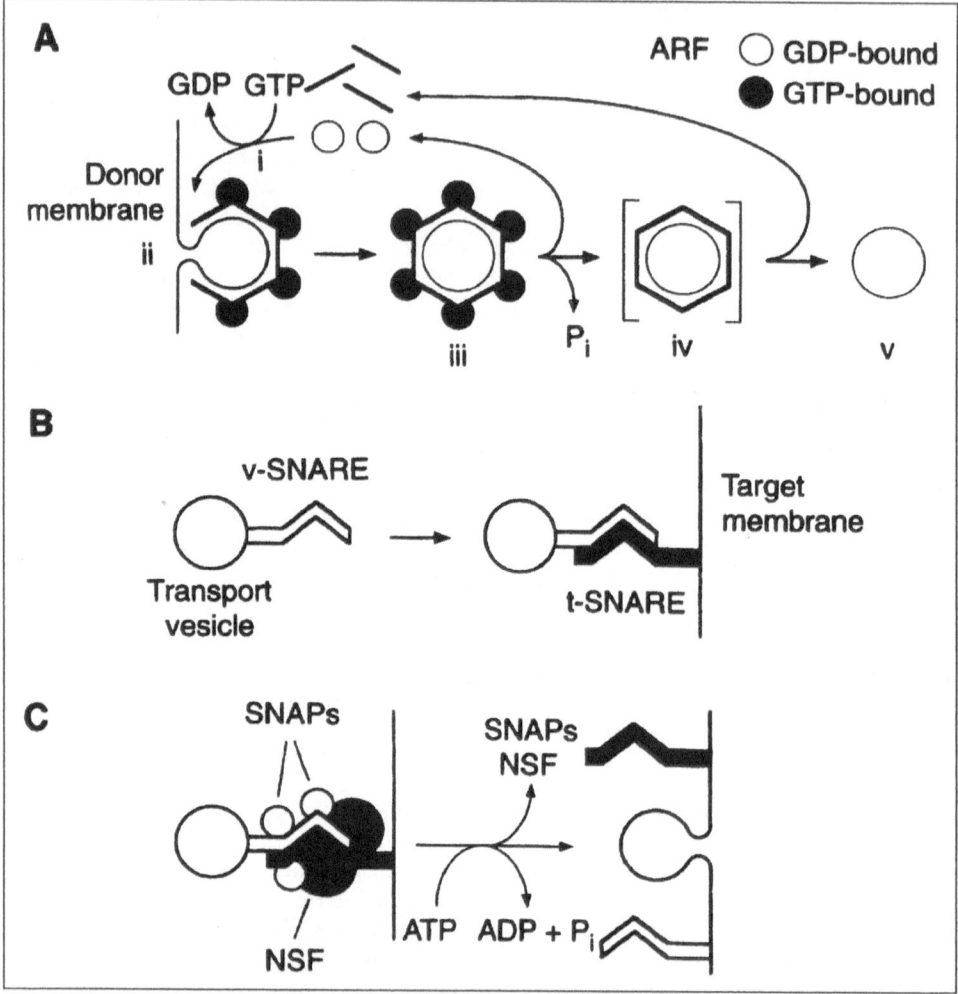

Fig. 6.2. Models for budding and docking steps of vesicle shuttle. Figure from ref. 8, reprinted with permission from Rothman JE, Wieland FT, Science 1996; 272:227-234. ©1996 American Association for the Advancement of Science.

microtubule organizing center to the plasma membrane (1), providing struts to which the closely associated Golgi cisternae can be anchored and tracks along which vesicles moving from the Golgi to the plasma membrane can travel. The Golgi apparatus is relocalized in migrating cells, apparently to more efficiently direct newly synthesized membrane components to the leading edge (15). Studies with microtubule-perturbing drugs indicate that microtubules are essential for the localization of Golgi stacks in the cell, but not for maintaining the Golgi cisternae in stacks (5). In polarized epithelial cells whose microtubules are organized differently (chapter 3) (1), microtubule depolymerization blocks only secretion from the apical, not the basolateral surface (16). However, this observation does not rule

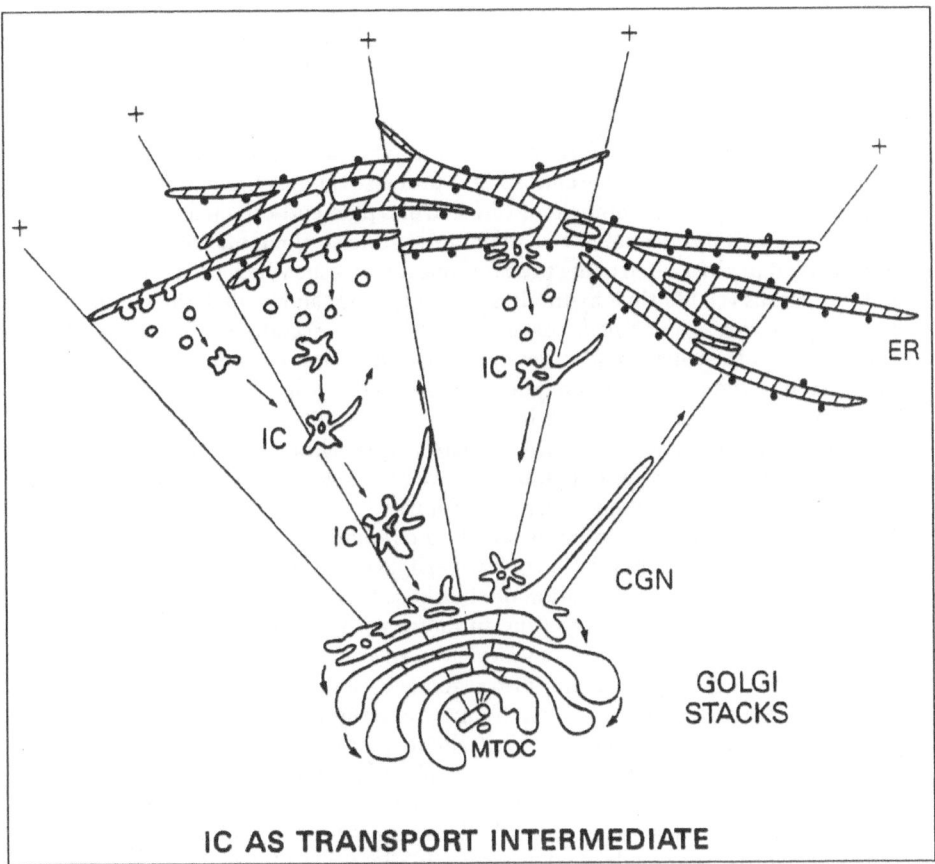

IC AS TRANSPORT INTERMEDIATE

Fig. 6.3. Model for extension of Golgi membranes along microtubule tracks. Figure from ref. 11 (Lippincott-Schwartz J. Trends Cell Biol. 1993; 3:81-88), reprinted with kind permission of Elsevier Science - NL, Sara Burgerhartstraat 25, 1055 KV Amsterdam, The Netherlands.

out the possibility of multiple basolateral pathways which are differentially sensitive to microtubule-depolymerizing agents.

Microtubules also appear to be important in the localization of and transport to lysosomes (5). Nocodozole depolymerization of microtubules causes relocalization of prelysosomal compartments and lysosomes from their juxtanuclear location. Likewise, transit of newly synthesized lysosomal proteins such as cathepsin D is inhibited by microtubule depolymerization (17). Microtubules have also been implicated in post-endocytic pathways, particularly in the transit from early to late endosomes (18), in which the cytoplasmic linking protein CLIP-170 may play an important role (2). In polarized epithelial cells, endocytosis occurs at both apical and basolateral surfaces, and the endocytosed materials become mixed in the late endosome in a microtubule-dependent process.

Role of Microtubule Motors

The orientation of both secretion (Golgi to plasma membrane) and endocytotic (plasma membrane to late endosomes) processes along cellular microtubules suggests that they involve minus-end-directed and plus-end-directed microtubule motors, respectively. Movements of membrane vesicles along microtubules have been verified in vitro, often using preparations from axons which can transport in both directions (1). Multiple methods and studies have shown that *kinesins catalyze plus-end-directed movements* of vesicles, as found in secretion and anterograde axonal transport, while *cytoplasmic dynein catalyzes minus-end-directed movements*, such as those of postendocytic pathways and retrograde axonal transport (19). The importance of the motors in intracellular motility raises questions about the regulation of their activities. Unfortunately, little is known about the specifics of this regulation. Two types of regulation have been suggested, modulating either motor activity or binding specificity and affinity.

Phosphorylation has been implicated in the modulation of kinesin motor activity (20). In vitro phosphorylation of kinesin light chain with PKA (21) or okadaic acid-induced hyperphosphorylation of kinesin-associated proteins (22) will both stimulate kinesin motor activity. Kinesin motor activity may be regulated additionally or instead by direct phosphorylation of proteins which associate with kinesin (20). Phosphorylation may also regulate dynein motor activity. Dynein is present on axonal vesicles moving in both the anterograde and retrograde directions (23), suggesting that the dynein on the former is inactive cargo. Metabolic labeling studies showed a lower degree of phosphorylation of the anterograde vesicle dynein, consistent with a phosphorylation-associated regulation of dynein activity (24). PKA phosphorylation of purified kinesin reduced its ability to bind synaptic vesicles to microtubules (25). However, inhibitors of PKA and other kinases did not alter axoplasmic transport of vesicles (26). Moreover, neither kinesin phosphorylation nor function appeared to be altered in cells lacking PKA. Phosphorylation of kinesin was observed in cultures of chick sympathetic neurons and PC12 cells on both heavy and light chains, though not on the motor domains, and on associated kinectin (see chapter 1) (27). These results suggest that phosphorylation regulates binding rather than motor activity. Thus, kinesin-mediated motility of vesicles may be regulated by multiple mechanisms.

Regulation of intracellular membrane movements via microtubule motor binding to vesicles is also suggested by studies on *dynactin* (28). Dynactin is a multimeric complex which associates with dynein and is postulated to provide a link between membranes and microtubules through the motor (29). Isolated dynactin contains 10 subunits (Table 6.2) (29), including a protofilament of an actin-related protein Arp1, apparently stabilized by capping proteins (Fig. 6.4). A projecting sidearm of the protofilament, composed of proteins related to the *Glued* gene product (see chapter 1), appears to be associated with the dynein. This association is apparently stabilized by a p50 subunit. Overexpression of p50 results in partial disassembly of the complex and yields cells with mislocalized Golgi, endosomes and lysosomes and distorted mitotic spindles, confirming the importance of dynein in both organelle localization and spindle assembly (29). Several mechanisms have been suggested for how dynactin may link dynein to membranes (29, 30). The most in-

Table 6.2. Dynactin subunit composition

Subunit	M_r	Stoichiometry
p150^{Glued}	145 000	1
p135^{Glued}	135 000	1
p62	53 086	1
p50	44 800	4
Arp1	42 616	9
β-actin	42 000	1
Capping protein α subunit	32 900	1
Capping protein β$_2$ subunit	30 613	1
p27	27 000	1
p24	24 000	1

Modified from ref. 29 (Schroer TA, Bingham JB, Gill SR. Trends Cell Biol. 1996; 6: 212-215), reprinted with kind permission of Elsevier Science - NL, Sara Burgerhartstraat 25, 1055 KV Amsterdam, The Netherlands.

triguing is that the membrane-associated, kinesin-binding protein kinectin (chapter 1) also serves as a dynactin link. This mechanism could allow a common regulation of both retrograde and anterograde transport and possibly account for reports of coordination between the two directions of membrane traffic (30). Thus, the regulation of kinectin behavior becomes very important.

One problem with understanding the control of the cytoskeleton in intracellular vesicle movements is separating the vesicle movements from membrane processes, such as budding and fusion. An important in vitro system for studying vesicle movements without these complications is *isolated axoplasm*, in which *both retrograde and anterograde movements of vesicles can be observed on microtubules* (31). An intriguing system in vivo involves observation of pigment granules of fish scale chromatophore cells (1). The obvious advantage of these cells is that the granules remain intact during their translocations, removing the complication of membrane changes. Movement of the granules is under adrenergic control through G-protein-linked receptors (chapter 2). Dispersion of the granules in melanophores occurs in reponse to β-adrenergic agonists. Aggregation is an α$_2$-adrenergic response involving cAMP (32, 33). In some species of fish, granule movements are regulated both by cAMP-dependent phosphorylations and by calcium-dependent phosphatases. Phosphorylation of microtubule-associated proteins has been implicated (1), but these need to be defined.

Much less is known about microfilament-dependent trafficking in cells. One signficant example is the PDGF-induced movement of c-Src from its site of synthesis to membrane ruffles in fibroblasts (34). This process is mediated by Rac and inhibited by cytochalasin D. Interestingly, the oncogenic analog v-Src is translocated to focal adhesions, probably along stress fibers by a similar process which is Rho dependent. Previous studies had shown that localization of the oncogene was critical to its transforming activity. The mechanism and regulation of the translocations are uncertain.

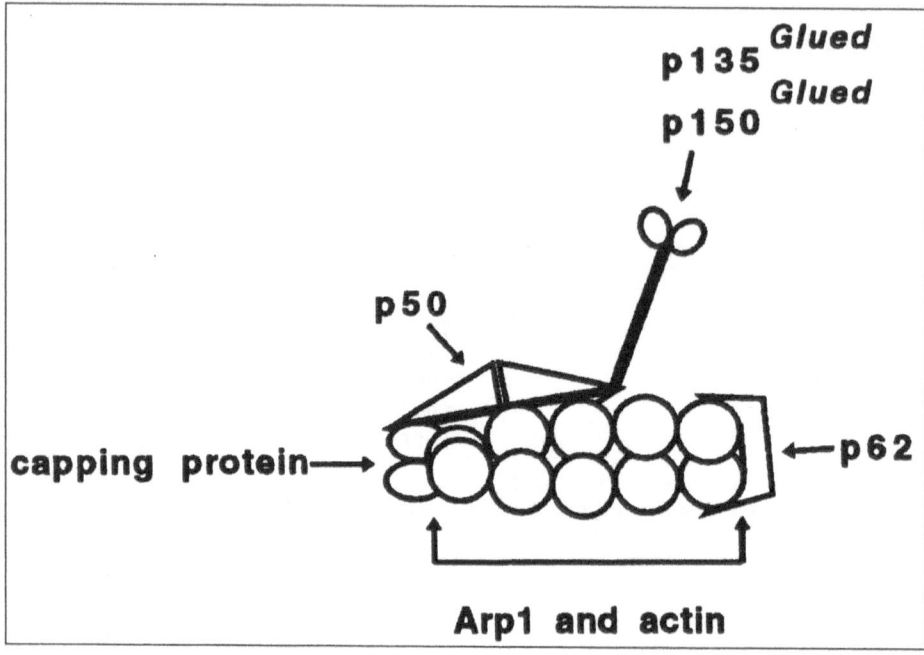

Fig. 6.4. Model for dynactin structure. Figure from ref. 29 (Schroer TA, Bingham JB, Gill SR. Trends Cell Biol. 1996; 6:212-215), reprinted with kind permission of Elsevier Science - NL, Sara Burgerhartstraat 25, 1055 KV Amsterdam, The Netherlands.

Exocytosis

Secretion is observed in many types of eukaryotic cells and is usually defined by two classes: constitutive and regulated. *Regulated secretion* involves the collection of secretory granules beneath the membrane of the secretory cell to be released in response to an external signal (35). However, recent studies demonstrating the commonality of the proteins between the two (36) suggest that the regulated secretion practiced by different cell types is simply a more complex variant of the constitutive form. By this view, regulated secretion differs primarily by the imposition of hierarchies of control steps on exocytosis which are specific for different cell types (37). For example, the basic components and mechanisms of the simple constitutive secretion in yeast are found in the highly elaborate, regulated neurotransmitter release which is controlled both temporally and spatially. The fundamental steps are described in Figure 6.5 for calcium-regulated exocytosis (38) along with common components. The exocytotic mechanism is simply an elaboration of the targeting and fusion process used generally in membrane trafficking, exemplified in the model in Figure 6.2B/C. Table 6.3 (37) describes variations observed in a number of systems exhibiting regulated secretion. Products for secretion can be packaged either in secretory vesicles or in dense-cored granules (37, 39). Exocytosis may involve membrane vesicles derived either from the secretory pathway via the Golgi and trans-Golgi network or from the recycling pathway via sorting endosomes.

The observations cited above indicate that an important feature of regulated secretion is a blockage of exocytosis. One model proposes that this block is imposed

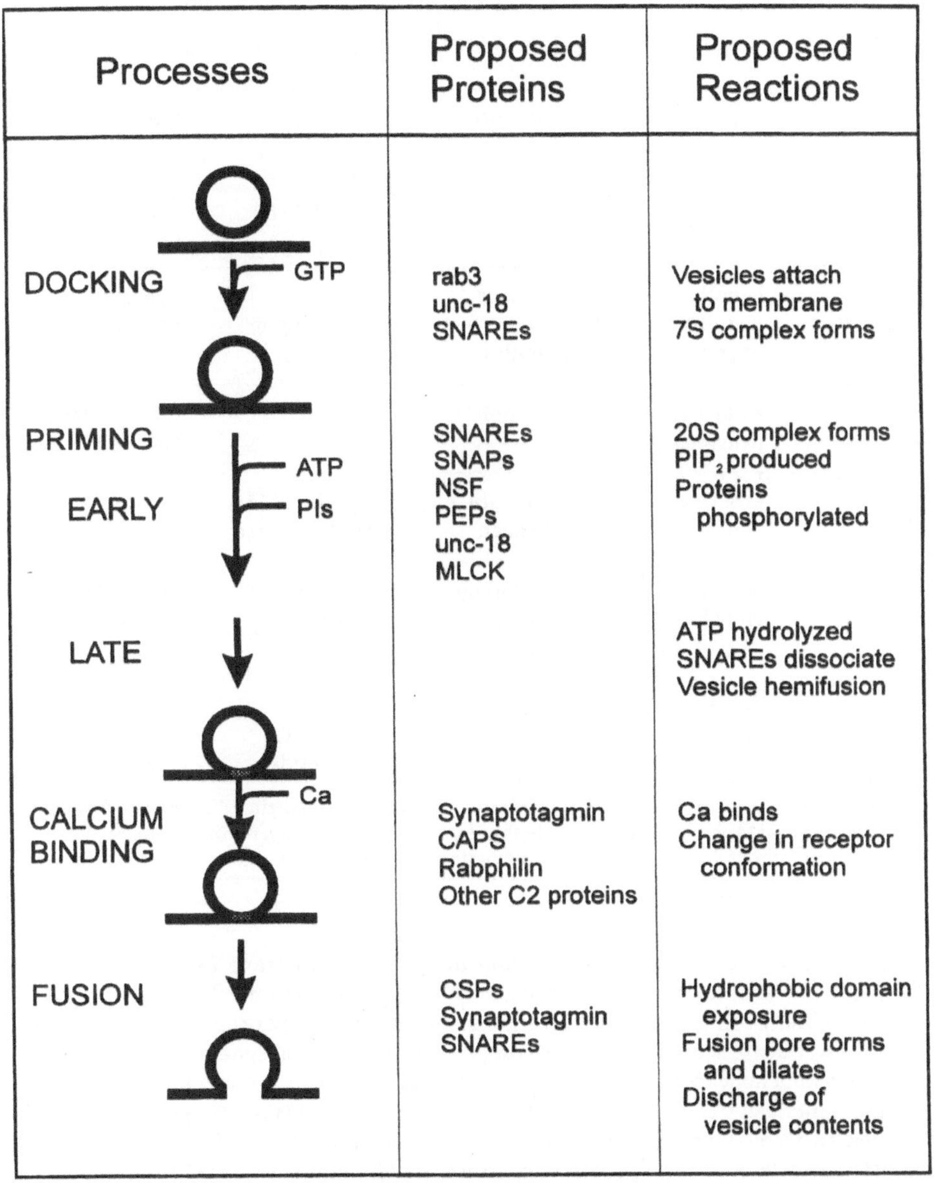

Processes	Proposed Proteins	Proposed Reactions
DOCKING ⊢ GTP	rab3 unc-18 SNAREs	Vesicles attach to membrane 7S complex forms
PRIMING EARLY ⊢ ATP ⊢ PIs	SNAREs SNAPs NSF PEPs unc-18 MLCK	20S complex forms PIP$_2$ produced Proteins phosphorylated
LATE		ATP hydrolyzed SNAREs dissociate Vesicle hemifusion
CALCIUM BINDING ⊢ Ca	Synaptotagmin CAPS Rabphilin Other C2 proteins	Ca binds Change in receptor conformation
FUSION	CSPs Synaptotagmin SNAREs	Hydrophobic domain exposure Fusion pore forms and dilates Discharge of vesicle contents

Fig. 6.5. Model for stages and components involved in calcium-mediated exocytosis. Figure from ref. 38 (Augustine GJ, Burns ME, DeBello WM, Pettit DL, Schweizer FE. Annu Rev Pharmacol Toxicol 1996; 36: 659-701), with permission, from the Annual Review of Pharmacology and Toxicology, Volume 9, 1993, by Annual Reviews Inc.

Table 6.3. Properties of some systems exhibiting regulated secretion

Secretory system	Source	Organelle	Contents	Stimulus and signal
Neuronal system	brain vesicle	synaptic	ACh, glu, gly	depolarization, Ca^{2+}
Neuro-secretory system	hypothalamo-neurohypo-physial system	posterior pituitary granules	VP, OT neurophysin	depolarization, Ca^{2+}, ATP
Endocrine gland	endocrine pancreas	β-granule	insulin	β-adrenergic, Ca^{2+}, ATP
Exocrine gland	exocrine pancreas	pancreatic zymogen	trypsin, RNAse, DNAse, lipase	Ca^{2+}, ATP
Unicellular systems	bone marrow and blood	neutrophils, eosinophils, basophils	lysozyme, hydrolases	depolarization, Ca^{2+}, complement, cAMP, cGMP
	blood platelet	5-HT granule	5-HT, ATP, nucleotides	thrombin, Ca^{2+}, ADP

Abbreviations: ACh, acetylcholine; HT, hydroxytryptamine; VP, vasopressin; OT, oxytocin. Abstracted from ref. 37 (Linial M, Parnas D. Biochim Biophys Acta 1996; 1286: 117-152), reprinted with kind permission of Elsevier Science - NL, Sara Burgerhartstraat 25, 1055 KV Amsterdam, The Netherlands.

by *cortical microfilaments* underlying the plasma membrane (40) as shown in Figure 6.6 (40) for non-neuronal cells. This type of secretion exhibits several distinct kinetic components, indicating that multiple steps can occur over a period of time to release secretory granule components. In contrast, synaptic vesicle secretion which appears to be part of a recycling pathway (35) is much faster and simpler kinetically. The importance of cortical microfilaments in exocytosis has been demonstrated by studies on permeabilized pancreatic acinar cells (41). Treatment of these cells with appropriate concentrations of β-thymosin or a gelsolin fragment led to apical actin depolymerization and exocytosis which were calcium independent. Blocking the actin depolymerization with phalloidin also blocked exocytosis. However, excess amounts of depolymerizing agents inhibited exocytosis, suggesting that a minimal microfilament cytoskeleton is required for exocytosis. Previous studies have shown that exocytosis from these cells involves two phases, the first phase mediated by calcium, the second mediated by PKC. The results with microfilament-depolymerizing agents suggest that limited microfilament depolymerization is sufficient to replace the first phase, which presumably requires calcium-dependent breakdown of the cortical cytoskeleton. These results also suggest that the submembrane microfilaments act as a clamp on exocytosis by retarding

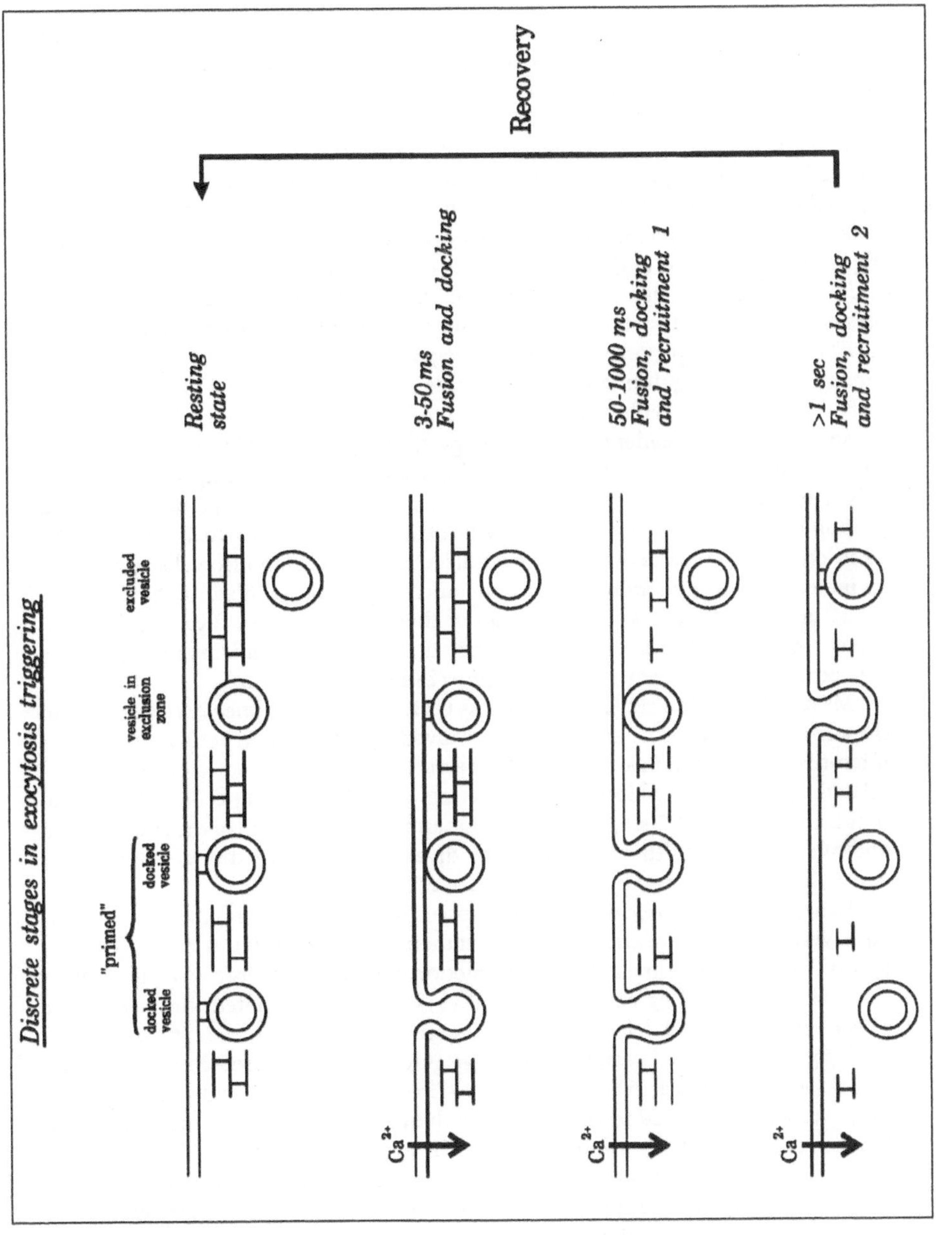

Fig. 6.6. Model for stages of exocytosis in non-neuronal cells. Figure from ref. 40 (Burgoyne RD, Morgan A. Biochem. J. 1993; 293: 305-316), courtesy of The Biochemical Society and Portland Press.

the fusion of primed granules with the plasma membrane (40). In contrast, the inhibition of exocytosis by more extensive actin filament depolymerization indicates that microfilaments are required for an earlier step in the secretion process, perhaps in bringing the granules to the sites for exocytosis (9).

Calcium acts as the triggering agent for exocytosis in many cell types, implicating calcium-binding proteins in the regulatory steps. One of the first calcium-binding proteins to be identified in the stimulation of exocytosis was *annexin II*, also called calpactin or lipocortin II, a member of the calcium and phospholipid-binding family of annexins (42). Annexin II is usually found as a tetramer of two 36 kDa and two 11 kDa subunits. Significantly, annexin II can aggregate chromaffin granules at 1 µM Ca²⁺ and sustains catecholamine release in response to a Ca²⁺ pulse in an ATP-dependent process. Granule aggregation is sensitive to phosphorylation, though the exact requirements remain unclear. Annexin II can be phosphorylated by both Ser/Thr and Tyr kinases. Annexin II associates with and bundles microfilaments in a Ca²⁺-dependent manner, suggesting a possible role in regulating the structure of cortical microfilaments during exocytosis. Interestingly, another signaling protein implicated in regulated exocytosis is from the 14-3-3 family described in chapter 2, which has one domain with substantial homology to annexin II. The addition of 14-3-3 isoforms to permeabilized adrenal chromaffin cells resulted in an increased catecholamine release concomitant with reorganization of the cortical actin network (43). The association of these proteins with PKC and their involvement in priming of exocytosis suggests possible mechanisms for regulating exocytosis by microfilament perturbations induced by calcium.

Mast cell secretion can be triggered by GTP. Both *heterotrimeric G-proteins* and the small G-proteins Rac and Rho are involved (44). G-protein activation induces microfilament reorganization as well as secretion. Although the two events are correlated, studies with inhibitors on secreting and nonsecreting cells indicate that they are independent. Using mutants of Rac and Rho, the control of the cytoskeleton and of secretion was shown to occur by divergent, parallel pathways, apparently controlled by a common upstream regulator.

Neurotransmitter release at the synapse is so fast that the exocytosis must involve vesicles which are already docked at their release sites (Fig. 6.7) (40). Consistent with this supposition, only a small fraction of vesicles undergo exocytosis upon arrival of the signaling action potential (45). Calcium mediated fusion of those presumably docked vesicles releases their contents, but further exocytosis requires docking of additional vesicles, a reserve pool, which are crosslinked within a cytoskeletal matrix rich in F-actin and spectrin (40). In the most widely cited model, these vesicles are linked to the matrix by the protein synapsin, a microfilament-binding protein (37). Phosphorylation of the synapsin by calmodulin-dependent kinase alters the vesicle association with actin to permit it to dock at the membrane. However, mice with both synapsin genes knocked out were still viable and did not exhibit drastic changes in their neurological parameters. Thus, it is difficult to envision that synapsins are necessary for synaptic transmission. An alternative method of regulating the cytoskeletal block to synaptic vesicle exocytosis is suggested by the identification of a 145 kDa cytosolic factor which can reconstitute calcium-activated secretion in neuroendocrine cells (46). This protein has binding domains for both tropomyosin and PIP₂ (47). Antibodies against the tropomyosin-binding domain blocked its stimulation of secretion and its interaction with F-actin. How the protein might perturb microfilaments as a contribution to secretion remains uncertain.

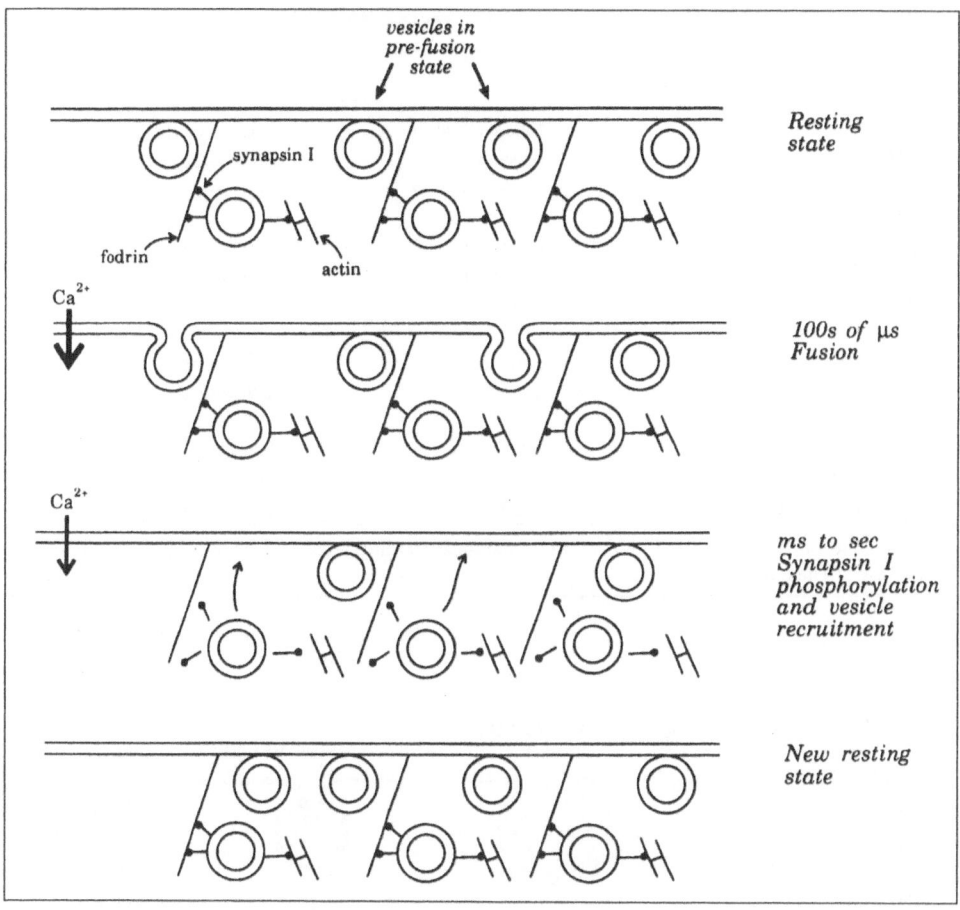

Fig. 6.7. Model for stages of exocytosis at synapse. Figure from ref. 40 (Burgoyne RD, Morgan A. Biochem. J. 1993; 293: 305-316), courtesy of The Biochemical Society and Portland Press.

Glutamate exocytosis from synaptic nerve terminals appears to involve a different control mechanism. Calcium-dependent release of glutamate is significantly enhanced by phorbol esters which activate PKC isoforms. Moreover, PKCε appears to play a role in sustaining glutamate exocytosis. Extraction and binding studies indicate that PKCε is tightly bound to F-actin in an active form (48). How this binding contributes to exocytosis is still uncertain.

Endocytosis

Endocytosis is the process used by virtually all cells to internalize external nutrients and to modulate expression of cell surface molecules. Two broad classes of endocytosis have been described, *pinocytosis* and *phagocytosis* (39). Both phagocytosis and pinocytosis can also be characterized as receptor-mediated and nonreceptor-mediated. Four different types of pinocytosis have been described (Table 6.4 and Fig. 6.8) (39, 49) which can be differentiated by

Table 6.4. Types of endocytosis

Type of endocytosis	Size of vesicle or vacuole	Intermediate organelle
Phagocytosis		
Receptor-mediated	0.1-10 µm	Phagosome
Non-receptor-mediated	0.1-10 µm	Vacuole
Pinocytosis		
Clathrin-dependent		
Receptor-mediated	100 nm	Endosome
Synaptic vesicle	50 nm	Endosome
Clathrin-independent		
Potocytosis	60 nm	Caveolae
Micropinocytosis	100 nm	Micropinosome
Macropinocytosis	0.5-6 µm	Macropinosome

Modified from ref. 39 (Liu J-P, Robinson PJ. Dynamin and endocytosis. Endocrin Rev 1995; 16: 590-607), courtesy of The Endocrine Society.

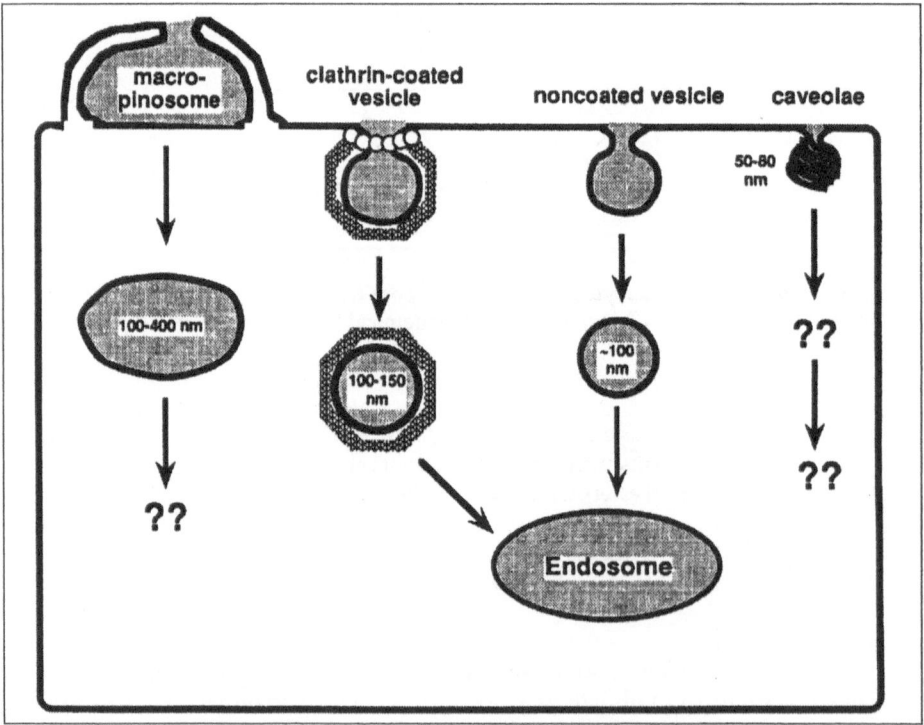

Fig. 6.8. Multiple pinocytotic pathways. Figure from ref. 49 (Lamaze C, Schmid SL. The emergence of clathrin-independent pinocytic pathways. Curr Opin Cell Biol 1995; 7:573-580), courtesy of Current Biology Ltd.

pharmacological treatments (49). In most cases, *receptor-mediated pinocytosis involves clathrin-dependent membrane invaginations.* Nonreceptor-mediated pinocytosis requires clathrin-independent mechanisms such as that involving caveolae.

Receptor-Mediated

In receptor-mediated, clathrin-dependent endocytosis, receptors and their ligands are concentrated in plasma membrane invaginations, called *coated pits*, containing clathrin and adaptor proteins (*APs*, also called adaptins) (50). This mechanism restricts the size of the invaginations and subsequent endocytic vesicles to that of the clathrin coat (85-110 nm diameter). Coat formation is initiated by the binding of APs to sites, apparently adaptin receptors on the plasma membrane (50). The APs can then bind receptor cytoplasmic domains to concentrate them into the nascent coated pits. Two general classes of receptors are internalized by this mechanism: receptor kinases, such as the EGF receptor, and "cargo receptors" such as the transferrin receptor and LDL receptor. In the case of receptor kinases, ligand binding to the receptor extracellular domain may enhance the binding of the receptors to APs (51). The two classes of receptors appear to have different internalization signals in their cytoplasmic domains, but they are co-concentrated in the same coated pits.

APs mediate the assembly of clathrin into a planar lattice at the plasma membrane cytoplasmic surface. Clathrin contains three copies each of heavy and light chains to form a three-pronged structure, the *triskelion*, which can associate into networks on the membrane. The microtubule-binding protein *dynamin* associates with the lattices which reorganize to produce membrane invaginations (52). Dynamin is related to both heterotrimeric and small G-proteins, but has a substantially higher GTPase activity (53). A model for the role of dynamin in coated pit formation posits that dynamin binds to the lattice in its GDP state. Binding of GTP triggers redistribution of the dynamin and assembly into helical rings which form the neck constrictions of the coated pits, the collars observed by electron microscopy (53). GTP hydrolysis mediates a second structural change to tighten the collar for membrane fusion. Release of the dynamin from the membrane after GTP hydrolysis allows it to be recycled for additional budding events. *In this mechanism dynamin acts as a GTP-driven mechanochemical motor, similar to myosin, kinesin or dynein, rather than as a G-protein switch.*

Although dynamin was originally isolated based on its association with microtubules from brain extracts, no evidence for a dynamin-microtubule interaction has been observed in vivo. How dynamin is associated with the clathrin lattice and plasma membrane is still unclear, but transfection experiments indicate multiple regions that target the protein to either clathrin-coated or uncoated membrane domains. Binding to clathrin-coated pits requires a 9 amino acid C-terminal motif containing a potential SH3 binding site (54). The SH3 binding site could also provide linkages to submembrane signaling elements which might regulate endocytosis or subsequent internalization events. SH3-containing proteins which bind dynamin and stimulate its GTPase activity include Grb2, PI3K and PLCγ (52). Interestingly, microinjection of a Grb2 SH2 domain peptide blocks EGF receptor internalization, indicating a role for the adaptor protein in endocytosis and providing a possible link between receptor downregulation and mitogenic signaling (55). In addition, dynamin has been shown to have a site for PKC phosphorylation near its C-terminus, suggesting that its function may be regulated by phosphorylation and dephosphorylation (50).

Microfilament Involvement

A role for microfilaments in endocytosis has been demonstrated in budding yeast (56) and at the apical, but not the basolateral surfaces of polarized mammalian cells (57). Studies on budding yeast have provided important insights into the endocytosis process because of the power of yeast genetics to identify proteins involved. Binding of the yeast pheromone-factor to its receptor activates receptor hyperphosphorylation, stimulates a signaling cascade through MAPK and induces receptor internalization (56). Neither phosphorylation nor signaling is required for endocytosis, but conditional mutants indicate that clathrin and actin, but not tubulin, are necessary for internalization. Furthermore, mutants in the actin-bundling protein fimbrin (Fig. 1.1) are also defective for α-factor endocytosis (58). Microfilaments in budding yeast occur in two locales, described in chapter 3, as cables extending through the cytoplasm and as cortical patches at the plasma membrane. Several endocytosis mutants have been shown to have altered localization of the cortical patches (59). However, other cytoskeletal mutants including mutations in myosin V, profilin and tropomyosin, have delocalized actin patches, but normal endocytosis. Cortical patches consist of microfilaments wound around plasma membrane invaginations (60). Perhaps endocytosis defective mutants alter the structure as well as the localization of the invaginations. Regardless, a characterization of the roles of these mutants should provide invaluable insights for understanding endocytosis.

Although the specifics of the microfilament involvement in yeast endocytosis remain to be defined, recent studies definitively implicated type I myosins (61, 62) previously discussed in chapters 1, 3 and 4. Deletion of the budding yeast *MYO3* gene caused no apparent phenotypic change. However, a synthetic lethality screen identified a *MYO5* as a second myosin I gene (61). The double mutant exhibited pronounced phenotypic changes, including slower growth, altered morphology, reduced actin cables, actin patch disorganization, reduced invertase secretion and defective fluid phase endocytosis. Myo5p was localized to actin patches, a site of filament nucleation. In a second study, a double deletion of the myosin I genes gave slow-growing mutants which can be complemented with a *MYO5* temperature-sensitive allele and are defective in endocytosis (62). In contrast, the secretory pathway was apparently unaffected. Thus, myosin I appears to be critical for endocytosis and is probably involved in other processes regulating cytoskeleton organization and contributing to cell growth.

Although microtubules appear not to be involved in the invagination step of endocytosis, they may play important roles in subsequent endocytotic pathways, including the segregation of receptor and ligand for recycling and degradation, respectively (63). After receptor-mediated endocytosis the internalized glycoprotein asialoorosomucoid (ASOR) and its receptor are both in vesicles bound to microtubules. ATP treatments release ASOR-containing vesicles, but not the receptor-containing vesicles. Microtubule binding of the ASOR-containing vesicles correlates with binding to dynein, but not kinesin. Furthermore, released vesicles can be immunoprecipitated with anti-dynein (63). These studies provide a model by which dynein binding to ligand-containing vesicles can move them along microtubules toward the lysosomal compartment, even perhaps provide a motive force for pulling the sorting endosome apart as part of the segregation process (Fig. 6.9).

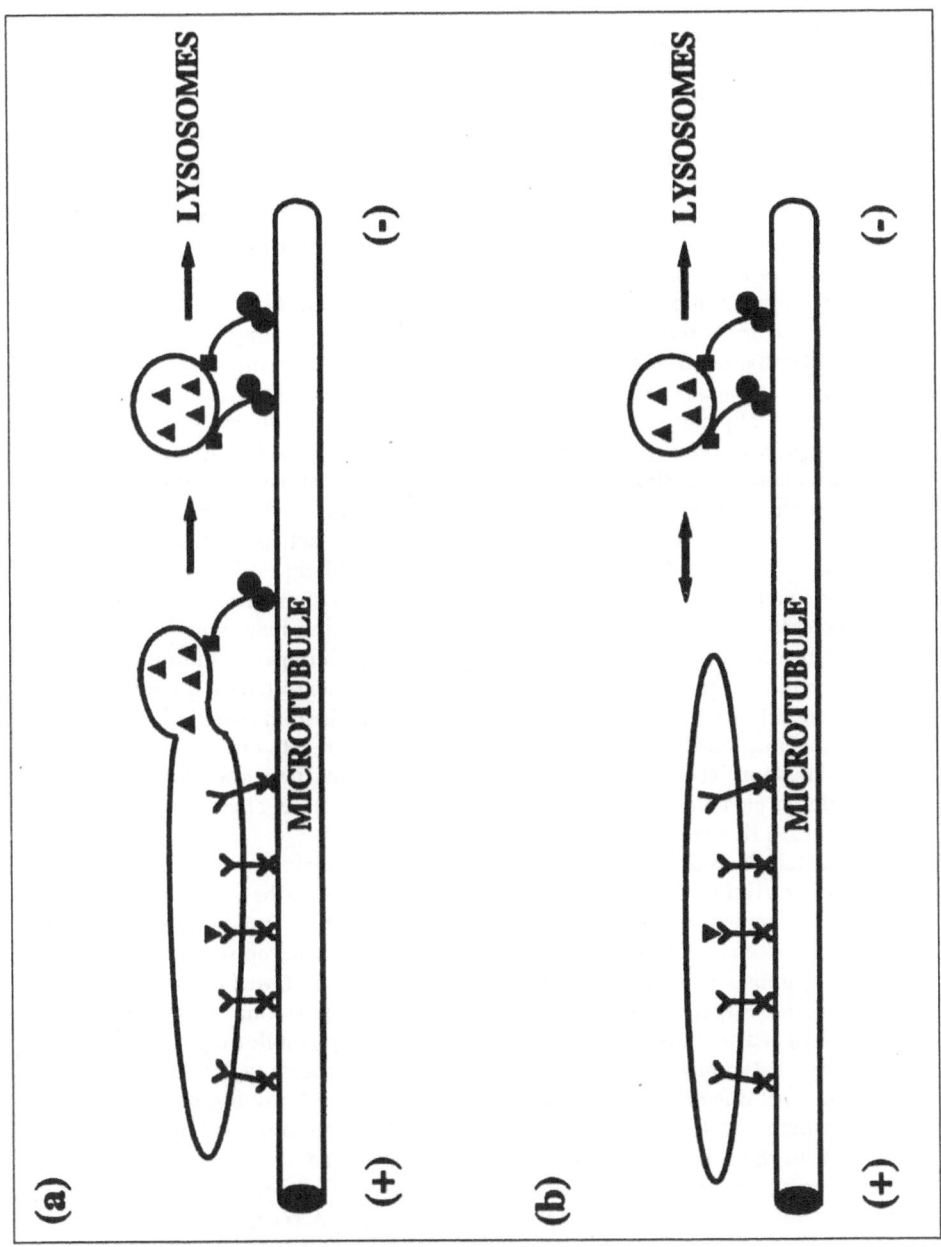

Fig. 6.9. Models for role of cytoplasmic dynein in endosome fracturing and sorting and in vesicle movement along microtubules. Figure from ref. 63 (Oda H, Stockert RJ, Collins C, Wang H, Novikoff PM, Satir P, Wolkoff AW. J Biol Chem 1995; 270: 15242-15249), courtesy of the American Society of Biochemistry and Molecular Biology.

Receptor Down-Regulation

One of the important functions of receptor-mediated endocytosis is the ligand-dependent regulation of the cellular content of signaling receptors, such the tyrosine kinase and G-protein-linked receptors. In some cases the regulation of endocytosis appears more stringent for signaling receptors than "cargo" receptors. For example, ligand-induced recruitment of the EGF receptor into coated pits requires a rate-limiting cytosolic component not necessary for transferrin receptor (64) and is inhibited by genistein, a tyrosine kinase inhibitor. Kinase-deficient EGF receptors are more slowly recruited into coated pits, but this deficiency can be corrected by adding soluble EGF receptor kinase domain (65). An intriguing aspect of receptor internalization is its relationship to signaling, particularly the possibility that signaling through receptors requires their internalization. This hypothesis appears not to be correct. Internalization defective mutants of the EGF receptor actually have a higher transforming activity than their wild-type counterparts (66). Moreover, other members of the EGF receptor family (ErbB-2 and -4) which are active in signaling processes appear to be endocytosis-impaired (67). However, these observations do not indicate that the internalized receptor cannot participate in signaling, only that it is not necessary. In fact, functional EGF receptor kinase has been demonstrated in endosomes (68), and cells with defects in endocytosis appear to differentially activate different signaling pathways of the EGF receptor (69), suggesting that endocytic trafficking may modulate specific signals. Insulin receptor (IR) is also down-regulated after ligand binding. Interestingly, insulin receptor in some cells appears to be concentrated on microvilli (70) and must migrate into coated pits for internalization. Insulin stimulation induces tyrosine phosphorylation of annexin II, which, in contrast to other insulin-stimulated phosphorylations, is blocked by inhibition of internalization (71). IR internalization is also blocked by agents which inhibit IR kinase. These results suggest that annexin II may play a role in the endocytosis of IR.

G protein-linked (serpentine) receptors are also regulated by endocytotic pathways. Interestingly, different receptors follow different paths (72). For example, the β2 adrenergic receptor (β2AR) can be sequestered in a process that involves cycling between the plasma membranes and endosomes. Receptor desensitization occurs by phosphorylation by specific receptor kinases, followed by binding of the inhibitor β-arrestin to the phosphorylated form. Internalization requires agonist-dependent redistribution of the receptor into coated pits followed by agonist-independent internalization in a dynamin-dependent process (73). Experiments in the same cell line demonstrated a different fate for the thrombin receptor (TR). After stimulation, the TR is similarly internalized into endosomes, but its pathway diverges to shunt it to the lysosomes for receptor degradation. The TR is then replaced on the cell surface from a naive receptor pool of newly synthesized receptors. Internalization of the angiotensin II type 1A receptor, another typical G-protein-linked receptor, does not normally require dynamin, though it can be shifted to the dynamin pathway by overexpression of β-arrestin (73). Thus, multiple pathways contribute to regulation of the G-protein-linked receptors, depending on the receptor and cell type and context.

Caveolae

Caveolae are small, flask-shaped invaginations of the plasma membrane involved in endocytosis and transcytosis in a number of cell types such as adipocytes, smooth muscle cells and endothelial cells (74, 75). Caveolae are enriched in a number

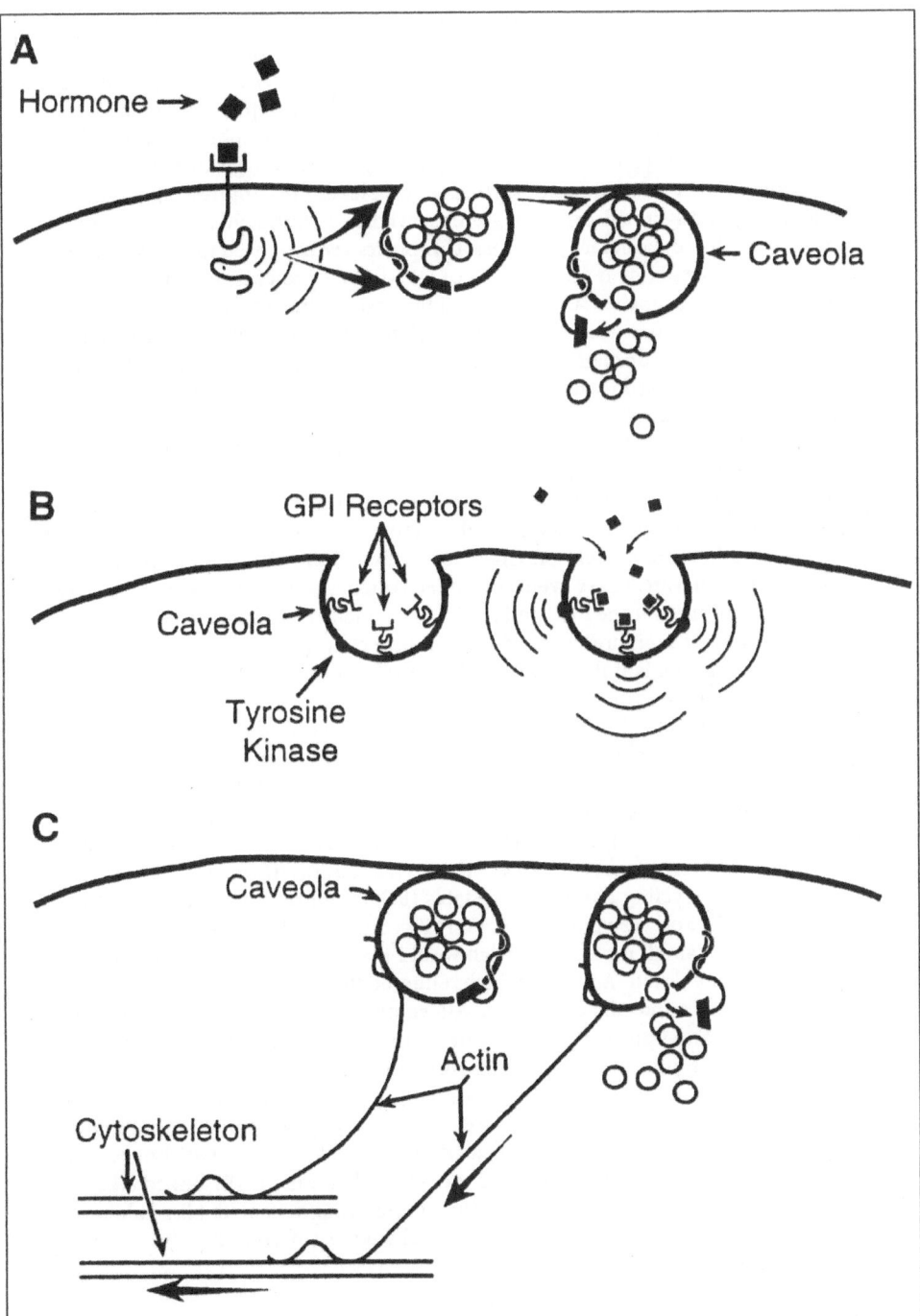

Fig. 6.10. Model for endocytosis and signaling by caveolae. Figure from ref. 81, courtesy of Dr. Richard Anderson.

of types of molecules including GPI-anchored proteins, sphingomyelin, ganglio-sides, PKC, IP$_3$ receptors, G-protein-coupled receptors, heterotrimeric G-proteins, nonreceptor tyrosine kinases and caveolin. *Caveolin* is a 21 kDa integral membrane protein associated with the membrane through a hairpin domain. It is isolated as a detergent-insoluble complex containing glycolipids, sphingomyelin, cholesterol, GPI-linked proteins and signaling components, suggesting that it may play a role in the organization of signaling complexes (76). Interestingly, similar complexes of GPI-linked proteins are observed in lymphocytes which contain neither caveolin or caveolae (77). These complexes are proposed to be involved in microfilament-dependent endocytosis of GPI-linked proteins. How they are related to caveolae is unclear. Caveolin itself has a 20 amino acid membrane proximal domain which has been shown to associate with G-protein subunits, H-Ras and c-Src (78). This domain is also involved in the oligomerization of caveolin and has been proposed to act as a scaffolding domain for organizing complexes of signaling molecules.

Endocytosis via either the lymphocyte complexes or caveolae is regulated by PKC (77, 79). PKA was also implicated in the case of the lymphocyte complexes (77). Dissociation of caveolae from the plasma membrane into cytoplasmic vesicles and reasociation of the vesicles with the membrane were correlated with PKC phosphorylation and protein phosphatase dephosphorylation, respectively, of a 90 kDa PKC substrate. These events occurred concomitantly with association and dissociation of membrane PKC and could be mimicked by histamine as a regulator of PKC and of membrane diacylglycerol (see chapter 2). How PKC associates with the membrane is unknown, but PKC binding proteins (RACKs) have been demonstrated to be associated with caveolae (79).

Caveolae differ from other membrane internalization domains in that they generally remain near the plasma membrane after pinching off and do not fuse with endosomes. Thus, they appear as either invaginated pits (caveolae) or as plasmalemmal vesicles of about 50 nm, both of whose cytoplasmic surfaces are decorated with a coat of filaments (80). The caveolae and vesicles are proposed to transfer small molecules between the extracellular fluid and cytoplasm (*potocytosis*) by cycling among three states (81): membrane-open (MO), vesicular- closed (VC) and vesicular-open (VO) (Fig. 6.10). The MO state has been proposed to collect small molecules which would then be transported from the cell surface into the VC state. A trigger which rendered these vesicles permeable would then release them into the cytoplasm. Alternatively, they could be stored in the VC form until an appropriate signal caused their fusion with the plasma membrane for release into the extracellular millieu. It is tempting to speculate that cortical microfilaments play a role in maintaining the localization of these vesicles and may be involved in triggering the release of their contents into the cytoplasm (81). Microfilaments have been implicated in the endocytosis of GPI-linked protein domains in lymphocytes (77) and the internalization of GPI-linked alkaline phosphatase by caveolae (82). Association of microfilaments with plasmalemmal vesicles has been suggested by morphological studies (83), but their function remains unclear.

Macropinocytosis

Macropinocytosis is a type of endocytosis accompanying *cell surface ruffling*, usually at the margins of spread cells (84). Macropinosomes are characterized by

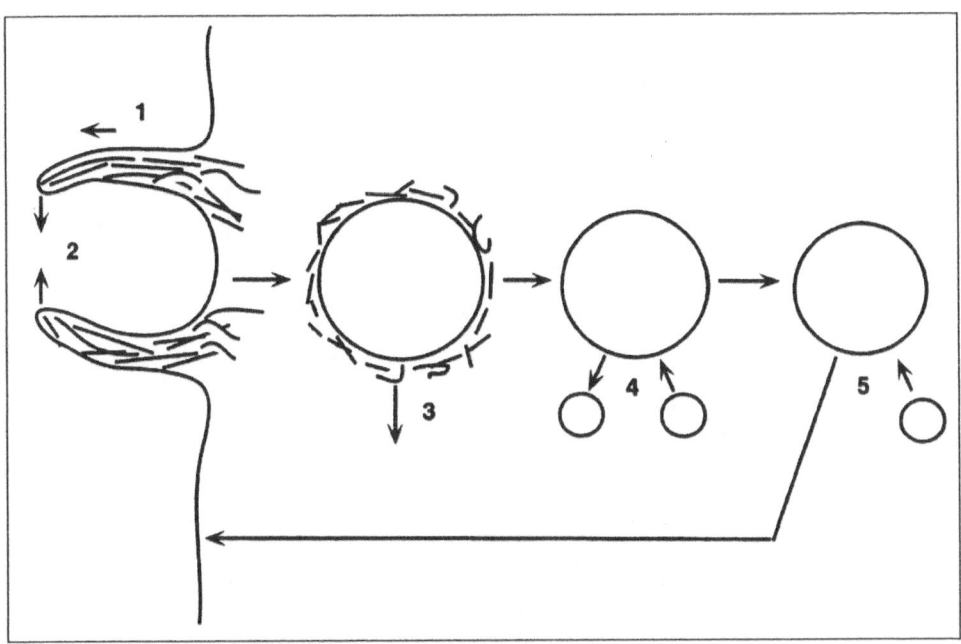

Fig. 6.11. Model for macropinocytosis. Figure from ref. 84 (Swanson JA, Watts C. Trends Cell Biol 1995; 5: 424-428), reprinted with kind permission of Elsevier Science - NL, Sara Burgerhartstraat 25, 1055 KV Amsterdam, The Netherlands.

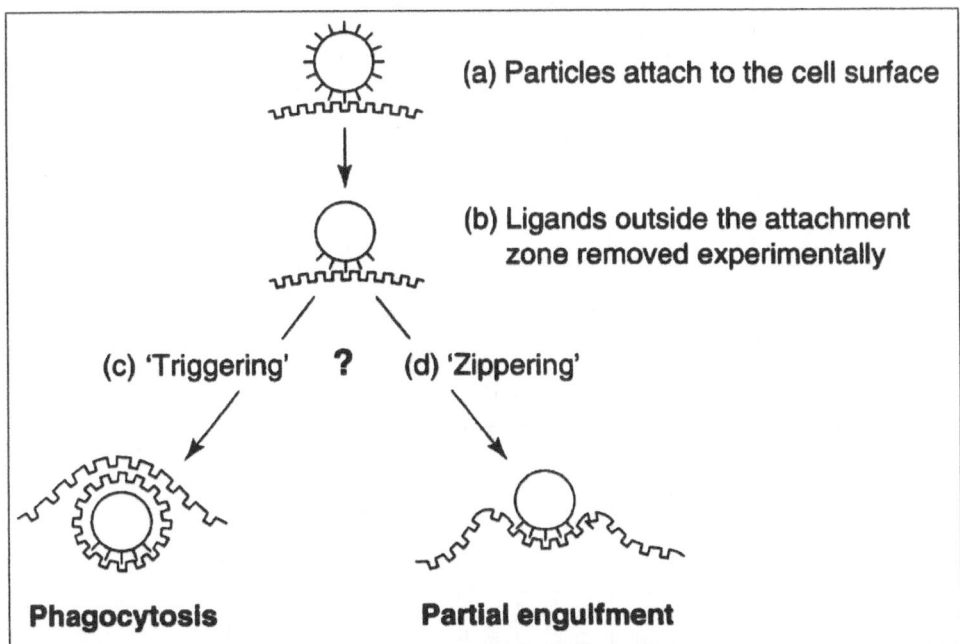

Fig. 6.12. Alternative models for phagocytosis. Figure from ref. 88 (Swanson JA, Baer SC. Trends Cell Biol 1995; 5: 89-93), reprinted with kind permission of Elsevier Science - NL, Sara Burgerhartstraat 25, 1055 KV Amsterdam, The Netherlands.

Table 6.5 Phagocytic receptors

Receptors (CD)	M_r	Gene family	Distribution	Ligand(s)
FcγRI (CD 64)	75	Ig	Monocytes, M-φ, IFN-stimulated PMN	Monomeric IgG
FcγRII (CD 32)	40	Ig	Leukocytes; platelets	Immune complexes
FcγRIII (CD 16)	50–70	Ig	PMN (PI-linked), Mφ, NK cells	Immune complexes
CR1 (CD 35)	160–220	RCA	Phagocytes, erythrocytes, B lymphocytes	C3b
CR3 (CD 11b/CD 18)	α 165 β 95	Integrin	Phagocytes, NK cells	iC3b, fibrinogen
CR4 (CD 11c/CD 18)	α 150 β 95	Integrin	Mφ, Dendritic Cells	Fibrinogen, iC3b
Mannose receptor	175	Lectin	Mφ	Mannose/fucose terminal residues

Abbreviations: Ig, immunoglobulin superfamily; RCA, regulators of complement activation; M-Φ, macrophages; PMN, polymorphonuclear neutrophils. Modified from ref. 90 (Brown EJ. BioEssays 1995; 17:109–117), courtesy of ICSU Press.

their location, lack of discernible coat, heterogeneity and large size (up to 5 microns). This size provides an ideal mechanism for nonspecific uptake of fluid phase solutes, although the advantage of such uptake to cells is not entirely clear (84). Macropinocytosis, which requires actin polymerization resembles phagocytosis without an inducing particle. Alternatively, macropinocytosis can be viewed as the fusion of membrane ruffles to the adjacent cell surface to enclose extracellular fluid. It is stimulated by agents which stimulate ruffling, such as cytokines, growth factors, phorbol esters or microinjection of oncogenic Ras. As previously mentioned, a central element in the signaling pathway for ruffling is the small G-protein Rac (85). Thus, each of these inducing agents presumably activates Rac, as described in chapter 4, which is proposed to regulate assembly of actin nucleating mechanisms at the plasma membrane. Simplistically, one can look at this process as a variant of the protrusive activity involved in motility. Adhesion of the protrusion leads to directed motility. Protrusions which do not adhere sweep back over the cell surface. When they fuse with membrane they form a macropinosome which encloses extracellular fluid. Successive stages in the internalization of the macropinosome are shown in Figure 6.11 (84) and include the removal of the microfilament coat and interaction with other endosomal compartments. The role of the cytoskeleton in these processes is unclear, but it is interesting to note that double myosin I mutants of *Dictyostelium* which exhibit conditional defects in fluid phase pinocytosis are still able to extend processes (86).

Phagocytosis

Phagocytosis is the cellular uptake of relatively large particles with two general requirements:

1) an interaction between molecules on the surface of the particle and receptors on the cell; and

2) actin polymerization (87).

In mammals, most phagocytosis involves "professional phagocytes", such as polymorphonuclear phagocytes, monocytes and macrophages. These cells provide a primary defense mechanism against invading microbes and tissue damage and accumulate at sites of infection, inflammation and tissue damage. Thus, they are also specialized to perform other defensive tasks such as the production of cytotoxic and degradative molecules. Other cells, nonprofessional phagocytes, are also able to internalize particles, though more selectively and less efficiently (87). Two different mechanisms have been proposed to describe phagocytosis: zippering and triggering (Fig. 6.12) (88). In the former engulfment of the particle proceeds by progressive binding of ligand sites on the surface of the particle by cellular receptors, similar to the progressive adhesion of a migrating cell to adhesion sites on the substratum. In contrast, the triggering mechanism involves only a limited initial contact of cell and particle to initiate engulfment of the particle. In fact, the triggering mechanism may not be a phagocytic response. Instead, it could be described as macropinocytosis induced by a particle which triggered increased ruffling (88). This analysis highlights one of the important aspects of the classic phagocytic response. Ligand-receptor interaction initiates only a local response in the cytoskeleton in the early phase of activation.

The key to phagocytosis is the recognition of ligand on the target particle by cellular receptors, some of which are listed in Table 6.5. Thus, the receptors must have evolved to recognize surface components of the particles or host proteins deposited on the particles such as opsonins. Receptor requirements for phagocytosis

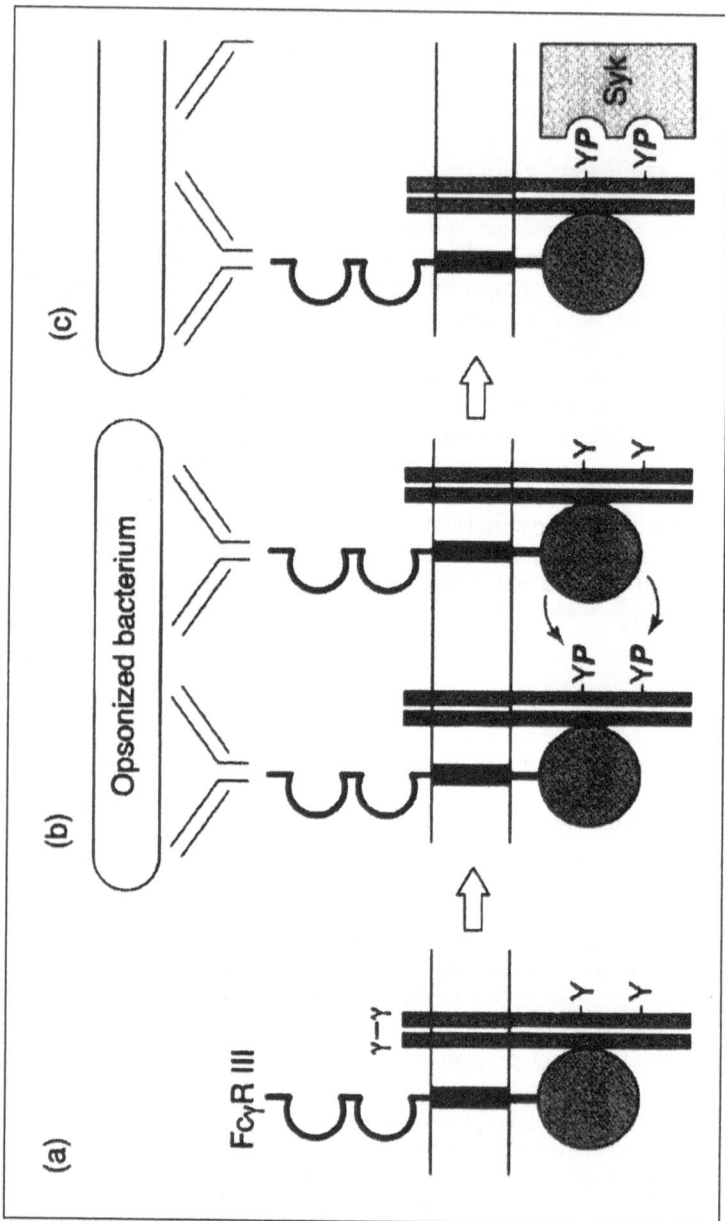

Fig. 6.13. Model for early signaling events of Fcγ receptors in macrophages. Figure from ref. 93 (Greenberg S. Trends Cell Biol 1995; 5: 93-99), reprinted with kind permission of Elsevier Science - NL, Sara Burgerhartstraat 25, 1055 KV Amsterdam, The Netherlands.

are quite varied and complex. Integrins are able to bind specific components on the surfaces of pathogens, though only a subset of these are internalized (89). Integrin-mediated phagocytosis is modulated by the affinity of the integrin-ligand interaction. High affinity associations induced phagocytosis at low integrin levels. Lower affinity ligands required higher integrin levels for phagocytosis. Ligand-receptor interactions can be modulated by other cell surface molecules. Leukocyte Mac1 (CD11b/CD18, CR3), a β2-integrin, associates with Fcγ receptors to enhance phagocytosis of targets coated with Ig opsonins (90). Extracellular matrix components can also stimulate phagocytosis. For example, integrin-mediated phagocytosis involving α3β1 and α6β1 integrins is stimulated by basement membrane entactin and laminin, respectively. Phagocytosis via an incompletely characterized β3 integrin on PMN and monocytes involves association with both the extracellular matrix and an integral membrane protein (integrin-associated protein, IAP) which is a member of the immunoglobulin superfamily. IAP is proposed to be a calcium channel which could regulate subsequent events in the phagocytic process.

Cortical microfilament rearrangements are necessary for phagocytosis. The requirement for actin polymerization is indicated by the ability of cytochalasin D to inhibit the process. As previously noted, polymerization requires nucleation sites and uncapping processes, but no specfic components have been identified with phagocytic mechanisms. Microfilament crosslinking may play a role in phagocytosis since the fimbrin analog L-plastin is phosphorylated in response to FcγRII adhesion and phagocytosis (91). Adhesion resulted in a translocation of the plastin to podosomes leukocyte adhesion sites. Phosphorylation was not inhibited by cytochalasin D, suggesting that localization and phosphorylation were independent events. How the cytoskeleton reorganization is regulated is unclear. Studies of Fcγ receptor-mediated phagocytosis of opsonized particles have shown changes in calcium, PKC activity and tyrosine phosphorylation, but none of these appear to be absolutely required (90). Since phagocytosis involves actin rearrangements similar to those observed in cell motility (chapter 4), one explanation is that different activating pathways could contribute to phagocytosis under different circumstances, thus obviating a requirement for any single stimulating event. Consistent with this hypothesis is the fact that complement- and IgG-opsonized particles are phagocytosed differently by macrophages (92). Phosphotyrosine-containing proteins and cytoskeletal proteins, such as F-actin, paxillin, α-actinin and vinculin, are distributed in punctate structures on the phagosome surface for complement-opsonized particles, but uniformly distributed for IgG-opsonized particles. Formation of the foci in the former case requires PKC. In contrast, ingestion of IgG beads was blocked by tyrosine kinase inhibitors, but phagocytosis of complement-opsonized beads was not.

Phagocytosis via FcγRII resembles T cell activation (93). These receptors contain a tyrosine activation motif (TAM) which becomes tyrosine phosphorylated probably by a member of the Src family by clustering of the receptors. Mutations in these sequences or tyrosine kinase inhibitors blocks this mechanism of phagocytosis. Phosphorylated TAMs can bind the tyrosine kinase Syk for recruitment to the membrane (Fig. 6.13) (93). How Syk is involved in subsequent events is unclear, though it has been shown to bind PI3K in platelets. PI3K is implicated in phagocytosis through inhibition studies with wortmannin. PI3K is associated with the regulation of isoforms of PKC, a component of ligated FcγR complexes, and of Rac, which is involved in microfilament regulation. Other lipid-associated enzymes which have been implicated in cytoskeleton changes may also be involved in phagocytosis, including phospholipase A2, whose action produces arachidonic acid, a factor in pseudopod formation (93). PKA and casein kinase II have been observed in FcγR complexes, supporting the possibility of multiple signaling pathways associated with responses leading to phagocytosis. How these signaling elements contribute to microfilament reorganization and the spatial restriction of the signal is unclear.

Summary

Microtubules are important for establishing and maintaining the localization of intracellular organelles and as tracks for trafficking of membrane vesicles between cellular organelles including the plasma membrane. Mechanisms and constituents involved in membrane budding and membrane fusion appear to be common for all types of membrane trafficking. Most intracellular movements of vesicles require that the microtubule motors kinesin and dynein form complexes which bind to both vesicles and microtubules. The motors are regulated by phosphorylation, though the molecular mechanisms for regulation remain unclear. Secretion can occur by either constitutive or regulated mechanisms. Microfilaments are proposed to clamp secretion by association with vesicles near the plasma membrane, releasing the vesicles in response to signals which activate microfilament depolymerization and/or vesicle release. The rate and extent of exocytosis is determined by the vesicle-filament association and dissociation. Different types of exocytotic cells respond to different signals, but calcium and phosphorylation are commonly implicated in exocytotic mechanisms.

The term endocytosis encompasses several processes by which materials are taken into cells. The best-studied is clathrin-mediated endocytosis, in which receptors and their ligands are concentrated in coated pits. The GTPase dynamin appears to act as a motor for constricting the pits to form vesicles. Dynamin can bind a number of signaling molecules containing SH3 domains which may provide a link between signaling and endocytosis. Actomyosin motors have also been implicated in endocytosis, but their specific role has not been determined. Caveolae are used for uptake of some molecules. These plasma membrane invaginations also contain a number of signaling molecules, but their role is still unclear. Macropinocytosis and phagocytosis are similar in the size of vesicle involved and the requirement for microfilaments. The former appears to be a consequence of the ruffling process, and is thus activated by agents which stimulate cell surface receptors and regulated by Rac. Phagocytosis is triggered by particle binding to specific receptors to activate intracellular signaling elements. Though a number of pathways have been implicated, none appears to be universal, suggesting redundancy in the mechanisms which can trigger particle engulfment.

References

1. Stebbings H. Microtubule-based intracellular transport of organelles. In: Hesketh HE, Pryme IF, eds. Treatise on the Cytoskeleton, Greenwich, CT: JAI Press, 1996:113-140.
2. Pickard JE, Kreis TE. CLIPs for organelle-microtubule interactions. Trends Cell Biol 1996; 6:178-183.
3. Sheetz MP, Dai J. Modulation of membrane dynamics and cell motility by membrane tension. Trends Cell Biol 1996; 6:85-89.
4. Goodson HV, Valetti C, Kreis TE. Motors and membrane traffic. Curr Opin Cell Biol 1997; 9:18-28.

5. Saraste J, Thyberg J. Function of microtubules in protein secretion and organization of the Golgi complex. In: Hesketh HE, Pryme IF, eds. Treatise on the Cytoskeleton, Greenwich, CT: JAI Press, 1996:239-273.

6. Harter C, Wieland F. The secretory pathway: mechanisms of protein sorting and transport. Biochim Biophys Acta 1996; 1286:75-93.

7. Schekman R, Orci L. Coat proteins and vesicle budding. Science 1996; 271: 1526-1533.

8. Rothman JE, Wieland FT. Protein sorting by transport vesicles. Science 1996; 272:227-234.

9. Fath KR, Burgess DR. Golgi-derived vesicles from developing epithelial cells bind actin filaments and possess myosin-I as a cytoplasmically oriented peripheral membrane protein. J Cell Biol 1993; 120: 117-127.

10. Mizuno M, Singer SJ. A possible role for stable microtubules in intra-cellular transport from the endoplasmic reticulum to the Golgi apparatus. J Cell Sci 1994; 107:1321-1331.

11. Lippincott-Schwartz J. Bidirectional membrane traffic between the endoplasmic reticulum and Golgi apparatus. Trends Cell Biol 1993; 3:81-88.

12. Lippincott-Schwartz J, Donaldson JG, Schweizer A, Berger EG, Hauri H-P, Yan LC, Klausner RD. Microtubule-dependent retrograde transport of proteins into the ER in the presence of brefeldin A suggests an ER recycling pathway. Cell 1990; 60:821-836.

13. Lippincott-Schwartz J, Cole NB. Roles for microtubules and kinesin in membrane traffic between the endoplasmic reticulum and the Golgi complex. Biochem Soc Trans 1995; 23:544-548.

14. Burgess TL, Kelly RB. Constitutive and regulated secretion of proteins. Annu Rev Cell Biol 1987; 3:243-293.

15. Singer SJ, Kupfer A. The directed migration of eukaryotic cells. Annu Rev Cell Biol 1986; 2:337-365.

16. Drenckhahn D, Jons T, Puschel B, Schmitz F. Role of the cytoskeleton in the development of epithelial polarity. In: Hesketh HE, Pryme IF, eds. Treatise on the Cytoskeleton, Greenwich, CT: JAI Press, 1996:141-165.

17. Scheel J, Matteoni R, Ludwig T, Hoflack B, Kreis TE. Microtubule depolymerization inhibits transport of cathepsin D from the Golgi apparatus to lysosomes. J Cell Sci 1990; 96:711-720.

18. Prydz K, Hovland KS, Orsen I. The role of microtubules in apical and basolateral endocytosis in epithelial Madin-Darby canine kidney (MDCK) cells. Biochem Soc Trans 1995; 23:53-56.

19. Holzbaur ELF, Vallee RB. Dyneins: molecular structure and cellular function. Annu Rev Cell Biol 1994; 10:339-372.

20. Haimo LT. Regulation of kinesin-directed movements. Trends Cell Biol 1995; 5:165-168.

21. Matthies HJG, Miller RJ, Palfrey HC. Calmodulin binding to and cAMP-dependent phosphorylation of kinesin light chains modulate kinesin ATPase activity. J Biol Chem 1993; 268:11176-11187.

22. McIlvain JM Jr, Burkhardt JK, Hamm-Alvarez S, Argon Y, Sheetz MP. Regulation of kinesin activity by phosphorylation of kinesin-associated proteins. J Biol Chem 1994; 269:19176-19182.

23. Hirokawa N, Sato-Yoshitake R, Yoshida T, Kawashima T. Brain dynein (MAP 1C) localizes on both anterogradely and retrogradely transported membranous organelles in vivo. J Cell Biol 1990; 111: 1027-1037.

24. Dillman JF III, Pfister KK. Differential phosphorylation in vivo of cytoplasmic dynein associated with anterogradely moving organelles. J Cell Biol 1994; 127: 1671-1681.

25. Sato-Yoshitake R, Yorifugi H, Inagaki M, Hirokawa N. The phosphorylation of kinesin regulates its binding to synaptic vesicles. J Biol Chem 1992; 267: 23930-23936.
26. Brady ST. A kinesin medley: biochemical and functional heterogeneity. Trends Cell Biol 1995; 5:159-164.
27. Hollenbeck PJ. Phosphorylation of neuronal kinesin heavy and light chains in vivo. J Neurochem 1993; 60:2265-2275.
28. Thaler CD, Haimo LT. Microtubules and microtubule motors: mechanisms of regulation. Int Rev Cytol 1996; 164: 269-327.
29. Schroer TA, Bingham JB, Gill SR. Actin-related protein 1 and cytoplasmic dynein-based motility—what's the connection? Trends Cell Biol 1996; 6: 212-215.
30. Vallee RB, Sheetz MP. Targeting of motor proteins. Science 1996; 271: 1539-1544.
31. Vale RD, Schnapp BJ, Reese TS, Sheetz MP. Organelle, bead, and microtubule translocations promoted by soluble factors from the squid giant axon. Cell 1985; 40:559-569.
32. Thaler CD, Haimo LT. Regulation of organelle transport in melanophores by calcineurin. J Cell Biol 1990; 111: 1939-1948.
33. Thaler CD, Haimo LT. Control of organelle transport in melanophores: Regulation of Ca^{2+} and cAMP levels. Cell Motil Cytoskel 1992; 22:175-184.
34. Fincham VJ, Unlu M, Brunton VG, Pitts JD, Wyke JA, Frame MC. Translocation of Src kinase to the cell periphery is mediated by the actin cytoskeleton under the control of the Rho family of small G-proteins. J Cell Biol 1996; 135: 1551-1564.
35. Kelly RB. Secretory granule and synaptic vesicle formation. Curr Opin Cell Biol 1991; 3:654-660.
36. Bennett MK, Scheller RH. The molecular machinery for secretion is conserved from yeast to neurons. Proc Natl Acad Sci USA 1993; 90:2559-2563.
37. Linial M, Parnas D. Deciphering neuronal secretion: tools of the trade. Biochim Biophys Acta 1996; 1286: 117-152.
38. Augustine GJ, Burns ME, DeBello WM, Pettit DL, Schweizer FE. Exocytosis: proteins and perturbations. Annu Rev Pharmacol Toxicol 1996; 36:659-701.
39. Liu J-P, Robinson PJ. Dynamin and endocytosis. Endocrin Rev 1995; 16: 590-607.
40. Burgoyne RD, Morgan A. Regulated exocytosis. Biochem J 1993; 293:305-316.
41. Muallem S, Kwiatkowska K, Xu X, Yin HL. Actin filament disassembly is a sufficient final trigger for exocytosis in nonexcitable cells. J Cell Biol 1995; 128: 589-598.
42. Raynal P, Pollard HB. Annexins: the problem of assessing the biological role for a gene family of multifunctional calcium- and phospholipid-binding proteins. Biochim Biophys Acta 1994; 1197: 63-93.
43. Roth D, Burgoyne RD. Stimulation of catecholamine secretion from adrenal chromaffin cells by 14-3-3 proteins is due to reorganisation of the cortical actin network. FEBS Lett 1995; 374:77-81.
44. Norman JC, Price LS, Ridley AJ, Koffer A. The small GTP-binding proteins, Rac and Rho, regulate cytoskeletal organization and exocytosis in mast cells by parallel pathways. Mol Biol Cell 1996; 7: 1429-1442.
45. Valtorta F, Benfenati F. Protein phosphorylation and the control of exocytosis in neurons. Ann New York Acad Sci 1994; 710:347-355.
46. Walent JH, Porter BW, Martin TFJ. A novel 145 kd branin cytosolic protien reconstitutes Ca^{2+}-regulated secretion in permeable neuroendocrine cells. Cell 1992; 70:765-775.
47. Seethaler G, Tooze S, Shields D. Fusion and confusion in the secretory pathway. Trends Cell Biol 1996; 6:239-242.

48. Prekeris R, Mayhew MW, Cooper JB, Terrian DM. Identification and localization of an actin-binding motif that is unique to the epsilon isoform of protein kinase C and participates in the regulation of synaptic function. J Cell Biol 1996; 132:77-90.

49. Lamaze C, Schmid SL. The emergence of clathrin-independent pinocytic pathways. Curr Opin Cell Biol 1995; 7:573-580.

50. Robinson MS. The role of clathrin, adaptors and dynamin in endocytosis. Curr Opin Cell Biol 1994; 6:538-544.

51. Sorkin A, Waters CM. Endocytosis of growth factor receptors. BioEssays 1993; 15:375-382.

52. Damke H. Dynamin and receptor-mediated endocytosis. FEBS Let 1996; 389: 48-51.

53. Warnock DE, Schmid SL. Dynamin GTPase, a force-generating molecular switch. BioEssays 1996; 18:885-893.

54. Shpetner HS, Herskovits JS, Vallee RB. A binding site for SH3 domains targets dynamin to coated pits. J Biol Chem 1996; 271:13-16.

55. Wang Z, Moran MF. Requirement for the adaptor protein GRB2 in EGF receptor endocytosis. Science 1996; 272: 1935-1939.

56. Riezman H. Yeast endocytosis. Trends Cell Biol 1993; 3:273-277.

57. Gottlieb TA, Ivanov IE, Adesnik M, Sabatini DD. Actin microfilaments play a critical role in endocytosis at the apical but not the basolateral surface of polarized epithelial cells. J Cell Biol 1993; 120:695-710.

58. Kubler E, Riezman H. Actin and fimbrin are required for the internalization step of endocytosis in yeast. EMBO J 1993; 12:2855-2862.

59. Munn AL, Stevenson BJ, Geli MI, Riezman H. *end5, end6* and *end7*: Mutations that cause actin delocalization and block the internalization step of endocytosis in *Saccharomyces cerevisiae*. Mol Biol Cell 1995; 6:1721-1742.

60. Mulholland J, Preuss D, Moon A, Wong A, Drubin D. Ultrastructure of the yeast actin cytoskeleton and its association with the plasma membrane. J Cell Biol 1994; 125:381-391.

61. Goodson HV, Anderson BL, Warrick HM, Pon LA, Spudich JA. Synthetic lethality screen identifies a novel yeast myosin I gene (*MYO5*): myosin I proteins are required for polarization of the actin cytoskeleton. J Cell Biol 1996; 133: 1277-1291.

62. Geli MI, Riezman H. Role of type I myosin in receptor-mediated endocytosis in yeast. Science 1996; 272:533-535.

63. Oda H, Stockert RJ, Collins C, Wang H, Novikoff PM, Satir P, Wolkoff AW. Interaction of the microtubule cytoskeleton with endocytic vesicles and cytoplasmic dynein in cultured rat hepatocytes. J Biol Chem 1995; 270:15242-15249.

64. Lamaze C, Schmid SL. Recruitment of epidermal growth factor receptor and transferrin receptors in vitro into coated pits: differing biochemical requirements. Mol Biol Cell 1993; 4:715-727.

65. Lamaze C, Schmid SL. Recruitment of epidermal growth factor receptor into coated pits requires their activated tyrosine kinase. J Cell Biol 1995; 129:47-54.

66. Wells A, Welsh JB, Lazar CS, Wiley HS, Gill GN, Rosenfeld MG. Ligand-induced transformation by a noninternalizing epidermal growth factor receptor. Science 1990; 247:962-964.

67. Baulida J, Kraus MH, Alimandi M, Di Fiore PP, Carpenter G. All ErbB receptors other than the epidermal growth factor receptor are endocytosis impaired. J Biol Chem 1996; 271:5251-5257.

68. Cohen S, Fava R. Internalization of functional epidermal growth factor:receptor tyrosine kinase complexes in A431 cells. J Biol Chem 1985; 260:12351-12358.

69. Viera AV, Lamaze C, Schmid SL. Control of EGF receptor signaling by clathrin-mediated endocytosis. Science 1996; 274:2086-2089.

70. Carpentier JL, Paccaud JP. Molecular and cellular biology of insulin-receptor internalization. Ann NY Acad Sci 1994; 733: 266-278.

71. Biener Y, Feinstein R, Mayak M, Kaburagi Y, Kadowaki T, Zick Y. Annexin II is a novel player in insulin signal transduction. Possible association between anexin II phosphorylation and insulin receptor internalization. J Biol Chem 1996; 271:29489-29496.

72. Rasenick MM, Caron MG, Dolphin AC, Kobilka BK, Schultz G. In: Pharmacological Sciences: Perspectives for Research and Therapy in the Late 1990s. Cuello AC, Collier B, eds., Basel, Switzerland: Burkhauser Verlag, 1995:91-103.

73. Zhang J, Ferguson SSG, Barak LS, Menard L, Caron MG. Dynamin and β-arrestin reveal distinct mechanisms for G-protein-coupled receptor internalization. J Biol Chem 1996; 271:18302-18305.

74. Anderson RGW. Plasmalemmal caveolae and GPI-anchored membrane proteins. Curr Opin Cell Biol 1993; 5:647-652.

75. Parton RG, Simons K. Digging into caveolae. Science 1995; 269:1398-1399.

76. Lisanti MP, Scherer PE, Tang Z, Sargiacomo M. Caveolae, caveolin and caveolin-rich membrane domains: a signalling hypothesis. Trends Cell Biol 1994; 4:231-235.

77. Deckert M, Ticchioni M, Bernard A. Endocytosis of GPI-anchored proteins in human lymphocytes: role of glycolipid-based domains, actin cytoskeleton, and protein kinases. J Cell Biol 1996; 133: 791-799.

78. Li S, Couet J, Lisanti MP. Src tyrosine kinases, G^{alpha} subunits, and H-Ras share a common membrane-anchored scaffolding protein, caveolin. J Biol Chem 1996; 271:29182-29190.

79. Smart EJ, Ying Y-S, Anderson RGW. Hormonal regulation of caveolae internalization. J Cell Biol 1995; 131:929-938.

80. Anderson RGW, Kamen BA, Rothberg KG, Lacey SW. Potocytosis: sequestration and transport of small molecules by caveolae. Science 1992; 255:410-411.

81. Anderson RGW. Caveolae: where incoming and outgoing messengers meet. Proc Natl Acad Sci USA 1993; 90: 10909-10913.

82. Parton RG, Joggerst B, Simons K. Regulated internalization of caveolae. J Cell Biol 1994; 127:1199-1215.

83. Izumi T, Shibata Y, Yamamoto T. Striped structures on the cytoplasmic surface membranes of the endothelial vesicles of the rat aorta revealed by quick-freeze, deep-etching replicas. Anat Rec 1988; 220:225-232.

84. Swanson JA, Watts C. Macropinocytosis. Trends Cell Biol 1995; 5:424-428.

85. Ridley AJ. Membrane ruffling and signal transduction. BioEssays 1994; 16:321-327.

86. Novak KD, Peterson MD, Reedy MC, Titus MA. *Dictyostelium* myosin I double mutants exhibit conditional defects in pinocytosis. J Cell Biol 1995; 131: 1205-1221.

87. Rabinovitch M. Professional and non-professional phagocytes: an introduction. Trends Cell Biol 1995; 5:86-88.

88. Swanson JA, Baer SC. Phagocytosis by zippers and triggers. Trends Cell Biol 1995; 5:89-93.

89. Isberg RR, Nhieu GTV. The mechanism of phagocytic uptake promoted by invasin-integrin interaction. Trends Cell Biol 1995; 5:120-124.

90. Brown EJ. Phagocytosis. BioEssays 1995; 17:109-117.

91. Jones SL, Brown EJ. FcγRII-mediated adhesion and phagocytosis induce L-plastin phosphorylation in human neutrophils. J Biol Chem 1996; 271: 14623-14630.

92. Allen LA, Aderem A. Molecular definition of distinct cytoskeletal structures involved in complement- and Fc receptor-mediated phagocytosis in macrophages. J Exper Med 1996; 184:627-637.

93. Greenberg S. Signal transduction of phagocytosis. Trends Cell Biol 1995; 5: 93-99.

94. De Camilli P, Emr SD, McPherson PS, Novick P. Phosphoinositides as regulators in membrane traffic. Science 1996; 271:1533-1539.

CHAPTER 7

Early Development and Neoplasia

Introduction

M ost of the studies cited in earlier chapters deal with molecular mechanisms of basic cellular processes of relatively homogeneous cell populations. Much of what is significant in biomedical science is concerned with complex, organized cell mixtures: tissues and organs. Thus, it is important to see how the principles we have previously defined apply to such systems. To address some of the questions involved in more complex systems, we have chosen to briefly discuss two topics, early development and neoplasia. In our view these represent two important aspects of multicellularity:
1) the initial and presumably simplest form of the organization of cells; and
2) the failure or reversal of the normal organization.

To avoid writing another monograph, we will focus these discussions narrowly on three topics of early development, fertilization, compaction and cavitation, gastrulation, and on aspects of oncogenes and tumor suppressors in neoplasia related to earlier topics in this book.

Early Development

Fertilization and Egg Activation

The first cell-cell interaction in the life history of the multicellular animal occurs when sperm meets egg. Both of these are unique and remarkable cells. Like many aspects of early development, the cells vary widely among different species of animals, but they share some common features. Essentially, the sperm is a motile nucleus, a wanderer with the capability of forced entry, driven in most species by a flagellum at the tail of a simple but highly structured body. In contrast, the egg in many species can simplistically be considered a tiny blueprint for an animal with its primary body plan encripted into a spatial code in the polarized cell structure. In many species, the egg is huge compared to the sperm and somatic cells and has the nutrients as well as the macromolecular constituents necessary to carry the embryo through the first stages of early development with minimal new synthesis of nucleic acid or protein.

Contact of the sperm with the egg is mediated by the acrosome, a membrane-bound organelle at the sperm tip which can be likened to the exploding tip on the whaler's harpoon. Interaction of the sperm with the extracellular coat of the egg triggers a *reciprocal activation* of the sperm and the egg. In the *activation of the sperm,* an entry mechanism, the acrosomal reaction of the sperm, is initated. The acrosome contains digestive enzymes capable of dissolving the extracellular coat

Signaling and the Cytoskeleton, by Kermit L. Carraway, Coralie A. Carothers Carraway and Kermit L. Carraway III. © 1998 Springer-Verlag and R.G. Landes Company.

of the egg which are released upon sperm-egg contact. The release mechanism includes calcium influx and cytoplasmic alkalinization, followed by the activation of membrane phospholipases, indicating that the appropriate ion channels and membrane enzymes must be located at the sperm tip in the highly structured sperm morphology. For sperm activation in mammals, the signaling ligand is the glycoprotein ZP3 of the egg extracellular coat which binds and clusters G-protein-linked receptors in the sperm to stimulate a G-protein-activated PLC and promote calcium release from intracellular stores. In some species such as sea urchins, the cytoplasmic alkalinization triggers a rapid actin polymerization to extend an acrosomal process. Concomitantly, this mechanism may trigger an activation of flagellar dynein ATPase to increase sperm motility and push the sperm through the dissolving extracellular egg coat. Studies of echinoderm sperm acrosomal process extension have provided one of the important models for actin polymerization and the formation of protrusions (1). Prior to activation, unpolymerized actin is present in short filaments complexed with profilin and spectrin in a cup structure beneath the acrosome. The increase in calcium is proposed to induce formation of phosphoinositides which can release actin from the complex and trigger polymerization to push the membrane forward. However, how the actin adds to the growing filament proximal to the membrane remains a question, as discussed in chapters 1 and 4.

These combined actions of the acrosomal reaction result in contact of the plasma membranes of the sperm and egg and consequent *egg activation*. The most immediate event in the sea urchin egg response is plasma membrane depolarization to establish a fast block to polyspermy (2) which would otherwise lead to abnormal development and death of the embryo. Within 30 sec tyrosine phosphorylation, a calcium spike and cell contraction are observed and a calcium wave appears at approximately 30 sec as a key event in egg activation. In fact, eggs have been used as one of the model systems for studying calcium-dependent signaling (3). The calcium increase appears to be a response to IP_3 produced via a G-protein-linked phosphoinositidase. Phosphoinositol is abundant in the unfertilized egg and is converted to PIP and PIP_2. Cleavage of the latter yields diacylglycerol as well as IP_3, providing the requirements for activating PKC. PKC in turn stimulates a Na^+/H^+ exchanger which promotes cytoplasmic alkalinization.

Concomitant with these signaling events are pronounced changes in the egg cell surface and cortical cytoskeleton (1). The unfertilized egg surface is covered with short microvilli overlying a thin cortical actin meshwork in which numerous secretion granules are entrapped. With the rise in calcium comes an initial breakdown of microfilaments and the exocytosis of massive numbers of cortical granules from beneath the plasma membrane as described in chapter 6, to form the extracellular fertilization envelope (hyaline membrane), the slow block to polyspermy. This is followed by rapid polymerization of submembrane actin into meshworks of ruffle-like protrusions which are converted into longer microvilli. Subsequently, surface membrane is endocytosed into the cell to form cortical, acidic vesicles with a reduction in microvilllus length. Although the specific molecular mechanisms for these changes remain unclear, studies on sea urchin egg cytoskeletal proteins (Table 7.1) provide some insights into the general scheme. The presence of severing/capping proteins suggests that they are involved in the initial filament breakdown by barbed end capping followed by phosphoinositide-stimulated release and repolymerization into meshworks. This polymerization appears to be dependent on cytoplasmic alkalinization (4). The meshworks are then con-

Table 7.1. Sea urchin egg proteins associated with actin filaments

Protein	M_r (K_d)	Relative quantity
Filamin	~260	+
Spectrin	240/235	+++
Myosin	205	++
α-Actinin	100	+++
Fascin	58	+++
Actin	45	+++
Tropomyosin	33	++
Actolinkin	20	ND

Modified from ref. 1 (Bonder EM, Fishkind DJ. Curr Top Devel Biol 1995; 31: 101-137), courtesy of Academic Press.

verted into loose filament bundles by dimeric filament binding proteins, such as spectrin and villin, followed by formation of tight, highly ordered microvillous cores by the association of the monomeric crosslinking protein fascin. A myosin I may be involved in the formation of the microvilli, as proposed for intestinal microvilli (5). Interestingly, tropomyosin is abundant in cortical cytoskeleton preparations, suggesting it may play a role in organizing or stabilizing the filaments.

Granule exocytosis and *plasma membrane endocytosis* are important for remodeling the cell surface while conserving cellular membrane for the multiple cell divisions to follow fertilization. In the unfertilized egg, spectrin is observed as cross-bridges between the plasma membrane and cortical granules and between cortical granules, suggesting a role in the secretory vesicle clamp and granule release (see chapter 6). After fertilization spectrin is associated with both microvillar microfilaments and with the acidic cortical vesicles, where it may play dual roles (1). Dynamin has been discovered in association with cortical granules in unfertilized sea urchin eggs, as well as in the cytoplasm and the cortical acidic vesicles after fertilization. How endocytosis in these cells is regulated is uncertain. However, PKC is implicated in other cells, and phorbol ester treatment of eggs will induce endocytosis without cortical granule exocytosis.

Cleavage, Compaction and Cavitation

The purpose of egg activation, of course, is to trigger mitosis and cell division. Multicellularity arises by a variety of cleavage mechanisms ranging from the radial, holoblastic (complete) cleavage of the echinoderm egg to the superficial (incomplete) cleavage of the eggs of some insects such as *Drosophila* (2). In the latter, the egg undergoes extensive nuclear division to form a structured syncytium which is then divided into cells by cytoskeletal and membrane reorganizations (6). This syncitial ordering represents the initial stage in the polarization of the fly embryo and formation of its body plan. The type of cleavage observed for each species depends on the evolutionary history of the organism, the amount and location of nutritional material (yolk) in the egg and the orientation and timing of the formation of the mitotic spindle. In most animals, except mammals, the egg contains the proteins and RNA necessary for all of the early cleavage stages. Thus, transcription is delayed until later. As a consequence of this restricted synthetic capability, cleavage occurs in these species without substantial increases in cytoplasmic volume, the egg dividing into increasingly smaller cells. In contrast, mammalian eggs are small and fertilized internally. Thus, their later development is dependent on maternal factors which are not supplied to externally fertilized eggs. Mammals therefore generally initiate RNA synthesis at an earlier stage, the timing dependent on the species. For example, fertilized mouse eggs initiate a cAMP-dependent transcription as early as the 2 cell stage (7).

Another crucial difference in early development between mammals and other species is *compaction* at the eight cell stage. Mammalian cells undergo a rotational, holoblastic cleavage to the eight cell stage to produce an embryo of loosely associated cells. Following this third cleavage, the cells rearrange into a compact structure, maximizing cell-cell contacts and forming free and apposed plasma membrane regions (8). The free regions develop into apical domains which can be distinguished from the basolateral domains by the presence of cytochalasin-resistant microfilaments. The *key event* in compaction appears to be a *calcium-dependent cell adhesion* mediated by the redistribution of E-cadherin to intercellular contact sites (8). Adhesion is followed by the establishment of gap junction intercellular communication sites, polarization of intracellular components, including microtubules and microfilaments and the appearance of tight junctions which maintain discrete apical and basolateral domains (9). These processes are discussed in more detail in chapter 3. With the establishment of polarity, the next division (to 16 cells) initiates formation of the *morula*, the first morphological differentiative event in mammalian embryogenesis to produce the unpolarized, pluripotential inner cell mass which becomes the embryo. At this stage the outer cell layer has formed an epithelium (see chapter 3), called the *trophoblast* or *trophectoderm*, which is involved in the *implantation* step necessary for maternal sustenance of the mammalian embryo.

This epithelial polarization is necessary for the second major event of preimplantation development, *cavitation*. The cavity within the trophoblast, the *blastocoel*, expands by osmotic accumulation of water due to ion concentration. Sodium ions are proposed to enter trophoblastic cells via apically localized channels or cotransporters, such as the Na^+,H^+-exchanger. They are then pumped into the blastocoel by a basolaterally-localized Na,K-ATPase which attains its polarized distribution at the time of cavitation (9) (see chapter 3). In contrast, Cl^- movement into the blastocoel occurs by paracellular transport. Its concentration within the blastocoel depends on the development of the tight junctions which remain leaky until the 32 cell stage. Thus, cavitation is a progressive event which requires the development of polarity (see chapter 3), the localization of specific membrane components, undoubtedly with the participation of the polarized cytoskeleton, and the maturation of the epithelium, particularly the junctional complexes. During cavitation the cells of the blastocoel are remodeled by *apoptosis*. A two signal mechanism has been proposed for this process (10). The first signal (*paracrine* soluble ligand) comes from the outer layer of *endoderm* of the blastocyst to induce apoptosis in the *inner ectodermal cells* and create the cavity. At the same time the columnar cells that line the cavity are protected from apoptosis by association with the basement membrane (*insoluble ligand*, described in chapter 2). This dual signaling mechanism involves two proposed modulators for apoptosis, described in chapter

5, and is postulated to be a general process for *tube morphogenesis*, a common developmental mechanism for organogenesis.

To understand the events of early development in the mammal, one needs additional knowledge of the regulatory factors involved. However, such studies are complicated by the limited amounts of material available for such analyses and have relied primarily on pharmacological effects and immunolocalization. For example, the PKC activator phorbol esters have been reported in separate studies to stimulate both compaction (11) and cavitation (12). Whether these effects occur in vivo and what might be the physiological regulator(s) are unclear. Mouse preimplantation development is also regulated by growth factors through receptor tyrosine kinases (13). Insulin-like growth factor I/II receptor has been implicated in cell proliferation, while the EGF receptor is associated with trophectoderm development. Ras and Raf have also been implicated, providing a possible connection not only to the EGF receptor, but also to PKC, which can activate Raf in some systems. The only ligand detected for the EGF receptor in preimplantation embryos is TGF-α. However, EGF is found in oviductal fluid. Since the EGF receptor appears to be located both at the apical and basolateral surfaces of the trophectoderm, it may be involved in two different types of functions. For example, EGF activation of the apical receptor may contribute to growth and survival of the embryos, while TGF-α produced by the embryo stimulates blastocoel expansion (13). How the receptors are localized remains an important question. A role for the cytoskeleton is likely. Undoubtedly, other receptors and growth factors will be found to contribute. Much more work is needed to decipher this aspect of early development.

Body Plan Development

Theoretically, body plan development can proceed by two extreme models: (1) mosaic development, entirely specified by localized determinants; and (2) regulative development in which localized determinants in the egg are irrelevant and developmental progress is determined by interactions between cells, whether direct or indirect. *Xenopus* early development provides a well-studied example in which an interplay of both models can be observed. As previously mentioned, transcript localization in the egg may play an important role in the specification of the embryonic (and adult) body plan. Definition of the body plan of *Xenopus* begins with fertilization, when the egg cortex undergoes a microtubule-dependent, approx. 30° rotation relative to the core of the egg, creating a specialized region opposite the sperm entry site called the gray crescent, which defines the *dorsal* side of the egg. Although the components involved in this determination are still being delineated, the Wnt signaling pathway, particularly β-catenin, has been implicated by localization studies (14). Wnt acts as a ligand for a serpentine-type receptor called Frizzled, one of a group of cell surface receptors whose regulation of gene expression appears to be interrelated (Fig. 7.1) (15). These include cadherin (chapter 3), Notch (chapter 1) and Patched. Wnt binding to Frizzled in *Xenopus* leads to the development of *ventral* body structures such as skeletal muscle. However, Wnt binding can be blocked by a protein related to the N-terminal extracellular domain of Frizzled, called Frzb (16, 17), leading to the development of dorsal structures such as the head. Since β-catenin is a key modulator of both gene expression and cell-cell interactions, a change in its function in the Frizzled pathway could lead to multiple effects. The importance of the β-catenin pathway is further indicated by blocking glycogen synthase kinase-3 which regulates degradation of β-catenin by phosphorylation. Inhibition of the kinase leads to the formation of two-headed

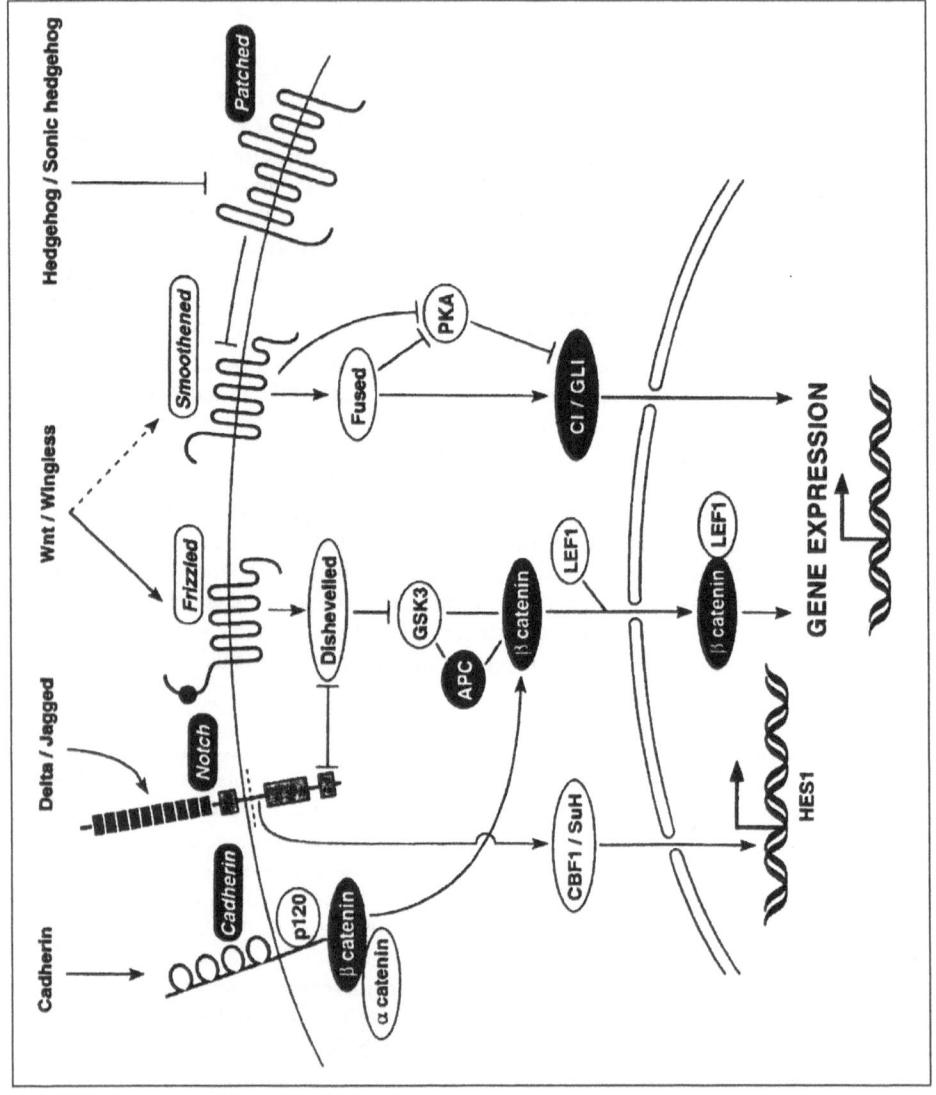

Fig. 7.1. Developmental signaling pathways involved in cancer. Figure from ref. 20 (Hunter T. Cell 1997; 88: 333-346), courtesy of Cell Press.

tadpoles (18). In mammals, the role of the β-catenin pathway is suggested by studies showing that its overexpression may contribute to colorectal cancer (19, 20).

One of the transcripts localized to the vegetal region of the *Xenopus* oocyte is Vg1, which encodes a member of the *TGF-β family* of growth factors. A two-step cytoskeletal model for localization of Vg1 implicates microtubules and microfilaments in transport and anchoring, respectively, of its mRNA (21, 22). The cortical rotation described above further defines the dorsoventral axis of the frog via an inductive mechanism because the region containing Vg1 transcript is rotated into the area near the gray crescent. This localization is maintained during cleavage to produce cells which contain the Vg1 transcript. Translation and proteolytic cleavage produce an active factor which is secreted to act by a paracrine mechanism on adjacent cells (23). These cells differentiate to form the blastula blastopore which is the site of the cell movements to initiate gastrulation. Thus, two putative cytoskeletal functions have been implicated in regulation of egg development: cortical rotation and Vg1 localization. In the case of Vg1 localization, the 3' end of the transcript appears to contain the binding domain (24), and cytokeratins have been implicated (25). Although the molecular details of both the rotation and transcript localization are still unclear, the importance of the cytoskeleton to several signaling events regulating these early stages of development should be obvious.

Gastrulation

Gastrulation is the process by which the basic multilayered body plan of the animal is established. Although gastrulation is a highly divergent process among different species of animals, in all species it involves migrations of sets of cells of the blastocyst to new positions to establish the three basic types of embryonic tissues: *endoderm*, primarily responsible for forming the gut and internal organs; *ectoderm*, which forms the epidermis; and *mesoderm*, which forms muscle and connective tissue. The key to establishing the body plan in animals is *positional information present in the egg* (26). This information may be present as *specifically localized proteins or transcripts*. Events subsequent to fertilization, but particularly gastrulation, develop this blueprint into the functional animal. These events can be described in simple terms for species such as the sea urchin. In higher organisms the ultimate body structure and processes for achieving and maintaining that structure become much more complex, though the basic mechanisms of cellular behavior are the same.

Sea urchin eggs, like those of most animals, exhibit a polarized structure in which the pigmented cytoplasm distinguishes the animal from the vegetal pole. Sea urchin eggs also exhibit radial, holoblastic cleavage so that distinct regions of the egg are segregated into the different cells produced by the cleavages. For example, the vegetal pole of the egg becomes incorporated into cells known as micromeres, while the animal pole is incorporated into the mesomeres. This segregation process continues through the formation of the blastula, a single layer of 1000-2000 cells surrounding a central cavity, the blastocoel. After hatching of the blastula from its fertilization envelope, the cells of its vegetal side (micromeres) begin to flatten, dissociate from each other and migrate into the blastocoel, moving along and probing the blastocoel surface with filopodia until they localize in regions for the formation of the urchin larval skeleton (27). Thus, the first stage in sea urchin gastrulation requires changes in cell adhesion from the extracellular envelope (hyaline layer formed by cortical granule secretion at fertilization), motility on the basal lamina of the blastocoel and adhesion to specific cellular sites at the blastocoel surface. *Filopodia* appear to be critical cellular components in the migration process (26), while fibronectin fibrils are important in the basal laminae to which the cells adhere during migration (28).

The next stage of urchin gastrulation involves invagination of the remaining cells at the vegetal plate to form the archenteron (primitive gut) extending about one-third of the distance through the blastocoel. The opening of the invagination is called the *blastopore*. The invagination requires only vegetal cells and is dependent on binding of the cells to the hyaline layer for cell shape changes (29). During a second stage of archenteron formation, the invaginated cells extend by flattening and migration to extend by about 3-fold. In some species of urchins, mesenchymal cells associate with the tip of the archenteron and extend filopodia which connect with the blastocoel wall at specific sites which are committed to become the ventral side of the organism (30). Again, filopodial extensions play a role in the recognition processes necessary for establishing specific body structures. How these morphological, adhesive and motility changes are regulated largely remains to be determined. In other species, gastrulation is more complex. Gastrulation in the frog is regulated by the Wnt pathway and Vg1, described above, among other factors. How they contribute to the cellular shape changes and migrations that are involved in gastrulation is under investigation. What is clear is that the complex operations of development follow the same basic principles of adhesion, morphogenesis and motility that we have described in previous chapters. Multicellularity simply imposes a larger set of instructions which must be integrated.

Neoplasia

The discovery of *src* as the first oncogene and its identification as a tyrosine kinase ushered in a new era in cancer biology by providing a molecular mechanism for oncogenesis (31). Studies of oncogenic viruses and of the transformation of cultured cells with DNA from tumor cells led to the identification of many oncogenes. With the observations of proto-oncogenes as normal cellular products of signaling pathways, often involved in regulating cell proliferation, an attractive theory of cancer formation arose. Damage to DNA activated cellular proto-oncogenes and caused dominant genetic effects in cells, leading to their uncontrolled proliferation. Although this theory still plays an important role in cancer research, it has proved to be too simplistic to explain human cancer. Earlier studies had shown that fusion of malignant cells with nonmalignant cells created primarily nontumorigenic hybrids, a result inconsistent with dominant oncogenic mutations (32). More recently, observations on hereditary retinoblastoma suggested a requirement for two genetic changes, only one of which was inherited (33). Moreover, investigations of genetic changes in human tumors found that most of these could not be explained by oncogenes, but indicated a prevalence of cytogenetic abnormalities such as chromosome deletions, suggesting the importance of the *loss of regulatory functions*. Finally, analyses of the prevalence of cancer with age indicated that *formation of a diagnosable tumor requires 5-6 separate genetic events* (34).

These combined observations led to the concept of *tumor suppressor genes*, which was validated with the transfer of chromosomes from normal human cells into tumor cells (35). Subsequent genetic analyses in colon cancer have identified 5-6 genes involved in the formation of colon tumors whose mutations can occur in different combinations and sequences (36). Three of these changes are due to losses of tumor suppressors, p53, APC (adenomatous polyposis coli) and DCC (deleted in colon cancer). Thus, *human cancers result from a combination of dominant mutations of oncogenes to promote cell proliferation and mutations which produce a loss of function of tumor suppressors which regulate cell function*. Since cancer is defined by invasiveness rather than simply cell proliferation, a goal in cancer research is to discover those properties of cancer cells which contribute to invasiveness, including cell proliferation, in order to develop therapeutic approaches which can block their effects. Because cancer development requires so many steps, attaining this goal would seem simple. One *caveat* is that *multiple pathways may lead to the same cellular phenotype* (see chapter 8), complicating the development of drugs to block that phenotype. Furthermore, the *genetic instability of tumors creates opportunities for the selection of cells expressing multiple neoplastic phenotypes*, particularly cells whose DNA repair mechanisms have been compromised. An additional *caveat* in developing treatments is that many of these oncogenic processes closely resemble normal cell functions. Finally, replacing the lost functions of tumor suppressors is much more complicated than blocking oncogenic functions.

Oncogenes

Oncogene research has outpaced tumor supressor research in part because there is a simple assay for oncogenes, stimulation of cell growth in culture. Furthermore, many oncogenes were identified from oncogenic viruses isolated from animal sources. Oncogene research played a major role in the early development of signaling elements and pathways for cell proliferation. The *caveat* to such studies is that the cell systems used are not normal cells. Thus, even oncogene products which cause test cells to become tumorigenic may never be found to contribute to human tumors in which *cellular and tissue context* may override the effects of the oncogene. However, by combining studies of oncogenes and their products with genetic analyses of development of simple organisms, enormous advances have been made in understanding critical aspects of cell regulation, which might ultimately be the most important contribution of oncogene research.

Tyrosine kinases were the first class of signaling molecules identified as oncogenes. As mentioned in earlier chapters, tyrosine kinases also play important roles in many of the cellular activities involved in tumor progression, including cell division, cell shape changes, adhesion and cell motility. Such observations raise the question of how mutations of tyrosine kinase genes contribute to their activation and their resultant role in cell proliferation and oncogenicity. At the molecular level, at least *three models can be proposed for the activation of protein kinases: relief from intramolecular inhibition, dimerization and re(mis)localization*; these are not mutually exclusive. Viral Src exhibits relief from intramolecular inhibition since its mutation results in a truncation of the C-terminus (31). This C-terminal peptide contains a tyrosine residue whose phosphorylation leads to an intramolecular interaction with the Src SH2 domain. This interaction is proposed to block both the kinase activity and cytoskeletal association of c-Src (37), reducing both activities which appear to contribute to the oncogenicity of the kinase. Intramolecular inhibition of c-Src can also be relieved by dephosphorylation of the inhibitory C-terminal tyrosine residue by specific phosphatases.

We have recently proposed a mechanism for c-Src activation and mislocalization via interactions of its SH3 domain which may be important in tumor progression. c-Src associates with a large, microfilament-associated signal transduction particle (STP) containing $p185^{neu}$/ErbB2 (38, 39) in the highly aggressive 13762 ascites rat mammary adenocarcinoma. ErbB2 is constitutively activated in these cells and assembles the Ras-to-MAP kinase mitogenic pathway (40). The STP also contains p58^{gag}, a retroviral Gag precursor homolog (41) obtained from these malignant mammary ascites cells. p58^{gag} is truncated at its C-terminus and does not have the nucleic acid-binding sequence. Purified p58 binds microfilaments and phospholipid (42) and has been implicated in cell surface stability and xenotransplantability of sublines expressing it (41-43). p58^{gag}, which is tyrosine-phosphorylated in the tumor cells (38), binds in in vitro studies to the SH3 domain of c-Src via a specific polyproline motif and is phosphorylated by the activated kinase. Transfection of p58 into cultured cells causes the reorganization of microfilaments (41) and relocalization of c-Src (44). We have proposed that p58^{gag} localizes the kinase to the constitutvely activated ErbB2-containing signal transduction particle, possibly playing a role in co-activation of the pathway (38).

PKC (a Ser/Thr kinase) activation can also be attributed to concomitant relief of intramolecular inhibition and relocalization. In this case, the tumor promoter phorbol ester (or the natural activator diacylglycerol) releases the inhibition concomitantly with translocation of the enzyme to the plasma membrane (45).

In contrast, receptor tyrosine kinases are believed to be activated through dimerization mechanisms (15), though these may also involve a concomitant relief of intramolecular inhibition in some cases. Either homo- or heterodimerization may contribute to kinase activation (46). A primary effect is dimer cross-phosphorylation to create tyrosine phosphate binding sites for SH2 domains to build up stable signaling complexes at the membrane (47). Oncogenic activation of the receptor tyrosine kinase Neu results from mutation of a residue in its transmembrane domain, facilitating dimerization (48), though this may not be important in human cancer. The functions of receptor tyrosine kinases may also be regulated by localization (47). Interestingly, some receptor kinases have been shown to bind to microfilaments. The EGF receptor can directly bind via a cytoplasmic domain sequence similar to one in profilin (49). Significantly, EGF receptors in some cells are present in both low and high affinity forms, but only the high affinity cytoskeleton-associated receptor is required for activation of the signal transduction cascade for mitogenesis (50). Neu has also been shown to be associated with microfilaments (39), though its association may be indirect.

Activation of tyrosine kinases by concomitant dimerization and cytoskeletal localization has been suggested for a member of the Ets family of transcription factors, TEL. A fusion protein of TEL and the β form of the PDGF receptor is found as a chromosome translocation associated with chronic myelomonocytic leukemia (51). A similar fusion protein of TEL and Abl is found in acute myeloid leukemia (52). The TEL-Abl fusion product is transforming, constitutively tyrosine phosphorylated and localized to the cytoskeleton. A TEL-Abl kinase negative mutant is not transforming or phosphorylated, but still cytoskeleton-associated. All of these phenomena are dependent on homo-oligomerization of the fusion protein via a TEL helix-loop-helix domain, suggesting that oligomerization contributes to the Abl activation and cytoskeletal localization. However, the interactions involved in oligomerization might also free the Abl actin binding domain (37) and catalytic site from intramolecular restraints. Whatever the mechanism, both activation and relocalization (from the nucleus) of Abl appear to be important to its oncogenic effects. The role of localization of Abl has been clarified in studies of the *bcr-abl* oncogene (53). Cell transformation results in both growth factor and anchorage independence, but Bcr-Abl can abrogate only the anchorage dependence (54). However, a recombinant Gag-Bcr-Abl protein can also induce growth factor independence. Interestingly, Bcr-Abl can transform hematopoietic cells, but not fibroblasts; Gag-Bcr-Abl is required for transformation of the latter (55). The likely explanation is that the myristoylated Gag protein derived from a retroviral precursor protein, can translocate Abl to the plasma membrane where its signal is needed for mitogenesis. This rationale is consistent with previous observations of the importance of membrane localization of the mitogenic signal (47).

Abl is a complex protein containing SH2, SH3, microfilament-binding and nuclear localization domains or motifs (56). Its primary normal function appears to be to regulate transcription in the nucleus by phosphorylation of Rb. Thus, mislocalization of the protein could lead to aberration in cell cycle regulation. Such changes might be induced by agents in the cytoplasm which directly bind to Abl or agents whose association with Abl changes its conformation to release binding sites such as the microfilament-binding domain which can contribute to relocalization. We have shown that c-Abl as well as c-Src is associated with the $p185^{neu}$/ErbB2-containing signal transduction particle in the 13762 ascites cells (38). In these cells, a large fraction of c-Abl is found in the microvilli (57), possibly bound to an endogenous retroviral Gag protein analog. $p58^{gag}$ has polyproline motifs similar to those known to bind to the SH3 domain of c-Abl (58), and in vitro studies showed that p58 bound via a specific polyproline motif to the c-Abl SH3 domain (57). The bound $p58^{gag}$ was also phosphorylated by the Abl. We have proposed that c-Abl mislocalization, as well as c-Src relocalization might be involved in the co-activation of the highly aggressive tumor cells. The expression of $p58^{gag}$ in the tumor is proposed to contribute to its malignant phenotype, specifically the xeno-transplantability of the particular Gag-expressing variant (41). $p58^{gag}$ directly associates with microfilaments (38, 42), as do many viral capsid precursors (59). Since most retroviral Gag precursor proteins contain similar proline-rich motifs to those found in $p58^{gag}$, we have proposed that this association with Abl and other SH3-containing proteins may play a role in maturation or pathogenesis of the retroviruses (44, 57).

One of the most prevalent oncogenes in human cancer is *ras*. Mutated Ras is maintained in its GTP-bound state and can activate multiple cellular pathways. Not surprisingly, Ras effectors which modulate nucleotide binding may also contribute to oncogenesis. Mutation of the Ras guanine nucleotide activating protein neurofibromin has been implicated in tumorigenesis (15). Neurofibromin, by reducing GTP-Ras levels, acts as a tumor suppressor (60). This example illustrates the yin-yang nature of oncogene-tumor suppressor couples in regulating cellular activities and transformation. Ras activation of the MAPK cascade of the growth factor-stimulated mitogenic pathway involves the proto-oncogene Raf (Chap. 2). In this case, activated Ras appears to serve as a recruiter to localize Raf at the plasma membrane (61, 62) where it can be activated and bound to a membrane-cytoskeletal site (Fig. 2.1) (47). The activator at this site is unknown; a number of activators of Raf have been proposed, suggesting that there may be multiple and alternative activation mechanisms, depending on the cellular context. Interestingly, *at least two pathways appear to be required for cell transformation via Ras.* One of these most likely includes PI3K (63). However, overexpression of strongly activated Raf is sufficient for transformation of some cells, possibly by inducing production of autocrine growth factors (64).

Ras is a potent effector of changes in the cytoskeleton, particularly microfilaments. Rho and Rac, which are involved in microfilament reorganizations (chapters 1, 4, 5 and 6), are required for transformation by Ras, further implicating the cytoskeleton as a contributor to transformation. Rac and cdc42 bind and activate a Pak Ser/Thr kinase to control the JNK (Jun kinase) and p38 MAPK pathways (53). This mechanism links these Rho family small G-proteins to another prominent class of oncogenes, the transcription factor genes. The activity of Jun as a transcription factor is regulated by phosphorylation, at least in part via this Rac-regulated MAPK pathway. Recently, Pak has been observed to bind the adaptor Nck, providing a link to tyrosine kinase receptors, which also bind Nck (65). Since Nck is reported to bind WASP, the proposed mediator between microfilaments and cdc42, the Nck may be acting as a scaffolding protein or integrator for complex formation regulating multiple pathways. A number of GEFs for the Rho family, including Dbl, Vav and Tiam1, have been shown to possess fibroblast transforming activity. Tiam1 is of particular interest because it has been implicated in Rac-dependent membrane ruffling, invasion and metastasis (66). The fact that transforming GEFs may regulate multiple Rho family members suggests that the putative cytoskeletal modifications are not highly specific. Instead, cytoskeletal reorganizations may be an integrated part of the transformation process, as suggested by the involvement of the cytoskeleton in cell cycle control.

One of the mechanisms for blocking aberrant cell behavior, as found in neoplasia, is apoptosis. As noted in chapter 5, apoptosis in some cell types occurs with a loss of adhesion. Moreover, apoptosis induces general morphological and cytoskeletal changes in addition to its effects on the nucleus (67). The key event in apoptosis is the activation of a family of proteolytic enzymes (caspases; also known as ICE-like proteases) which act on a multitude of cellular substrates (68). Among the substrates for these is a protein, Gas2, whose cleavage appears to cause dramatic microfilament reorganizations (67). Other cytoskeletal substrates are the nuclear envelope lamins and spectrin, which could contribute to changes in the nucleus and cortical actin filaments, respectively. As expected from the variety of apoptotic initiators described in chapter 5, caspase activation can be induced by several mechanisms. One of particular relevance to cell morphology involves the small G-protein-mediated kinase PAK2 (69). However in this case, PAK is activated by proteolytic cleavage of the enzyme itself, rather than by the GTPase. This effect suggests the mechanism involves the release of an inhibitory factor, either intramolecular or intermolecular. The activated PAK then affects the actomyosin of the apoptotic cell, possibly by acting on myosin heavy chain (70).

As noted above and in chapter 2, PI3K has been implicated in transformation via the Ras pathway and possibly by other mechanisms (71). It is proposed to act either directly as an adaptor protein or through its synthesis of phosphoinositides, as discussed in chapters 2 and 5. One suggested target for the phosphoinositides is the Ser kinase Akt, an oncoprotein which is also found as a product of the AKT8 murine retrovirus (15). Akt has a phosphoinositide-binding PH domain required for c-Akt activation. In contrast, v-Akt is a fusion protein with the Gag protein of the retrovirus. V-Akt is thus myristoylated, constitutively membrane associated and activated even without its PH domain. This mechanism exemplifies a common phenomenon. Many oncogenes are fusion proteins with Gag proteins, thus myristoylated and potentially membrane bound. An obvious exception is Src, but Src has its own myristoylation site. The exact role of myristoylation is unclear, but it may *serve to steer specific signaling components to complexes at membrane-microfilament sites*, where effects on the cytoskeleton can be integrated with incoming signals from extracellular ligands, either soluble or matrix-bound. Most importantly, these observations once again indicate how localization can contribute to signaling effects or defects.

Table 7.2. Some known or candidate tumor suppressor genes

Protein	Location	Function
APC	Cytoplasm?	?
DCC	Membrane	Cell adhesion
NF1	Cytoplasm	GTPase-activator
Merlin/schwanomin	Membrane skeleton?	Links membrane to cytoskeleton?
P53	Nucleus	Transcription factor
RB	Nucleus	Transcription factor
RET	Membrane	Receptor tyrosine kinase
Gelsolin	Cytoplasm, serum	Microfilament severing/ capping
α-Actinin	Cytoplasm	Microfilament crosslinking
p16^{INK4a}	Nucleus	CDK inhibitor
PTEN/MMAC	Cytoplasm?	Phosphatase?
Tropomyosin isoforms	Cytoplasm	Microfilament stabilization

Tumor Suppressors

Tumor suppressors have been much more difficult to identify because they are recessive and are not easily assayed. About a dozen tumor suppressors have been identified by positional cloning from families with hereditary forms of cancer. Since Knudson predicts a minimum of 50, based on hereditary cancers (33), many of these types must remain undescribed. A more rapid approach is to use *Drosophila* mutagenesis to create tumors and the power of *Drosophila* genetics to search for tumor suppressors (72). Finally, a different philosophical approach is to define components which should act as tumor suppressors, based on the characteristics of tumors. For example, tumors exhibit reduced cell-cell adhesiveness so components which are involved in cell-cell interactions may act as tumor suppressors. In fact, both E-cadherin and catenins, its cytoplasmic linkers to microfilaments, have been proposed to be tumor suppressors (see chapter 3). One can make a similar argument for integrins, cell cycle regulators, cytoskeleton regulators, some protein phosphatases, Ras GAPs, etc. In a sense, this approach is less concerned with finding tumor suppressor genes than with finding components and cellular processes which might block tumor progression. Table 7.2 lists some of the proteins proposed to have tumor suppressor function. Prominent among these is a growing list of cytoskeletal proteins, largely modulators of microfilament organization, again suggesting an important role for the cytoskeleton in neoplasia.

The two best-studied tumor suppressors are p53 and Rb, both of which are involved in regulating the cell cycle and can act as transcription factors. The role of Rb in progression through G1 was described in chapter 5 (see Fig. 5.5). Rb is the target for oncoproteins of several small DNA tumor viruses, notably the E1A proteins of adenovirus, large T antigen of polyomavirus and E7 protein of papillomavirus (73). The loss of Rb function prevents it from acting as an inhibitor of passage through G1 and lifts that restriction on proliferation. Mutants of tumor suppressor p53 are found in ≈60% of human tumors (74). Wild type p53 acts as a part of a sensing mechanism for DNA damage during progression through the cell cycle and for other signals of cellular distress (75). The p53 signal can be expressed in two ways, through activation of apoptosis or transcription activation. In the latter case, p53 induces expression of cell cycle inhibitors which act on cdk/cyclin kinases and the Rb pathway. Mutants of p53 lose their ability to activate apoptosis and suppress proliferation, though the effects may vary with different mutations. Some mutations can even promote tumorigenesis. This observation led to an early identification of p53 as an oncogene (75). Since p53 functions as a multimer, mutants which lose function, but maintain their capability for interaction may sequester wild type subunits and inhibit their suppressive effects.

Genetic analyses of colon tumor progression have revealed two additional tumor suppressors, APC and DCC, both of which were originally assumed to play a role in cell adhesion. APC was shown to bind Armadillo domains of β-catenin, implicating it in the cadherin cell-cell interaction complex described in chapter 3. The catenins are required for the linkage of cadherin to the cytoplasmic cytoskeleton and are involved in the regulation of cadherin function during tissue morphogenesis (8). However, β-catenin was later shown to be a component in the Wnt signaling pathway (15). The Wnt oncogene product is a cytokine which interacts with a cell serpentine receptor to induce gene expression via a pathway involving β-catenin (Fig. 7.1). In this mode, β-catenin directly modulates a transcription factor, while APC regulates degradation of β-catenin by binding to a site competitive for cadherin binding. Regulation of degradation is additionally dependent on phosphorylation by glycogen synthase kinase 3. β-catenin also interacts with the actin-bundling protein fascin at a site competitive with the cadherin binding site (76). Finally, APC associates with microtubules and localizes to microtubule-containing protrusions of migratory MDCK epithelial cells (77). *These combined results suggest that a multifunctional β-catenin may act as a signal integrator of proliferation (via the Wnt pathway) and cell morphology (via cadherin, microtubules and fascin).* Since β-catenin overexpression has been implicated in colon tumors (19, 20), cadherin may act as a tumor suppressor by sequestering β-catenin rather than by acting as a cell adhesion component.

Although APC may not be involved directly in cell-cell interactions, other cell adhesion molecules have also been implicated in tumor suppression. These include cadherin, integrins and immunoglobulin-like cell surface components (78) such as an NCAM-like (Ig) adhesion molecule and the product of the *Drosophila fat* gene, an analog of cadherin (72). Integrin and cadherin are major determinants of polarity and differentiated function in epithelial cells and cadherin of compaction during development (8). Disruption of cadherin's linkage to the cytoskeleton by whatever mechanism, prevents stable cell-cell interactions and the ability of epithelial tissue to perform its normal differentiated functions. For example, antibodies to cadherins perturbed morphogenesis of some embryonic organs in vitro (79). Other studies (80-82) suggested a role for E-cadherin in tumor suppression. Down-regulation of E-cadherin expression has been found to correlate with tumor progression in many breast carcinomas, but not in others (83). Numerous studies have found that loss of expression or function of catenins, particularly β-catenin, regardless of cadherin state, is correlated with tumor progression. Clearly, a role for loss of expression of cadherin and its associated cytoplasmic proteins can be postulated in tumor progression. However, the organization

Table 7.3. Integrins and the tumorigenic phenotype

Integrin type	Effect on cells and tissues
$\alpha5\beta1$	Reduced in tumor cells; overexpression suppresses growth and tumorigenesis; induces gas-1 and Bcl2 expression.
$\beta1c$	Diffuse cell surface distribution; inhibits cell cycle progression at G_1 phase if overexpressed.
$\alpha2\beta1$	Inhibits malignancy of a mammary carcinoma; enhances metastasis in a rhabdomyosarcoma cell line.
$\alpha3\beta1$	Expressed in most tumor cells, but not in a rhabdomyosarcoma cell line where its overexpression inhibits malignancy.
$\alpha v\beta5$	Suppresses anchorage independence and increases terminal differentiation of squamous cell carcinoma keratinocytes.
$\alpha6\beta4$ ($\alpha6$ variant)	Associated with malignant conversion of mouse skin keratinocytes.
$\alpha6$	Levels elevated in breast, hepatocellular and nonsmall-cell lung carcinoma, but identified as a potential tumor suppressor by PCR differential display of mammary epithelial cells.
$\alpha4\beta1$	Inhibits metastasis and homotypic aggregation when expressed on B16 melanoma and T-cell lymphoma but enhances metastasis to the bone and lungs of mice if transfected into CHO and K562 cells.
$\alpha v\beta3$	Essential for angiogenesis; colocalizes with metalloprotease MMP2 on invasive melanoma and angiogenic cells.

MMP2, matrix metalloproteinase 2. From ref. 79 (Ben Ze'ev A. Curr Opin Cell Biol 1997; 9:99–108), courtesy of Current Biology Ltd.

and regulation of cadherin complexes in tissues is complex, and other mechanisms for disruption of these complexes resulting in decreased cell-cell adhesiveness can be envisioned. The interplay between cadherin-mediated cell-cell adhesions and other ligand-receptor systems in a given cell will also be expected to contribute to (or detract from) the normal adhesive properties of cadherin. For example, the integrity of the cadherin-mediated junction is disrupted by soluble factors such as "scatter factor" from mesenchymal cells during certain stages of normal development. An aberrant intervention by these factors after tissue development has been proposed to lead to an epithelial-mesenchymal transition which may be linked to tumor progression (81).

The role of cell adhesion molecules as tumor suppressors is further complicated by the multiplicity of their interactions with different cell types and different ligands, as exemplified in Table 7.3 for the integrins (84). For example, $\alpha5\beta1$, which is reduced in tumor cells, can suppress tumor cell growth through its induction of *gas-1*, a growth arrest gene which blocks the transcription of immediate early genes. Furthermore, $\alpha5\beta1$ can also activate the Bcl-2 pathway to induce apoptosis. In contrast, $\alpha2\beta1$ inhibits mammary carcinoma, but enhances metastasis of a rhabdomyosarcoma cell line. Similarly, αvβ3 is required for angiogenesis, which is necessary for tumor growth in vivo. The disparate effects of different integrins can be rationalized by their involvement in different cell functions through different integrin signaling pathways or through their integration with pathways triggered by other agents such as growth factors, cell-cell adhesions or selectins (85). Many tumors are characterized by a loss of adhesiveness and a reduced anchorage dependence of proliferation (86), a complex effect on the cell cycle discussed in chapter 5. Integrins also play an important role in suppressing apoptosis, though the effects are both cell-type and integrin specific (87). An imbalance of the opposing roles of proliferation and apoptosis in tissue homeostasis can potentially be explained by the multiple roles of integrins and their signaling pathways.

Integrin binding can be modulated by inside-out signaling, discussed in chapter 4. Thus, cytoskeletal components of integrin signaling complexes or their associated microfilaments might be expected to act as tumor suppressors via their effects on adhesiveness or other effects on cell morphology or motility. Consistent with this idea are the observations of reduction of expression of a number of cytoskeletal proteins in cancer cells, including tropomyosin isoforms, gelsolin, α-actinin and vinculin (88). Changes in these components can at least partly account for the morphological and cytoskeletal changes which have been long associated with neoplastic transformation (86). Restoration of the levels of α-actinin and vinculin suppresses neoplastic behavior of such cells, as does overexpression of high molecular weight isoforms of tropomyosin (84). Finally, genetic analysis of neurofibromatosis 2, an inherited disposition to schwannomas and meningiomas, identified the tumor suppressor gene *merlin/schwanomin*, a member of the ezrin/radixin/moesin family of proteins involved in linking microfilaments to membranes (89). These studies provide further evidence for the importance of cytoskeletal complexes at the membrane in regulating complex cellular behaviors.

The possibility that phosphatases can act as tumor suppressors is supported by the recent cloning of a gene which is mutated in several types of cancer (90). Significantly, the gene contains a phosphatase-like sequence related to cdc14, a cell division control enzyme from yeast, and a sequence similar to a region of tensin, a microfilament-binding protein which is a component of focal adhesions. These results suggest that the protein may be specifically localized to membrane-microfilament interaction sites, where it could play an important role in regulating cell morphology.

Summary

Early development is noteworthy for its variety in different species, but this variety is typically achieved through variations on common molecular mechanisms. The first step in animal development is sperm-egg contact which initiates reciprocal activation of both cell types. Sperm activation triggers signaling events and cytoskeletal rearrangements necessary for penetrating the egg coat and fusing the sperm and egg membranes. Egg activation triggers an ordered series of signaling processes and cytoskeletal reorganizations to prepare the egg for cell division and subsequent development of the animal body plan. This development is controlled via specific localization of transcripts and regulatory proteins in the egg and through subsequent cell divisions. In mammals, a critical step in early development is a cell-adhesion dependent compaction at the eight cell stage. The subsequent differentiative step forms the blastocyst, consisting of the inner cell mass, which becomes the embryo, and the trophoblast, a polarized epithelium responsible for implantation. Expansion of the cavity within the trophoblast involves both fluid movements, driven by specifically localized ion pumps in the epithelium, and apoptosis of selected cells of the inner cell mass of the growing embryo. The development of the embryo body plan progresses with gastrulation, in which specific cell populations migrate from their original positions to form the three basic types of embryonic tissues, endoderm, mesoderm and ectoderm. These migrations depend on changes in both cell adhesions and protrusive activities. They are regulated by paracrine effects of cytokines produced by specifically localized cells.

Neoplasia results from mutations in genes encoding proteins which regulate specific activities of cells of normal tissues, particularly cell division, cell adhesion and cytoskeletal organization. Two types of genetic changes are important: those which activate proteins encoded by oncogenes to enhance proliferation and those which result in a loss of activity for proteins encoded by tumor suppressor genes involved in functions altered in neoplasia. Prominent oncogenic proteins include tyrosine kinases, Ras family members, adaptor proteins and transcription factors. Much less is known about tumor suppressors because of the difficulties in identifying them. The primary tumor suppressors identified are proteins involved in transcriptional regulation, cell adhesion, protein dephosphorylation and organization of the cytoskeleton at the cell membrane. The results obtained so far suggest that many tumor suppressors have yet to be identified.

References

1. Bonder EM, Fishkind DJ. Actin-membrane cytoskeletal dynamics in early sea urchin development. Curr Top Devel Biol 1995; 31:101-137.
2. Gilbert S. Developmental Biology, 3rd ed., Sinauer Associates, Inc., 1991.
3. Miyazaki S. Calcium signalling during mammalian fertilization. Ciba Foundation Symp 1995; 188:235-251.
4. Spudich A. Actin organization in the sea urchin egg cortex. Curr Topics Devel Biol 1992; 26:9-21.
5. Louvard D, Kedinger M, Hauri HP. The differentiating intestinal epithelial cell: establishment and maintenance of functions through interactions between cellular structures. Annu Rev Cell Biol 1992; 8:157-195.
6. Schejter ED, Wieschaus E. Functional elements of the cytoskeleton in the early *Drosophila* embryo. Annu Rev Cell Biol 1993; 9:67-99.
7. Schultz RM. Regulation of zygotic gene activation in the mouse. BioEssays 1993; 15:531-538.
8. Gumbiner BM. Cell adhesion: the molecular basis of tissue architecture and morphogenesis. Cell 1996; 84:345-357.
9. Watson AJ, Kidder GM, Schultz, GA. How to make a blastocyst. Biochem Cell Biol 1992; 70:849-855.
10. Coucouvanis E, Martin GR. Signals for death and survival: a two-site mechanism for cavitation in the vertebrate embryo. Cell 1995; 83:279-287.
11. Winkel GK, Ferguson JE, Takeichi M, Nuccitelli R. Activation of protein kinase C triggers premature compaction in the four-cell stage mouse embryo. Dev Biol 1990; 138:1-15.
12. Ohsugi M, Yamamura H. Differences in the effects of treatment of uncompacted and compacted mouse embryos with phorbol esters on pre- and postimplantation development. Differentiation 1993; 53:173-179.
13. Wiley LM, Adamson ED, Tsark EC. Epidermal growth factor receptor function in early mammalian development. BioEssays 1995; 17:839-846.

14. Rowning BA, Wells J, Wu M, Gerhart JC, Moon RT, Larabell CA. Microtubule-mediated transport of organelles and localization of β-catenin to the future dorsal side of *Xenopus* eggs. Proc Natl Acad Sci USA 1997; 94:1244-1229.

15. Hunter T. Oncoprotein networks. Cell 1997; 88:333-346.

16. Wang S, Krinks M, Lin K, Luyten FP, Moos M Jr. Frzb, a secreted protein expressed in the Spemann organizer, binds and inhibits Wnt-8. Cell 1997; 88: 757-766.

17. Leyns L, Bouwmeester, Kim S-H, Picccolo S, De Robertis DM. Frzb-1 is a secreted antagonist of Wnt signaling expressed in the Spemann organizer. Cell 1997; 88:747-756.

18. Hopkin K. Wnt world: a window on gastrulation, development, and cancer. J NIH Res 1997; 9:21-23.

19. Korinek V, Barker N, Morin PJ, van Wichen D, de Weger R, Kinzler KW, Vogelstein B, Clevers H. Constitutive transcriptional activation by a β-catenin-Tcf complex in APC$^{-/-}$ colon carcinoma. Science 1997; 275:1784-1787.

20. Morin PJ, Sparks AB, Korinek V, Barker N, Clevers H, Vogelstein B, Kinzler KW. Activation of β-catenin-Tcf signaling by mutations in β-catenin or APC. Science 1997; 275:1787-1790.

21. Yisraeli JK, Sokol S, Melton DA. A two-step model for the localization of maternal mRNA in *Xenopus* oocytes: involvement of microtubules and microfilaments in the translocation and anchoring of Vg1 mRNA. Development 1990; 108: 289-298.

22. Bassell G, Singer RH. mRNA and cytoskeletal filaments. Curr Opin Cell Biol 1997; 9:109-115.

23. Thomsen GH, Melton DA. Processed Vg1 protein is an axial mesoderm inducer in *Xenopus*. Cell 1993; 74:433-441.

24. Mowry KL, Melton DA. Vegetal messenger RNA localization directed by a 340nt RNA sequence element in *Xenopus* oocytes. Science 1992; 255:991-994.

25. Pondel MD, King ML. Localized maternal mRNA related to transforming growth factor beta mRNA is concentrated in a cytokeratin-enriched fraction from *Xenopus* oocytes. Proc Natl Acad Sci USA 1988; 85:7612-7616.

26. Bearer EL. Introduction. Curr. Topics Devel Biol 1992; 26:1-7.

27. Gustafson T, Wolpert L. Studies on the cellular basis of morphogenesis in sea urchin embryos: Directed movements of primary mesenchyme cells in normal and vegetalized larvae. Exp Cell Res 1961; 24:64-79.

28. Fink RD, McClay DR. Three cell recognition changes accompany the ingression of sea urchin primary mesenchyme cells. Dev Biol 1985; 107:66-74.

29. Gustafson T, Wolpert L. Cellular movement and contact in sea urchin morphogenesis. Biol Rev 1967; 42:442-498.

30. Hardin J, McClay DR. Target recognition by the archenteron during sea urchin gastrulation. Dev Biol 1990; 142:87-105.

31. Kellie S. Tyrosine Kinases and Neoplastic Transformation. Austin, Texas: R.G. Landes Co., 1994.

32. Harris H, Miller OJ, Klein G, Worst P, Tachibana T. Suppression of malignancy by cell fusion. Nature 1969; 223:363-368.

33. Knudson AG. Antioncogenes and human cancer. Proc Natl Acad Sci USA 1993; 90: 10914-10921.

34. Ruddon RW. Cancer Biology, 3rd ed., Oxford, UK: Oxford University Press, 1995.

35. Levine AJ. The tumor suppressor genes. Annu Rev Biochem 1993; 62:623-651.

36. Fearon ER, Vogelstein B. A genetic model for colorectal tumorigenesis. Cell 1990; 61:759-767.

37. Carraway CAC, Carraway KL. Interactions of membrane receptors and cell signaling systems with the cytoskeleton. In: Hesketh HE, Pryme IF, eds. Treatise on the Cytoskeleton. Vol. 2. Greenwich, CT: JAI Press, 1996:207-238.

38. Juang S-H, Carvajal ME, Whitney M, Liu Y, Carraway CAC. Tyrosine phosphorylation at the membrane-microfilament interface: A p185neu-associated signal transduction particle containing Src, Abl and phosphorylated p58, a membrane- and microfilament-associated retroviral Gag-like protein. Oncogene 1996; 12: 1033-1042.

39. Carraway CAC, Carvajal ME, Li Y, Carraway KL. Association of p185neu with microfilaments via a large glycoprotein complex in mammary carcinoma microvilli. Evidence for a microfilament-associated signal transduction particle. J Biol Chem 1993; 268:5582-5587.

40. Carraway CAC, Carvajal ME, Carraway KL. Association of the Ras/MAP kinase signal transduction pathway with microfilaments. Evidence for a p185neu-containing cell surface signal transduction particle linking the mitogenic pathway to a membrane-microfilament association site. Submitted.

41. Juang S-H, Huang J, Li Y, Salas PJI, Fregien N, Carraway CAC, Carraway KL. Molecular cloning and sequencing of a 58 kDa membrane-and microfilament-associated protein from ascites tumor cell microvilli with sequence similarities to retroviral Gag proteins. J Biol Chem 1994; 269:15067-15075.

42. Liu Y, Carraway KL, Carraway CAC. Isolation and characterization of a 58 kda membrane and microfilament-associated protein from ascites tumor cell microvilli. J Biol Chem 1989; 264: 1208-1214.

43. Howard SC, Hull SR, Huggins JW, Carraway CAC, Carraway KL. Relationship between xenotransplantability and cell surface properties of ascites sublines of a rat mammary adenocarcinoma. J Nat Cancer Inst 1982; 69:33-40.

44. Huang J, Li Y, Mayer B, Carraway KL, Carraway CAC. c-Src association with and phosphorylation of p58gag, a membrane- and microfilament-associated retroviral gag protein. Submitted.

45. Nishizuka Y. The role of protein kinase C in cell surface signal transduction and tumor promotion. Nature 1984; 308: 693-698.

46. Carraway KL III, Cantley LC. A neu acquaintance for erbB3 and erbB4: a role for receptor heterodimerization in growth signaling. Cell 1994; 78:5-8.

47. Carraway KL, Carraway CAC. Signaling, mitogenesis and the cytoskeleton: Where the action is. BioEssays 1995; 17:171-175.

48. Peles E, Yarden Y. Neu and its ligands: From an oncogene to neural factors. BioEssays 1993; 15:815-824.

49. den Hartigh JC, van Bergen en Henegouwen PMP, Verkleij AJ, Boonstra J. The EGF receptor is an actin-binding protein. J Cell Biol 1992; 119:349-355.

50. Defize LHK, Boonstra J, Meisenhelder J, Kruijer W, Tertoolen LGJ, Tilly BC, Hunter T, Van Bergen en Henegouwen PMP, Moolenaar WH, de Laat SW. Signal transduction by epidermal growth factor occurs through the subclass of high affinity receptors. J Cell Biol 1989; 107:939-949.

51. Golub TR, Barker GF, Lovett M, Gilliland DG. Fusion of PDGF receptor β to a novel ets-like gene, tel, in chronic myelomonocytic leukemia with 5(5;12) chromosomal translocation. Cell 1994; 77:307-316.

52. Golub TR, Goga A, Barker GF, Afar DEH, McLaughlin J, Bohlander SK, Rowley JD, Witte ON, Gilliland DG. Oligomerization of the ABL tyrosine kinase by the Ets protein TEL in human leukemia. Mol Cell Biol 1996; 16:4107-4116.

53. Gutkind JS, Vitale-Cross J. The pathway linking small GTP-binding proteins of the Rho family to cytoskeletal components and novel signaling kinase cascades. Sem Cell Devel Biol 1996; 7: 683-690.

54. Renshaw MW, McWhirter JR, Wang JYJ. The human leukemia oncogene *bcr-abl* abrogates the anchorage requirement but not the growth factor requirement for proliferation. Mol Cell Biol 1995; 15: 1286-1293.

55. Daley GQ, van Etten RA, Jackson PK, Bernards A, Baltimore D. Nonmyristoylated Abl proteins transform a factor-dependent hematopoietic cell line. Mol Cell Biol 1992; 12:1864-1871.
56. Wang JYJ. Abl tyrosine kinase in signal transduction and cell-cycle regulation. Curr Opin Genet Devel 1993; 3:35-43.
57. Huang J, Li Y, Mayer B, van Etten R, Carraway KL, Carraway CAC. Abl association with and phosphorylation of p58^{gag}, a membrane- and microfilament-associated retroviral gag protein. Submitted.
58. Cohen GB, Ren R, Baltimore D. Modular binding domains in signal transduction proteins. Cell 1995; 80:237-248.
59. Higley S, Way M. Actin and cell pathogenesis. Curr Opin Cell Biol 1997; 9: 62-69.
60. Rey I, Hall A. Tumour suppressors and the regulation of GTP-binding protein activity. Trends Cell Biol 1993; 3:39-42.
61. Leevers SJ, Paterson HF, Marshall CJ. Requirement for Ras in Raf activation is overcome by targeting Raf to the plasma membrane. Nature 1994; 369:411-414.
62. Stokoe D, Macdonald SG, Cadwallader K, Symons M, Hancock JF. Activation of Raf as a result of recruitment to the plasma membrane. Science 1994; 264: 1463-1467.
63. Rodriguez-Viciana P, Warne PH, Khwaja A, Marte BM, Pappin D, Das P, Waterfield MD, Ridley A, Downward J. Role of phosphoinositide 3-OH kinase in cell transformation and control of the actin cytoskeleton by Ras. Cell 1997; 89: 457-467.
64. McCarthy SA, Samuels ML, Pritchard CA, Abraham JA, McMahn M. Rapid induction of heparin-binding epidermal/diphtheria toxin receptor expression by Raf and Ras oncogenes. Genes Dev 1995; 9:1953-1964.
65. Galisteo ML, Chernoff J, Su Y-C, Skolnik EY, Schlessinger J. The adaptor protein Nck links receptor tyrosine kinases with the serine-threonine kinase Pak1. J Biol Chem 1996; 271:20997-21000.
66. Michiels F, Habets GGM, Stam JC, van der Kammen RA, Collard JG. A role for Rac in Tiam1-induced membrane ruffling and invasion. Nature 1995; 375: 338-340.
67. Brancolini C, Benedetti M, Schneider C. Microfilament reorganization during apoptosis: the role of Gas2, a possible substrate for ICE-like proteases. EMBO J 1995; 14:5179-5190.
68. Porter AG, Ng P, Janicke RU. Death substrates come alive. BioEssays 1997; 19: 501-507.
69. Rudel T, Bokoch GM. Membrane and morphological changes in apoptotic cells regulated by caspase-mediated activation of PAK2. Science 1997; 276:1571-1574.
70. Brzeska H, Korn ED. Regulation of class I and class II myosins by heavy chain phosphorylation. J Biol Chem 1996; 271: 16983-16986.
71. Cantley LC, Auger KR, Carpenter C, Duckworth B, Graziani A, Kapeller R, Soltoff S. Oncogenes and signal transduction. Cell 1991; 64:281-302.
72. Bryant PJ. Towards the cellular functions of tumour suppressors. Trends Cell Biol 1993; 3:31-35.
73. Helin K, Harlow E. The retinoblastoma protein as a transcriptional repressor. Trends Cell Biol 1993; 3:43-46.
74. Rotter V, Foord O, Navot N. In search of the functions of normal p53 protein. Trends Cell Biol 1993; 3:46-49.
75. Levine A. p53, the cellular gatekeeper for growth and division. Cell 1997; 88: 323-331.
76. Tao YS, Edwards RA, Tubb B, Wang S, Bryan J, McCrea PD. β-catenin associates with the actin-bundling protein fascin in a noncadherin complex. J Cell Biol 1996; 134:1271-1281.
77. Adams JC. Cell adhesion—spreading frontiers, intricate insights. Trends Cell Biol 1997; 7:107-111.

78. Hedrick L, Cho KR, Vogelstein B. Cell adhesion molecules as tumour suppressors. Trends Cell Biol 1993; 3:36-39.
79. Takeichi M. The cadherins: cell-cell adhesion molecules controlling animal morphogenesis. Development 1988; 102: 639-655.
80. Gumbiner BM, McCrea PD. Catenins as mediators of the cytoplasmic functions of cadherins. J Cell Sci Suppl 1993; 17: 155-158.
81. Birchmeier W, Hulsken J, Behrens J. E-cadherin as an invasion suppressor. Ciba Found Symp 1995; 189:124-136.
82. Bracke ME, Van Roy FM, Mareel MM. The E-cadherin/catenin complex in invasion and metastasis. Curr Top Microbiol Immunol 1996; 213:123-161.
83. Behrens J. Cell contacts, differentiation, and invasiveness of epithelial cells. Invasion Metastasis 1994-95; 14:61-70.
84. Ben Ze'ev A. Cytoskeletal and adhesion proteins as tumor suppressors. Curr Opin Cell Biol 1997; 9:99-108.
85. Zetter BR. Adhesion molecules in tumor metastasis. Semin Cancer Biol 1993; 4:219-29.
86. Nicolson GL. Transmembrane control of the receptors on normal and tumor cells. II. Surface changes associated with transformation and malignancy. Biochim Biophys Acta 1976; 457:57-108.
87. Meredith JE Jr, Schwartz MA. Integrins, adhesion and apoptosis. Trends Cell Biol 1997; 7:146-150.
88. Ben-Ze'ev A. The cytoskeleton in cancer cells. Biochim Biophys Acta 1985; 780: 197-212.
89. Tsukita S, Yonemura S, Tsukita S. ERM (ezrin/radixin/moesin) family: from cytoskeleton to signal transduction. Curr Opin Cell Biol 1997; 9:70-75.
90. Li J, Yen C, Liaw D, Podsypanina K, Bose S, Wang SI, Puc J, Miliaresis C, Rodgers L, McCombie R, Bigner SH, Giovanella BC, Ittmann M, Tycko B, Hibshoosh H, Wigler MH, Parsons R. PTEN, a putative protein tyrosine phosphatase gene mutated in human brain, breast and prostate cancer. Science 1997; 275: 1943-1947.

CHAPTER 8

Perspectives

In the Preface we discussed two questions which encompass most of the research on the cytoskeleton and signaling:

1) How are the structure and function of the cytoskeleton affected by external signals which impinge on the cell?
2) How does the cytoskeleton influence the cellular signaling processes which determine cell behaviors?

In the intervening chapters, we have used a functional approach to address the first question, examining many of the important cellular functions involving the cytoskeleton. From these analyses some general concepts concerning the cytoskeleton and its regulation emerge. First, although there are three types of cytoskeletal elements which can operate and be studied independently, their behavior and functions must be integrated in the cell. Second, the organization and function of the cytoskeleton is determined primarily by four processes: polymerization/depolymerization, crosslinking, linkage to membranes and motor protein activities. Third, these four processes contribute to varying degrees for the three types of filaments and for different cellular functions, and they must be regulated and integrated. Understanding the behavior and regulation of the cytoskeleton involves a hierarchical process: recognizing a specific behavior, identifying the cytoskeletal components involved, elucidating the mechanism(s) responsible for the cytoskeletal involvement and determining the regulatory aspects controlling the cytoskeletal involvement and integrating it into cellular behavior.

A good example of this investigational hierarchy can be seen in studies of the role of microfilaments in cell shape changes at the cell periphery, such as the extension of filopodia. This is a well-defined cell behavior which predominantly involves, if not exclusively, one type of filament, microfilaments. Actin polymerization is critical, but the mechanism by which actin monomers are added to the membrane-proximal end of growing microfilaments is still unclear and remains one of the most important questions in cytoskeleton research. The issue of which membrane components participate is largely unresolved. A myosin motor may be involved, but exactly which motor protein participates and how it functions remain uncertain. Filopodial extensions can be triggered by extracellular signals. The signaling pathway(s) involved requires the participation of the small G-protein cdc42. This protein may provide an integrating element for formation of filopodia by its linkage to actin polymerization via the Wiscott-Aldrich Syndrome protein (WASP). However, cdc42 has also been implicated in vesicle movements to the membrane. Both of these processes are critical to filopodial extensions. In summary, the microfilament-dependent protrusion of the cell surface is fundamental to both cell motility and some types of endocytosis, but our understanding of this process

Signaling and the Cytoskeleton, by Kermit L. Carraway, Coralie A. Carothers Carraway and Kermit L. Carraway III. © 1998 Springer-Verlag and R.G. Landes Company.

remains fragmentary. In that respect it provides a both a paradigm and a model for research on the response of the cytoskeleton to external signals. We have accumulated a large number of facts about the processes involved, but the overall picture is still very incomplete.

The second question, how the cytoskeleton influences signaling, has received much less attention. We have previously postulated that a critical element of signaling pathways is the location and organization of their components, and that the cytoskeleton plays an important role in that localization and organization (1). For example, many of the important signaling pathways which control cell behavior are localized in complexes at sites where the cytoskeleton associates with membranes. Some examples of these complexes are listed in Tables 3.1 and 3.2. Probably the most thoroughly studied of these complexes is the focal adhesion, which has been implicated in adhesion, motility, cell proliferation, apoptosis and gene expression. Judging from the temporal aspects of the formation of microfilaments at the focal adhesion, one might predict that their primary role there may be to stabilize the organization of the components, including the associated signaling elements. However, studies with cytochalasin suggest that microfilaments also play an earlier role in formation and function of the focal adhesion, though the nature of that role is unclear. One intriguing possibility is in translocation of components to the cell periphery, since a recent study has shown that v-Src, which is localized to focal adhesions, is moved from the perinuclear region of the cell to adhesion sites by a process mediated by microfilaments and regulated by Rho (2). Thus, the emerging paradigm is that *the cytoskeleton exerts one influence over signaling by providing an intracellular matrix on which signaling pathways are organized,* particularly at membrane-cytoskeleton interaction sites, and *exerts a second influence by providing tracks or guideways over which signaling elements can be moved within the cell.* This second role would be in addition to the organizing and integrating functions.

The question of the intracellular movement of signals remains one of the most perplexing in cell biology. Kinases, such as Abl, MAP kinase and Rsk, which are believed to activate transcription factors by phosphorylation in the nucleus, are themselves activated in the cytoplasm, often at the plasma membrane. Although the relocalization of these enzymes can be demonstrated microscopically (3), the mechanisms remain obscure. Since the consistency of the cytoplasm of most cells is not highly conducive to rapid diffusion of proteins, some facilitated mechanism of movement appears likely. Obviously, one possibility is that a membrane signaling complex is moved by vesicular transport to a perinuclear site, where the appropriate components are released for nuclear import, as described below.

In dealing with specific aspects of cell signaling, one can become overwhelmed with the amount of information impinging on cells and the variability of cellular responses. This complexity sometimes tempts the investigator to take too simplistic an approach toward analyzing cause-and-effect relationships between a signal and its cellular response. Attempts to understand how cells function must take into account three principles: *multiplicity, additivity* and *redundancy. Multiple signals* impinge on cells and different signals may yield the same apparent response. An example is the mitogenic stimulation of a cell by more than one growth factor/cytokine. On the other hand, the same ligand-receptor complex may trigger signaling pathways leading to different responses in different cells. For example, PI3K activation is involved in cell proliferation in some cell types, but not in others (4). It is obvious that *cellular context* must play a critical role in the final translation of any incoming signal. Further, different responses can be elicited when a single receptor has more than one ligand, as in the case of the EGF receptor (discussed in chapter 2). Binding of each ligand can activate a different signaling pathway, leading to a *multiplicity of cellular responses*. To add to the complexity of interpreting signaling, the comfortable paradigm of ligand binding to a single receptor type has been found too simplistic in the case of the ErbB receptors (5). Ligands for the ErbB receptors appear to prefer binding to receptor heterodimers than to homodimers, creating additional diversity or multiplicity. To complicate matters further, binding of a ligand to a specific receptor heterodimer can lead to different responses in different cell types. Heregulin/NDF stimulates either differentiation or proliferation in different mammary tumor cell lines (5), again emphasizing the importance of cell context.

Another layer of complexity is encountered when considering the manner in which cells respond to the multiplicity of signals impinging on á cell at any given time. The *responses elicited by the signals are not just additive*; the cellular response may be the result of *combinatorial effects*, integrating additive, synergistic and antagonistic effects. Combinatorial signaling occurs, for example, during mitogenic activation of a cell via a growth factor receptor. A cell growing either in a tissue or on a dish will have constraints already placed upon it by interactions with ECM and possibly with other cells. These constraints act to maintain a polarized cell morphology and to resist deformation, whereas a soluble factor such as a growth factor, may impose a directive leading to global reorganization. The cell must integrate these directives and respond in a manner which reflects an appropriate contribution from each ligand-receptor complex. An important aspect of integration is linkage of the signaling pathway initiated by ligand binding with pathways regulating cytoskeleton organization. Signaling and responses to signals may be *hierachical* as well as combinatorial, as described for focal adhesion formation (chapter 4). Epithelial cell polarization involves another type of hierarchical response in which cell-matrix adhesions precede cell-cell adhesions (chapter 3). Soluble ligands such as the hepatocyte growth factor (HGF) or "scatter factor", which binds the receptor c-Met and plays an important role in tissue remodeling during development, may occur only at certain stages of the hierarchy. The *sequential nature of cellular response to multiple signals* provides an important aspect of specificity and regulation. An example of the integration of pathways elicited by the engagement of c-Met by HGF to give the complex modulation of cell morphology, gene expression and mitogenesis of epithelial cells is shown in Figure 8.1 (6).

A further problem arises in understanding signaling when one makes the assumption that signaling pathways are linear. This assumption has led to the apparent presence of signaling components in multiple places in pathways. Removing the assumption of linearity and allowing parallel interacting pathways with multiple crossover points may reconcile such contradictions. One possible example is the interactions between the effects of the Rho family of proteins, Rho, Rac and cdc42, which are not easily explained by a linear scheme (7). The presence of feedback loops in pathways must also be considered. Two examples of feedback mechanisms which repress signals involve phosphorylation of a receptor by specific serine/threonine-specific kinases:

1) The tyrosine kinase activity of the EGF receptor is inhibited via phosphorylation by protein kinase C (8).

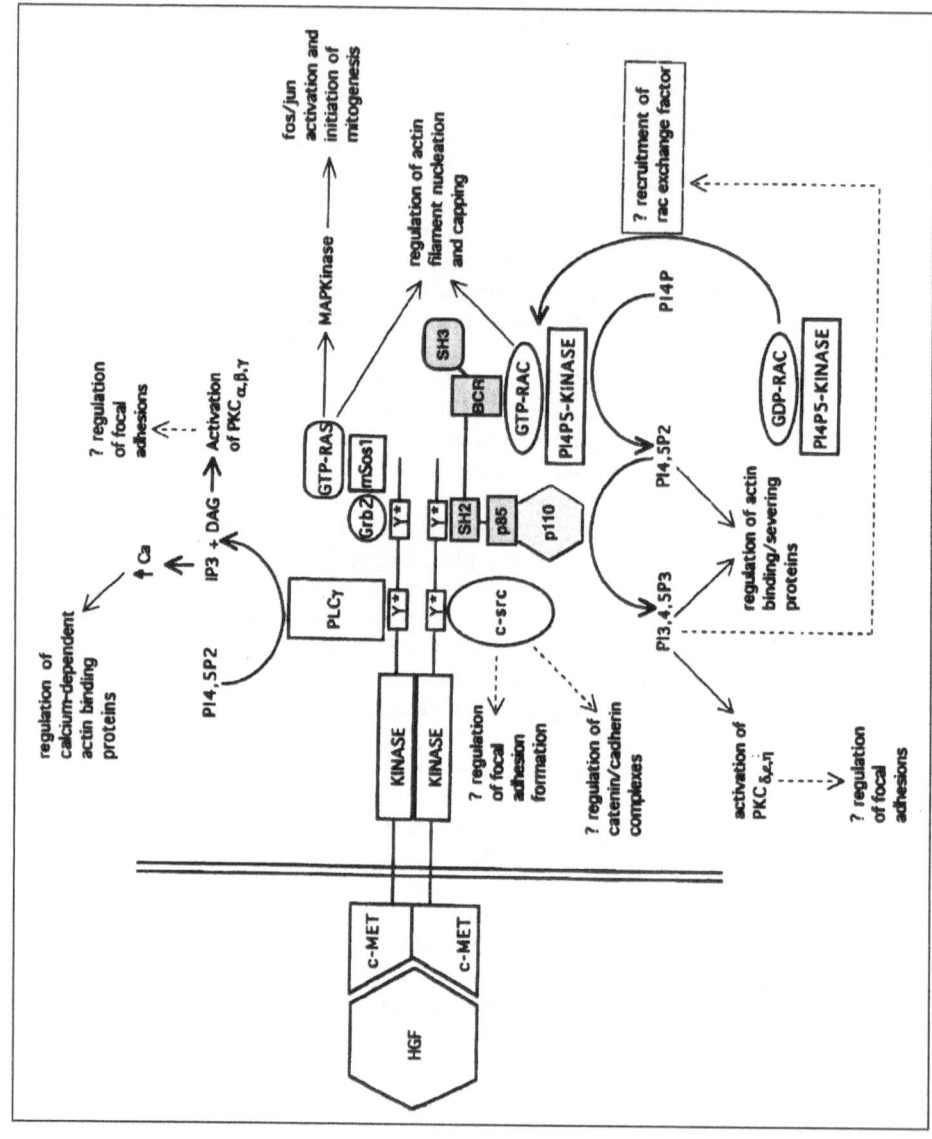

Fig. 8.1. Model for HGF effects on cell structure and functions. Figure from ref. 6 (Cantley LG. Am J Physiol 1996; 271: F103-13), courtesy of American Society for Physiology.

2) The β-adrenergic receptor's activity is modulated by phosphorylation by β-adrenergic kinase (βARK), leading to desensitization of the receptor to further stimulation (9).

An additional complexity in understanding signaling pathways is the *redundancy in pathways* for many functions, perhaps especially the most critical ones. A classic example arises in the knock-out experiments on Src family members. Gene disruptions of individual family members show relatively mild phenotypic reponses. Only in knockouts of multiple family members are substantial effects observed (10).

It is clear that attempts to provide simplistic answers to the critical questions in signaling must ultimately be frustrated. The quesion of importance is how the cell organizes and integrates all of these inputs and outputs. One proposal is that the cell operates primarily in the manner of a solid state system of interconnected membranes and cytoskeletal elements. *This framework provides the scaffolding for the organization of components of individual pathways*, but equally important, the *sites for association of components of different pathways for integration of signals*. A prototype for this model is the complex which integrates yeast budding and mating factor-dependent inhibition of cell cycle progression (see chapter 5), in which the protein Ste5 acts as a scaffold to bind several components of the pathways at the membrane-microfilament interface. Other proteins and complexes have been identified which may act as scaffolds in the formation of signaling complexes or the integration of different signaling pathways (Table 8.1).

An alternative mechanism to the formation of interactive complexes which facilitate signaling is the *sequestration of signaling components as a regulatory mechanism* (1, 11). Although there are few well defined examples of sequestration, binding domains for associations of proteins with other proteins and with lipids (described in chapter 2) provide molecular mechanisms for sequestration of important signaling proteins. For example, neuregulin/heregulin/ARIA is associated via a heparin-binding domain with ECM components at the muscle cell surface near its receptor and can be released by a triggering mechanism and proteolysis to bind to the receptor (12, 13). *Sequestration may act as a positive, as well as a negative mechanism*, if the signaling component is sequestrated near its future site of action. *Anchorage in the bilayer* by means of a lipid modification, as in the case of myristoylation of c-Src, is a very efficient method of concentrating potential signaling proteins near a receptor for facile recruitment after a signaling event. Proteins which do not have these modifications may still be targeted to the bilayer by associations with specific lipids. For example, associations of proteins with inositide membrane lipids are mediated by PH domains (14). These lipids may thus provide temporary sites for recruitment or sequestration of signaling components which could then be released by modification (phosphorylation or hydrolysis) of the lipids in response to a triggering mechanism. Negative sequestration has been proposed to play a role in apoptosis where membrane-associated CED-9 may bind and restrict the association of CED-4 with other proteins of the apoptotic pathway (15). However, no molecular mechanism for this sequestration interaction has been proposed.

Another logical mechanism for sequestration is association with cytoskeletal structures. Some mode of sequestration is clearly involved in the numerous examples of retention of transcription factors and other cytoplasmically localized proteins targeted for the nucleus, some of which are listed in Table 8.2. These proteins reside in the cytoplasm until a signaling event permits their entry into the

Table 8.1. Some putative scaffolding and integrator proteins of signaling complexes

Protein	Associated Signaling Source	Location/ Component(s)	Reference
Ste5	membrane complex	budding yeast	50
AKAPs	PKA	PSDs, others	51
ACKS	PKC	multiple	52
TMCgps	ascites microvilli (membrane cytoskeleton interface)	RTK-MAPK pathway Cadherin-catenins	11, 22, 53
PDZs	multiple	PSDs, others	54, 55
Dystroglycan	Grb2, NMJ complex	muscle NMJ, other	56
Catenins	Tyrosine kinases, transcription factors	epithelial, endothelial junctions	57
Nck	Pak, WASP, multiple tyrosine kinase receptors	cytoplasm, membrane?	58
Caveolin	sphingomyelin, Ras, Src, heterotrimeric G protein-α	plasma membranes	59
Pbs2p (MAPKK)	MAPK cascade	plasma membrane?	60

Table 8.2 *Cytoskeleton-linked signaling elements*

Signaling element	Function	Cytoskeletal element bound	Role of the interaction
MAP kinase cascade MEKK (Raf) MEK (MAPKK) MAPK	Transduction of multiple types of signals to nucleus, depending on cell context and signal	Microfilaments in yeast and ascites cells, others?	Integration of signaling pathways in yeast (24) and ascites microvill (11) Transduce signal to the cytoskeleton for initiation of morphological changes?
Glucocorticoid	Transcription factor	Microfilaments	Sequestration in cytoplasm
Heat shock proteins hsp27 hsp70	Confer thermal resistance to heat-stressed cells	Microfilaments	Acts as a capping protein, prevents polymerization
EF1α	Elongation factor, translation regulator	Microfilaments Microtubules	Regulates MF assembly, crosslinking, stability; severs MTs; regulates length and stability?
mRNA	Translation to protein	Microfilaments Intermediate filaments Microtubules?	Localization of message to specific cellular sites
Receptors of various types	Signal reception	Microfilaments, microtubules?	Localization of signal input, stabilization of complexes?
Ion channels	Ion transport	Microfilaments	Localization of transport activities and signals
G-proteins, heterotrimeric	Molecular switch	Microtubules	Localization of signal?

nucleus. For example, intracellular receptors of the steroid or retinoid class are known to be sequestered in association with an inhibitory complex prior to activation (discussed in chapter 2). Ligand binding frees the receptor from the inhibitor complex, providing the signal permitting nuclear localization. How and where such proteins are docked is unclear, and mechanisms for their transport from the cytoplasm to the nucleus have not been delineated. In the case of the glucocorticoid receptor, which is bound to HSP90 in the inactivated state, a role for microfilaments has been proposed for organizing the receptor/hsp90 complex and for sequestering it in the cytoplasm. Similarly, microtubules have been implicated in facilitating the movement of the activated ligand-receptor complex to the nucleus (17). Heat shock proteins, which are responsible for ordering a cell response to thermal and other types of environmental stress, appear to both modulate and be modulated by cytoskeletal structures. HSP70 is one of a class of actin-related proteins (ARPs) which comprise a much larger actin superfamily containing the actins, hsp/hsc70s, sugar kinases and several bacterial cell cycle proteins (18). HSP70, an F-actin capping protein (19), has been implicated in the linking of proteins to the actin filament network and in facilitation of protein transport to organelles (20). The mechanisms for its interaction with the cytoskeletal stuctures or their associated proteins are not understood, but the mode of association with target proteins for transport is thought to be electrostatic. HSP27 also acts as an actin capping protein and can inhibit actin polymerization. This effect is modulated by phosphorylation via a mitogen- and stress-activated pathway involving p38 MAP kinase (21). Clearly, one mode of action of heat shock proteins involves reorganization of the actin cytoskeleton. In the case of stress, this mechanism may provide an actin-based response to the new cellular conditions.

Activated kinases of the MAP kinase pathway, Rsk and the JAK/STAT pathways are also retained in the cytoplasm before their release or transport to the nucleus. Although these proteins have nuclear targeting sequences, the timing and mode of targeted transport must be tightly regulated. Little is known about mechanisms for their retention and eventual delivery. In the microvilli of the highly aggressive 13762 ascites, rat mammary cells Ras, Raf, MAPKK, MAPK and Rsk are sequestered at the membrane-microfilament interface by their stable interaction with a microfilament-associated signal transduction particle containing constitutively activated p185[neu]/ErbB2 (11, 22). The mechanisms for release from the constraints imposed by association with the membrane-microfilament interface site and for transport to the nucleus are presently unknown. *However, docking with cortical microfilaments as a sequestration mechanism may be a developing paradigm for cytoplasmic retention of activated kinases* prior to their release (by unknown mechanisms) for entry into the nucleus. Microfilaments have been implicated in the sequestration of Raf (1, 23) in mammalian cells and of MAPK and MEK analogs in yeast via STE5 and BEM1, an actin binding protein (24) (discussed in chapter 3).

Another question posed in chapter 2, concerns the rapid transit through the crowded cytoplasm of the activated signaling proteins. Simple diffusion cannot account for either the speed or the specificity in nuclear targeting. One possibility is that a membrane signaling complex is moved by vesicular transport to a perinuclear site, where the appropriate components such as MAPK and Rsk are released for nuclear import and other components of the signaling complex are either degraded or recycled to the plasma membrane. No evidence for an association with cytoskeletal structures or perinuclear sites has been reported for the cases described above, but perinuclear intermediate filaments and their associated proteins are well positioned to play a role. The major unresolved issue is how regulation of the cytoskeleton is involved in getting the plasma membrane complex components through the cytoplasm. One might envision a seqence of temporal and spatial events directly linking the activated receptor and its associated mitogenic pathway complex with a transit mechanism to the nucleus. The signal transduction complex containing MAPK and possibly Rsk is initially endocytosed and moved into the cytoplasm by cortical microfilaments. Transfer of the complex/vesicle to microtubules has to occur for transport using the microtubule motors into the perinuclear region (chapter 6), where it could be taken up by the nucleus via the nuclear transport mechanism (chapter 2). Specificity of the loading, unloading and transit steps is presumably regulated by phosphorylation, possibly involving MAPK or MARK (chapter 6), which can phosphorylate the ubiquitous, cytoplasmic MAP-4. All three cytoskeletal filament structures interact (discussed in chapter 1), and may be linked by MAPs, as well as other cytoskeletal crosslinkers. MAPs are substrates for the MAP kinases (ERKs) which have been suggested to have a role in microtubule re-organization (25). Thus, the cargo might participate in the transport mechanism as well as being transported. This hypothesis might be modified in a number of logical ways to give a similar linear pathway. The related questions of sequestration and nuclear targeting remain some of the most interesting of the major unaddressed areas in all of signal transduction and deserve some focused attention.

An intriguing story is developing on the involvement of the cytoskeleton in the localization and functions of EF1α, a member of a highly conserved family of translation-regulating proteins. EF1α is known to facilitate the GTP-dependent binding of aminoacyl-transfer RNA to ribosomes. However, EF1α is also the second most abundant protein after actin in eukaryotic cells, comprising 1-2% of the total protein in normal growing cells. The level of expression of this translation factor correlates with proliferation level, and is higher in actively proliferating normal cells in culture, in embryos and in tumor cells. Recent studies have demonstrated interactions with the cytoskeleton that are important in translation of mRNA and in the organization of the mitotic spindle microtubule-organizing centers (26). These findings suggest multiple functions for the EF1α and provide a rationale for the overexpression with respect to other components of the translation machinery. For example, EF1α binds, bundles and severs microtubules (27, 28) and binds and crosslinks microfilaments (29-31). It is found in purported mitotic spindle microtubule-organizing center complexes containing α-, β- and γ-tubulins and the heat shock protein HSC70 or actin, depending on the presence of ATP. These properties implicate EF1, not only in compartmentation of translation, but also in other microtubule- and microfilament-dependent functions, such as cell division. Its high cellular concentration and relatively high K_d assure that under the appropriate conditions EF1α will compete efficiently with other actin crosslinking proteins (32). EF1α-crosslinked microfilaments may provide a more conducive environment for some steps in the translation process.

Interestingly, EF1α co-localizes with mRNA, EF2 and ribosomes, all of which co-localize with microfilaments (26). These studies, as well as the demonstration of EF1α and EF2 binding to microfilaments and EF1α and polysome binding to microtubules, implicate these cytoskeletal structures in important regulatory roles in *localization of translation*. Much of the message in many cells is associated with cytoskeletal structures, and roles for cytoskeletal involvement in targeting and transport have been postulated (33, 34). Localization on microtubules is thought to facilitate active transport of RNA-containing granules and anchoring of mRNAs as well as translation. Spatial information for mRNA targeting is postulated to reside in specific localizing sequences on the 3' end (35) which interact with as yet uncharacterized sequences at target sites. Localization of message appears to be quite important for translation. For example, aminoacyl-tRNAs cannot be incorporated into protein in cell extracts, whereas even free amino acids can serve as protein precursors in the intact cell, suggesting an absolute requirement for a *channeling mechanism* for translation (36). Channeling, or direct transfer of metabolic intermediates sequentially from one enzyme to another has been known for many years to order and facilitate metabolic pathways (37). Channeling requires a mechanism for structural organization such as assembly into multi-enzyme complexes or association with membranes or with the cytoskeleton. The links between EF1α and both microfilaments and microtubules establishes the cytoskeleton as a major channeling effector in translation. A number of intriguing questions remain concerning mRNA localization to cytoskeletal structures and its cellular consequences:

1) What are the molecular mechanisms involved in the interactions of EF1α and EF2 with specific cytoskeletal structures or their associated proteins?

2) What processes control the temporal regulation of binding of the translation machinery proteins to cytoskeletal structures? Translation must be tightly coupled to cell cycle and/or signaling events. In both cases, phosphorylation is a major mechanism for regulation and might logically play a role in the regulation of protein-protein interactions.

3) What is the cellular significance of the reciprocal regulation of translation by the cytoskeleton and of the cytoskeleton by components of the translational machinery, such as the microtubule- and microfilament-modulating protein EF1α?

The answers to the last two questions are likely complex, whereas the question of how the translation machinery associations and other such integrative pathways are organized for their functions can be extrapolated from other systems. They involve specific binding domains and motifs. It is interesting to note that many signaling components have multiple binding domains, as well as functional activities. Multidomain proteins can be divided into two broad classes, those with and without functional activities (38), discussed in chapter 2. The latter are often referred to as adaptor proteins, a term coined to describe proteins with SH2 and SH3 domains and lacking any apparent active site (39) such as Grb2, Shc, Nck and Crk. Interestingly, the last of these is an oncogene even though it apparently lacks a functional activity further supporting the importance of the formation of multimeric complexes. Subsequent studies have shown that other domains in multidomain proteins serve similar functions. PH domains appear in many different types of signaling proteins, suggesting that targeting to the membrane bilayer is very important in their function. For example, the insulin substrate IRS1 has both PH and PTB domains in its N-terminal sequence (14). An unusual possibility is ErbB-3 which has little or no catalytic activity, but forms heterodimers with and is phosphorylated by other ErbB family members. The phosphotyrosine motifs on ErbB3 are different from those on ErbB2 and are capable of recruiting a different subset of SH2 domain-containing proteins to multimeric complexes for signaling (13). One of the enzymes bound by ErbB3, but not the other ErbBs, is PI3K, which has been implicated in mitogenesis. Thus ErbB3 may function more like IRS1 than like the other ErbB receptors. Likewise, the p85 regulatory subunit of PI3K has SH2 and SH3 domains (40), allowing it to form complexes with multiple proteins for specifying localization or substrate preferences (41). PI3K has an additional twist, exhibiting two activities, a protein kinase as well as the lipid kinase activity (42). Thus, it may be able to modify and modulate other membrane proteins as well as the lipids as part of its demonstrated multiplicity of actions (40, 41).

Many other active signaling proteins also have multiple binding domains in their active (catalytic) subunits. All of the Src family members have both SH2 and SH3 domains, as does Abl, which has in addition microfilament-binding and nuclear localization sequences (chapters 2 and 7). Another multidomain tyrosine kinase is FAK (chapter 4), which may play a structural as well as a catalytic role in signaling via integrins and other adhesive complexes (43). Many of the modulators of small G-proteins also appear to have multiple binding domains; PH domains are common in these important proteins. These protein-protein interaction sites likely reflect a requirement for their association with multimeric complexes (44, 45). For example, both Vav, a nucleotide exchange protein, and RasGAP, a GTPase-activating protein, have both SH2 and SH3 domains. Vav also has a PH domain and DAG/PE-binding domains, suggesting mechanisms for association with membranes. These examples provide only a part of the evidence for the importance of the multimeric complexes and the specificity of the associations which form them.

Another broad family of proteins important in signaling, those that transduce signal to the cytoskeleton, also have multiple binding domains and/or motifs. Some of these have lipid modifications, and others bind specific phospholipids (46). Thus both are targeted to the plasma membrane where they are ideally situated to interact with activated plasma membrane receptor complexes. Many of the membrane skeletal proteins are actin binding proteins and have actin binding domains. As expected from the different manners in which actin binding proteins may interact with actin, multiple actin binding domains are being identified. Although these sequences can be grouped into families of related domains, this lack of a readily identifiable actin binding sequence makes identification of actin binding domains in newly sequenced proteins difficult. It is likely that secondary and possibly tertiary structure may play a role in some if not all of these domains (for example, see the discussion of vinculin below). In addition to actin binding motifs, these proteins can have binding domains or motifs for other membrane skeletal proteins. Many of these multiple domain-containing proteins are known components of organization sites such as focal adhesions and thus can be considered to be a subclass of scaffolding proteins.

It is becoming apparent that regulation of and by membrane skeletal proteins may be somewhat more complex than other intracellular proteins. A good example of the multi-domain membrane skeletal proteins exhibiting this kind of complex modulation is vinculin (47), one of the major proteins found in contact sites. Vinculin is localized to focal contacts and to cell-cell adhesions sites in many animal cell types (48). Vinculin associates preferentially with acidic phospholipids, especially well with $PI4,5P_2$, and interacts with several structural proteins in addition to actin and with regulatory proteins, including α-actinin, talin, paxillin and protein kinases. Its structure consists of an N-terminal 90 kDa globular head region and a C-terminal 27 kDa rod-like domain, linked by a proline-rich region. The three-dimensional structure of vinculin appears to be conformationally regulated by an intramolecular head-to-tail interaction such that the binding domain for each ligand is not always available. Further, phosphorylation of specific sites by both tyrosine and serine/threonine-specific kinases is repressed in the same manner.

Vinculin can be conformationally altered in the presence of talin to allow phosphorylation; F-actin binding decreases phosphorylation. Further, binding of acidic phospholipids, particularly PI4,5$_2$, causes a conformational activation of vinculin, enhancing the binding of both F-actin and talin. This is one example of how the complex interactions of membrane skeletal proteins with lipids and with multiple proteins may modulate their functions in sites such as adhesions.

Ultimately, it is these complex organizations of membrane and cytoskeletal components which determine a cell's fate: proliferation, differentiation, apoptosis or quiescence. As described in chapter 5, the relevant decisions concerning these options to the cell are made in the G1 phase of the cell cycle. Cell adhesion is important in determining the result. Different integrins cause different responses, presumably by promoting the assembly of different adhesion complexes whose signals combine in specific ways with signals from soluble ligands. Organization of the actin cytoskeleton is critical to the progress through G1 for many cell types. However, transformed or neoplastic cells have lost key regulatory aspects of G1 progression. Thus, they have also lost their ability to undergo apoptosis and differentiation, instead proceeding through the proliferation phase in spite of contrary signals or due to a loss of some of those signals. Thus, each cell type appears to require a balance of informational inputs from its environment to determine its state by regulating not only biochemical changes in signaling pathways, but also their organization, contributed by the cytoskeleton and its interactions with membranes. Agents which contribute to that organization, such as phosphoinositides, adhesion molecules, small G-proteins and cytoskeletal proteins, are the key elements for cellular homeostasis.

In the last quarter of a century, many cellular and molecular concepts derived from many approaches and a wide diversity of tools have laid a basic foundation of understanding of cell function. Even the complexities of tumorigenesis and tumor progression have yielded to the onslaught of classical biochemical and cell biological approaches to advance our understanding of alterations in cell function resulting from oncogenic events. Yet some complex areas such as development have not yielded so readily to these classical approaches. In the past decade, an explosion in technological developments in molecular genetics has begun to make inroads into this difficult area. With the cloning of bacterial and yeast genomes, and human sequences becoming increasingly available, a new era is beginning to emerge in the study of biology. We are now in a position to use the information so laboriously gathered on proteins involved in basic cellular functions to identify proteins found by genetic methods to be critical in specific aspects of development. Key to our ability to link these convergent approaches is our emerging capability to predict protein homologies based on sequence. Whereas previously, emphasis has been placed on *overall* sequence similarities in assessment of protein *relatedness*, a major focus in the postgenome era will be the identification of *domains* or *motifs* to obtain hints about protein *functionality*.

An important concept emerges from the relatively recent recognition that many proteins involved in signaling and in membrane-cytoskeletal organization have multiple domains and motifs, many of which are involved in protein interaction with other proteins, lipids, nucleic acids or glycoconjugates. This beads-on-a-bracelet view of important regulatory proteins allows us to identify not only catalytic activities, but also potential interactions with other molecules in signaling pathways. Thus methodologies being developed for the facile identification of binding domains and of specific motifs to which they bind, such as the oriented peptide library (49) (chapter 2), are key to our ability to propose testable hypotheses relating to interactions and functions for newly identified gene products. By coupling these identifications with biochemical studies of protein associations in complexes and immunolocalization studies of proteins in cells and tissues, we should be able to greatly expand our comprehension of the organization of signaling complexes and their cytoskeletal associations in cellular functions.

To summarize our concepts about signaling and the cytoskeleton, we need to return to the two questions in the preface and the beginning of this chapter. Simplistically, the cytoskeleton as a complex entity regulates spatial aspects of cell behavior, i.e., cell organization, including the localization and organization of signaling elements. Similarly, cell signaling controls temporal aspects of cell behavior, including temporal aspects of cytoskeleton organization. Thus, these two fundamental elements of cell behavior are inextricably intertwined, just as the different filaments of the cytoskeleton or the different signaling pathways are integrated in the cell. This analysis justifies our basic paradigm for cell signaling introduced in the Preface. *The cell is more than the sum of its parts; it is the product of its parts integrated over time and space.*

References

1. Carraway KL, Carraway CAC. Signaling, mitogenesis and the cytoskeleton: where the action is. Bioessays 1995; 17:171-175.
2. Fincham VJ, Unlu M, Brunton VG, Pitts JD, Wyke JA, Frame MC. Translocation of Src kinase to the cell periphery is mediated by the actin cytoskeleton under the control of the Rho family of small G-proteins. J Cell Biol 1996; 135: 1551-1564.
3. Edwards DR. Cell signalling and the control of gene transcription. Trends Pharm Sci 1994; 15:239-244.
4. Carpenter CL, Cantley LC. Phosphoinositide 3-kinase and the regulation of cell growth. Biochim Biophys Acta 1996; 1288:M11-M16.
5. Carraway KL III, Cantley LC. A neu acquaintance for ErbB3 and ErbB4: a role for receptor heterodimerization in growth signaling. Cell 1994; 78:5-8.
6. Cantley LG. Growth factors and the kidney: regulation of epithelial cell movement and morphogenesis. Am J Physiol 1996; 271:F1103-13.
7. Lim L, Hall C, Monfries C. Regulation of actin cytoskeleton by Rho-family GTPases and their associated proteins. Sem Cell Devel Biol 1996; 7:699-706.
8. Yarden Y, Ullrich A. Growth factor receptor tyrosine kinases. Annu Rev Biochem 1988; 57:443-478.
9. Hausdorff WP, Caron M, Lefkowitz RJ. Turning off the signal: desensitization of β-adrenergic receptor function. FASEB J 1990; 4:2881-2889.
10. Brown MT, Cooper JA. Regulation, substrates and functions of src. Biochim Biophys Acta 1996; 1287:121-149.
11. Carraway CAC, Carraway KL. Interactions of membrane receptors and cell signalling systems with the cytoskeleton. In: Hesketh HE, Pryme IF, eds. Treatise on the Cytoskeleton, vol. 2, Role in Cell Physiology 1996; 207-238, Greenwich, CT: JAI Press Inc.
12. Loeb JA, Fischbach GD. ARIA can be released from extracellular matrix through cleavage of a heparin-binding domain. 1995; J Cell Biol 130:127-135.
13. Carraway KL, Carraway CAC, Carraway KL III. Roles of ErbB-3 and ErbB-4 in the physiology and pathology of the mammary gland. J Mammary Gland Biol Neoplasia 1997; 2:187-198.
14. Lemmon MA, Falasca M, Ferguson KM, Schlessinger J. Regulatory recruitment of signalling molecules to the cell membrane by pleckstrin-homology domains. Trends Cell Biol 1997; 7:237-242.

16. Wu D, Wallen HD, Nunez G. Interaction and regulation of subcellular localization of CED-4 by CED-9. Science 1997; 275:1126-1129.

17. Pratt WB, Sanchez ER, Bresnick EH, Meshinchi S, Scherrer LC, Dalman FC, Welsh MJ. Interaction of the glucocorticoid receptor with the Mr 90,000 heat shock protein: an evolving model of ligand-mediated receptor transformation and translocation. Cancer Res 1989; 49: 2222s-2229s.

18. Frankel S, Mooseker MS. The actin-related proteins. Curr Opin Cell Biol 1996; 8:30-37.

19. Weeds A, Maciver S. F-actin capping proteins. Curr Opin Cell Biol 1993; 5: 63-69.

20. Tsang TC. New model for 70 kDa heat-shock protein's potential mechanisms of function. FEBS Lett 1993; 323:1-3.

21. Landry J, Huot J. Modulation of actin dynamics during stress and physiological stimulation by a signaling pathway involving p38 MAP kinase and heat-shock protein 27. Biochem Cell Biol 1995; 73:703-707.

22. Carraway CAC, Carvajal ME, Carraway KL. Association of the Ras/MAP kinase signal transduction pathway with microfilaments. Evidence for a p185neu-containing cell surface signal transduction particle. Submitted.

23. Stokoe D, Macdonald SG, Cadwallader K, Symons M, Hancock JF. Activation of Raf as a result of recruitment to the plasma membrane. Science 1994; 264: 1463-1467.

24. Chant J. Cell polarity in yeast. Trends in Genetics 1994; 10:328-333.

25. Crews CM, Alessandrini A, Erikson RL. Erks: their fifteen minutes has arrived. Cell Growth Differ 1992; 3:135-42.

26. Condeelis J. Elongation factor 1α, translation and the cytoskeleton. Trends Biochem Sci 1995; 20:169-170.

27. Durso NA, Cyr RJ. A calmodulin-sensitive interaction between microtubules and a higher plant homolog of elongation factor-1. Plant Cell 1994; 6:893-905.

28. Shiina N, Gotoh Y, Kubomura N, Iwamatsu A, Nishida E. Microtubule severing by elongation factor. Science 1994; 266:282-285.

29. Dharmawardhane S, Demma M, Yang F, Condeelis J. Compartmentalization and actin binding properties of ABP-50: the elongation factor-1 of Dictyostelium. Cell Motil Cytoskeleton 1991; 20:279-288.

30. Kurasawa Y, Hanyu K, Watanabe Y, Numata O. F-actin bundling activity of Tetrahymena elongation factor1 is regulated by Ca2+/calmodulin. J Biochem (Tokyo). 1996; 119:791-798.

31. Yang F, Demma M, Warren V, Dharmawardhane S, Condeelis J. Identification of an actin-binding protein from Dictyostelium as elongation factor 1α. Nature 1990; 347:494-496.

32. Owen CH, DeRosier DJ, Condeelis J. Actin crosslinking protein EF-1α of *Dictyostelium discoideum* has a unique bonding rule that allows square-packed bundles. J Struct Biol 1992; 109:248-254.

33. Bassell G, Singer RH. mRNA and cytoskeletal filaments. Curr Opin Cell Biol 1997; 9:109-115.

34. Hesketh JE. Sorting of messenger RNAs in the cytoplasm: mRNA localization and the cytoskeleton. Exp Cell Res 1996; 225: 219-236.

35. Singer RH. The cytoskeleton and mRNA localization. Curr Opin Cell Biol 1992; 4:15-19.

36. Negrutskii BS, Deutscher MP. Channeling of aminoacyl-tRNA for protein synthesis in vivo. Proc Natl Acad Sci USA 1991; 88:4991-4995.

37. Srere, P. Complexes of sequential metabolic enzymes. Annu Rev Biochem 1987; 56:89-124.

38. Pawson T, Gish GD. SH2 and SH3 domains: from structure to function. Cell 1992; 71:359-362.

39. Mayer BJ, Baltimore D. Signalling through SH2 and SH3 domains. Trends Cell Biol 1993; 3:8-13.

40. Varticovski L, Harrison-Findik D, Keeler ML, Susa M. Role of PI 3-kinase in mitogenesis. Biochim Biophys Acta 1994; 1226:1-11.

41. Kapeller R, Cantley LC. Phosphatidylinositol 3-kinase. BioEssays 1994; 16: 565-576.

42. Abraham RT. Phosphatidylinositol 3-kinase related kinases. Curr Opin Immunol 1996; 8:412-418.

43. Richardson A, Parsons JT. Signal transduction through integrins: a central role for focal adhesion kinase? BioEssays 1995; 17:229-236.

44. Bokoch GM, Der CJ. Emerging concepts in the *Ras* superfamily of GTP-binding proteins. FASEB J 1993; 7:750-759.

45. Pronk GJ, Bos JL. The role of p21^{ras} in receptor tyrosine kinase signalling. Biochim Biophys Acta 1994; 1198: 131-147.

46. Isenberg G. Actin-binding protein-lipid interactions. Cell Motil Cytoskel 1991; 12:136-144.

47. Jockusch, BM, Rüdiger, M. Crosstalk between cell adhesion molecules: vinculin as a paradigm for regulation by conformation. Trends Cell Biol 1996; 6: 311-315.

48. Jockusch BM, Bubeck P, Giehl K, Kroemker M, Moschner J, Rothkegel M, Rüdiger M, Schlüter K, Stanke G, Winkler J. The molecular architecture of focal adhesions. Annu Rev Cell Dev Biol 11:379-416.

49. Songyang Z, Cantley LC. Recognition and specificity in protein tyrosine kinase-mediated signalling. Trends Biochem Sci 1995; 20:470-475.

50. Wittenberg C, Reed SI. Plugging it in: signaling circuits and the yeast cell cycle. Curr Opin Cell Biol 8:223-230.

51. Faux MC, Scott JD. Molecular glue: kinase anchoring and scaffold proteins. Cell 1996; 85:9-12.

52. Mochly-Rosen D. Localization of protein kinases by anchoring proteins: a theme in signal transduction. Science 1995; 268: 247-251.

53. Li Y, Carraway, KL, Carraway CAC. Molecular Associations of teh care glycoprotein complex from a microfilamnet-associated signal transduction particle from ascite tumor cell microvilli. Submitted.

54. Garner CC, Kindler S. Synaptic proteins and assembly of synaptic junctions. Trends Cell Biol 1996; 6:429-433.

55. Songyang Z, Fanning AS, Fu C, Xu J, Marfatia SM, Chishti AH, Crompton A, Chan AC, Anderson JM, Cantley LC. Recognition of unique carboxyl-terminal motifs by distinct PDZ domains. Science 1996; 275:73-77.

56. Henry MD, Campbell KP. Dystroglycan: an extracellular matrix receptor linked to the cytoskeleton. Curr Opin Cell Biol 1995; 8:625-631.

57. Ben-Ze'ev A. Cytoskeletal and adhesion proteins as tumor suppressors. Curr Opin Cell Biol 1997; 9:99-108.

58. Galisteo ML, Chernoff J, Su Y-C, Skolnik EY, Schlessinger J. The adaptor protein Nck links receptor tyrosine kinases with the serine-threonine kinase Pak1. J Biol Chem 1996; 271:20997-21000.

59. Lisanti MP, Scherer PE, Tang Z, Sargiacomo M. Caveolae, caveolin and caveolin-rich membrane domains: a signalling hypothesis. Trends Cell Biol 1994; 4:231-235.

60. Posas F, Saito H. Osmotic activation of the HOG MAPK pathway via Ste11p MAPKKK: scaffold role of Pbs2p MAPKK. Science 1997; 276:1702-1705.

Index